Microbial Biodegradation

From Omics to Function and Application

Edited by

Jerzy Długoński

Department of Industrial Microbiology and Biotechnology
Faculty of Biology and Environmental Protection
University of Łódź
Łódź
Poland

Caister Academic Press

Caister Academic Press
Norfolk, UK

www.caister.com

British Library Cataloguing-in-Publication Data
A catalogue record for this book is available from the British Library

ISBN: 978-1-910190-45-6 (paperback)
ISBN: 978-1-910190-46-3 (ebook)

Cover created based on design ideas of Andrzej Długoński and Jerzy Długoński

Contents

Contributors

Haris N. Abubackar
Chemical Engineering Laboratory
Faculty of Sciences
University of La Coruña
La Coruña
Spain

haris.nalakath@udc.es

Przemysław Bernat
Department of Industrial Microbiology and Biotechnology
Faculty of Biology and Environmental Protection
University of Łódź
Łódź
Poland

pbernat@biol.uni.lodz.pl

Carl E. Cerniglia
Division of Microbiology
National Center for Toxicological Research
Food and Drug Administration
Jefferson, AR
USA

carl.cerniglia@hhs.fda.gov

Jianmeng Chen
College of Biological and Environmental Engineering
Zhejiang University of Technology
Hangzhou
China

jchen@zjut.edu.cn

Jerzy Długoński
Department of Industrial Microbiology and Biotechnology
Faculty of Biology and Environmental Protection
University of Łódź
Łódź
Poland

jdlugo@biol.uni.lodz.pl

Łukasz Drewniak
Laboratory of Environmental Pollution Analysis
Faculty of Biology
University of Warsaw
Warsaw
Poland

ldrewniak@biol.uw.edu.pl

Łukasz Dziewit
Department of Bacterial Genetics
Faculty of Biology
University of Warsaw
Warsaw
Poland

ldziewit@biol.uw.edu.pl

Geoffrey M. Gadd
Geomicrobiology Group
School of Life Sciences
University of Dundee
Dundee
Scotland
UK; and
Laboratory of Environmental Pollution and Bioremediation
Xinjiang Institute of Ecology and Geography
Chinese Academy of Sciences
Urumqi
China

g.m.gadd@dundee.ac.uk

Aleksandra Góralczyk
Department of Industrial Microbiology and Biotechnology
Faculty of Biology and Environmental Protection
University of Łódź
Łódź
Poland

olagoralczyk92@tlen.pl

Katarzyna Hrynkiewicz
Department of Microbiology
Faculty of Biology and Environmental Protection
Nicolaus Copernicus University
Toruń
Poland

hrynk@umk.pl

Roksana Iwanicka-Nowicka
Laboratory of Systems Biology
Faculty of Biology
University of Warsaw; and
Laboratory of Microarray Analysis
Institute of Biochemistry and Biophysics
Polish Academy of Sciences
Warsaw
Poland

roxana@ibb.waw.pl

Anna Jasińska
Department of Industrial Microbiology and Biotechnology
Faculty of Biology and Environmental Protection
University of Łódź
Łódź
Poland

ajas@biol.uni.lodz.pl

Christian Kennes
Chemical Engineering Laboratory
Faculty of Sciences
University of La Coruña
La Coruña
Spain

kennes@udc.es

Seong-Jae Kim
Division of Microbiology
National Center for Toxicological Research
Food and Drug Administration
Jefferson, AR
USA

seong-jae.kim@fda.hhs.gov

Konrad Kowalski
Department of Experimental Oncology
Institute of Immunology and Experimental Therapy
Wrocław
Poland

konrad.kowalski@iitd.pan.wroc.pl

Ohgew Kweon
Division of Microbiology
National Center for Toxicological Research
Food and Drug Administration
Jefferson, AR
USA

oh-gew.kweon@fda.hhs.gov

Katarzyna Paraszkiewicz
Department of Industrial Microbiology and Biotechnology
Faculty of Biology and Environmental Protection
University of Łódź
Łódź
Poland

katapar@biol.uni.lodz.pl

Sylwia Różalska
Department of Industrial Microbiology and Biotechnology
Faculty of Biology and Environmental Protection
University of Łódź
Łódź
Poland

rozalska@biol.uni.lodz.pl

Mirosława Słaba
Department of Industrial Microbiology and Biotechnology
Faculty of Biology and Environmental Protection
University of Łódź
Łódź
Poland

mirsla@biol.uni.lodz.pl

John B. Sutherland
Division of Microbiology
National Center for Toxicological Research
Food and Drug Administration
Jefferson, AR
USA

john.sutherland@fda.hhs.gov

Rafał Szewczyk
Department of Industrial Microbiology and Biotechnology
Faculty of Biology and Environmental Protection
University of Łódź
Łódź
Poland

rszewcz@biol.uni.lodz.pl

Dariusz Trzmielak
Department of Marketing
Faculty of Management and Center for Transfer Technology
University of Łódź
Łódź
Poland

dtrzmielak@uni.lodz.pl

María C. Veiga
Chemical Engineering Laboratory
Faculty of Sciences
University of La Coruña
La Coruña
Spain

veiga@udc.es

Aleksandra Ziembińska-Buczyńska
Environmental Biotechnology Department
Faculty of Energy and Environmental Engineering
The Silesian University of Technology
Gliwice
Poland

aleksandra.ziembinska-buczynska@polsl.pl

Current Books of Interest

Cyanobacteria: Omics and Manipulation	2017
Foot and Mouth Disease Virus: Current Research and Emerging Trends	2017
Brain-eating Amoebae: Biology and Pathogenesis of *Naegleria fowleri*	2016
Staphylococcus: Genetics and Physiology	2016
Chloroplasts: Current Research and Future Trends	2016
Influenza: Current Research	2016
MALDI-TOF Mass Spectrometry in Microbiology	2016
Aspergillus and *Penicillium* in the Post-genomic Era	2016
Omics in Plant Disease Resistance	2016
Acidophiles: Life in Extremely Acidic Environments	2016
Climate Change and Microbial Ecology: Current Research and Future Trends	2016
Biofilms in Bioremediation: Current Research and Emerging Technologies	2016
Microalgae: Current Research and Applications	2016
Gas Plasma Sterilization in Microbiology: Theory, Applications, Pitfalls and New Perspectives	2016
Virus Evolution: Current Research and Future Directions	2016
Arboviruses: Molecular Biology, Evolution and Control	2016
Shigella: Molecular and Cellular Biology	2016
Aquatic Biofilms: Ecology, Water Quality and Wastewater Treatment	2016
Alphaviruses: Current Biology	2016
Thermophilic Microorganisms	2015
Flow Cytometry in Microbiology: Technology and Applications	2015
Probiotics and Prebiotics: Current Research and Future Trends	2015
Epigenetics: Current Research and Emerging Trends	2015
Corynebacterium glutamicum: From Systems Biology to Biotechnological Applications	2015
Advanced Vaccine Research Methods for the Decade of Vaccines	2015
Antifungals: From Genomics to Resistance and the Development of Novel Agents	2015
Bacteria-Plant Interactions: Advanced Research and Future Trends	2015
Aeromonas	2015
Antibiotics: Current Innovations and Future Trends	2015
Leishmania: Current Biology and Control	2015
Acanthamoeba: Biology and Pathogenesis (2nd edition)	2015
Microarrays: Current Technology, Innovations and Applications	2014
Metagenomics of the Microbial Nitrogen Cycle: Theory, Methods and Applications	2014

Full details at www.caister.com

Preface

Mankind has transformed and contaminated the environment in parallel with the progression of civilization. Earth pollution, on a scale so large that it threatens wildlife and humans, has been noticeable from the mid-19th century. This pollution has been a consequence of the significant increase in the total number of living humans on Earth and of rapid industrial development, which has produced an increasing number of products for the expanding civilization and has contributed to the progressive degradation of the natural environment. The possibility of applying microorganisms to remove pollutants was fully appreciated in the late 19th century. Since that time, the first artificial waste water treatment plants were constructed and have been constantly improved.

Currently, the elimination of pollution of natural origin (domestic waste, waste from the agro-food industry, etc.) is generally not much of a problem, from the viewpoint of the effectiveness and efficiency of the technologies used. A major problem, both from technological and scientific standpoints, is the microbial degradation of deleterious xenobiotics, especially those (such as numerous pesticides, pharmaceuticals or synthetic dyes) that are very toxic and difficult to metabolize by microbes. The main reasons are combined with a relatively limited knowledge regarding the mechanisms and factors that have a crucial effect on the course and yield of biodegradation processes carried out by complex microbial populations in contaminated areas. New valuable prospects for better elucidating these intricate relationships are formulated by the advances in modern science, especially 'omics' sciences such as: genomics, proteomics, lipidomics, metabolomics and other related branches.

The presented book involves panels of selected subjects related to biodegradation processes, with an emphasis on the possibility of applying recent achievements in the field of 'omics' sciences for the microbial removal of deleterious contaminants.

The first set of units (chapters 1–3) provides an introduction to the burgeoning area of metagenomic and metatranscriptomic techniques used in the research on microbial organic pollutant degradation and heavy metal elimination, as well as the characteristics of housekeeping genes applied in microbial identification and functional genes exploited as molecular markers in biodegradation studies.

The next set of chapters (4–6) describes the usefulness of an analytical technique – mass spectrometry (MS) – in studies on microbial detoxification and degradation of contaminants, with special attention paid to MS-omic (metabolomics, proteomics and lipidomics) challenges in this area of knowledge.

The panel of chapters 7–11 reviews the latest advances in the investigations of the pathways and mechanisms of microbial detoxification and degradation of the most deleterious contaminants: EDCs (endocrine disrupting compounds), PAHs (polycyclic aromatic hydrocarbons), VOCs (volatile organic compounds), dyes covering also both the updates of 'omics' sciences and the aspects of practical application.

The last section (chapters 12 and 13) is focused on heavy metal elimination by microorganisms and the possibilities of applying surface-active agents in numerous bioremediation processes.

In summary, this book involves an overview of the current state-of-the-art in the field of microbial biodegradation, and I believe that it will be useful to students, research scientists and consulting professionals in the spheres of environmental microbiology and biotechnology, as well as molecular biology.

I am deeply grateful to all contributing authors for their cooperation and patience. The efforts of Anna Jasińska and Milena Radzioch from my team in typing the index for this book and their attention to detail is much appreciated. This work was in part supported by University of Łódź and Faculty of Biology and Environmental Protection University of Łódź, Poland.

Jerzy Długoński

Organic Pollutants Degradation by Microorganisms – Genomics, Metagenomics and Metatranscriptomics Backgrounds

1

Sylwia Różalska and Roksana Iwanicka-Nowicka

Contents

Abstract

Environment pollution by organic compounds of anthropogenic origin is a major global problem. Microorganisms persistent in the environment play a crucial role in removal of pollutants, but until recently it was difficult to determine their exact functions in the elimination of contaminants. Here, the next-generation sequencing (NGS) techniques, which have had a remarkable impact on microbial studies and provided new insights into microbial communities, their biodiversity, and function, are discussed. This review also describes recent achievements in genomics, metagenomics, and metatranscriptomics, which have been implemented in studies on the microbial composition of environments contaminated with organic compounds. The importance of the new approaches for determination of the genes responsible for degradation of toxic compounds and metabolic genes is also discussed.

1.1 Introduction

Contamination of the environment with various anthropogenic pollutants is a major global problem. Development of novel technologies to restore contaminated environments remains a significant challenge and is an interesting alternative to chemical and physical methods. Microorganisms play a significant role as components degrading or transforming various pollutants and thus re-establish contaminated environments (de Menezes *et al.*, 2012). To effectively control the bioremediation processes, it is necessary to recognize which microorganisms are responsible for particular processes and to understand their role in the environment decontamination. However, it is estimated that only 1% of living microorganisms are culturable in laboratory conditions. Until very recently, culturability was a prerequisite for genome sequencing and for full access to the genetic components of individual organisms (Cowan *et al.*, 2015).

Advanced metagenomic or metatranscriptomic approaches are relatively new methods and are based on high-throughput sequencing, which can be used to directly sequence genomic DNA or cDNA. This methodology may overcome the problems of both culture-dependent methods and PCR. These approaches will open up great opportunities to explore the structure and functions of microbial communities of contaminated environments. These methods will also enable the discovery of novel environmental microorganisms or genes with important applications to biotechnology or bioremediation of contaminated environments (Cardoso *et al.*, 2011; Fang *et al.*, 2013; Cowan *et al.*, 2015).

In this chapter, we describe the application of genomic, metagenomic and metatranscriptomic techniques in recent studies on biodegradation of organic compounds and provide information on taxonomic abundance and abundance of catabolic genes during the degradation processes.

1.2 Genomic, metagenomic and metatranscriptomic techniques

Eleven years ago, next-generation sequencing (NGS) technologies appeared on the market and, since that time, they made a tremendous impact on many types of microbial studies. Amazing progress has been achieved in terms of speed, read length and throughput, along with a reduction in sequencing costs.

NGS-based analysis allows for finding answers to questions that could not be addressed before, as it enables a culture-independent sequencing of DNA or RNA collected from microbial communities for content analysis. The principles of different NGS platforms are described in dedicated reviews (e.g. Voelkerding *et al.*, 2009; Metzker, 2010; Glenn, 2011; Pareek *et al.*, 2011; Zhang *et al.*, 2011; Liu *et al.*, 2012; Shokralla *et al.*, 2012; Mardis, 2013; Morey *et al.*, 2013; Knief, 2014). Despite differences, most sequencing platforms use roughly similar protocols for library preparation, library amplification and the sequencing process (Fig. 1.1).

Library preparation: the input material for library preparation is either genomic DNA (gDNA), cDNA or RNA. Owing to size limitations for library molecules, nucleic acid has to be fragmented. This is usually achieved with physical methods, i.e. Bioruptor® Sonication System or Covaris Focused-ultrasonicator System (to obtain fragments for end-repair reaction), or enzymatically (blunt-ended DNA fragments are obtained).

Depending on the sequencing platform that is going to be used, the appropriate fragment size of a library is:

- for Illumina libraries – 150–300 bp for SE libraries and < 800 bp for PE libraries;
- for 454 libraries – approximately 1500 bp (to obtain long reads with a modal length of 700 bp);
- for Ion Torrent Personal Genome Machine (PGM) libraries – median insert size is 410 bp, 320 bp, 260 bp and 139 bp for a 400-base-read library, 300-base-read library, 200-base-read library and 100-base-read library, respectively.

The fragmented DNA is ligated to compatible, sequencing platform-specific, DNA adapters. To complete the linkage between the adapters and DNA inserts, ligation is followed by nick-repair (PGM). For other technologies (Illumina, 454 Roche), the sample is ready for loading, in most of the cases, right after ligation.

Adaptors play an important role. They:

- enable binding of the library of fragments to a surface – microbeads (454, Ion PGM) or a glass slide (Illumina);
- allow for amplification of the library by PCR – emulsion (454, Ion PGM) or bridge (Illumina);
- serve as priming sites for sequencing.

Usually, external fragment of the adaptor contains elements for the immobilization of library molecules on a solid surface, while the internal part – the priming site for sequencing. The adapter-ligated library can be size-selected for optimum length.

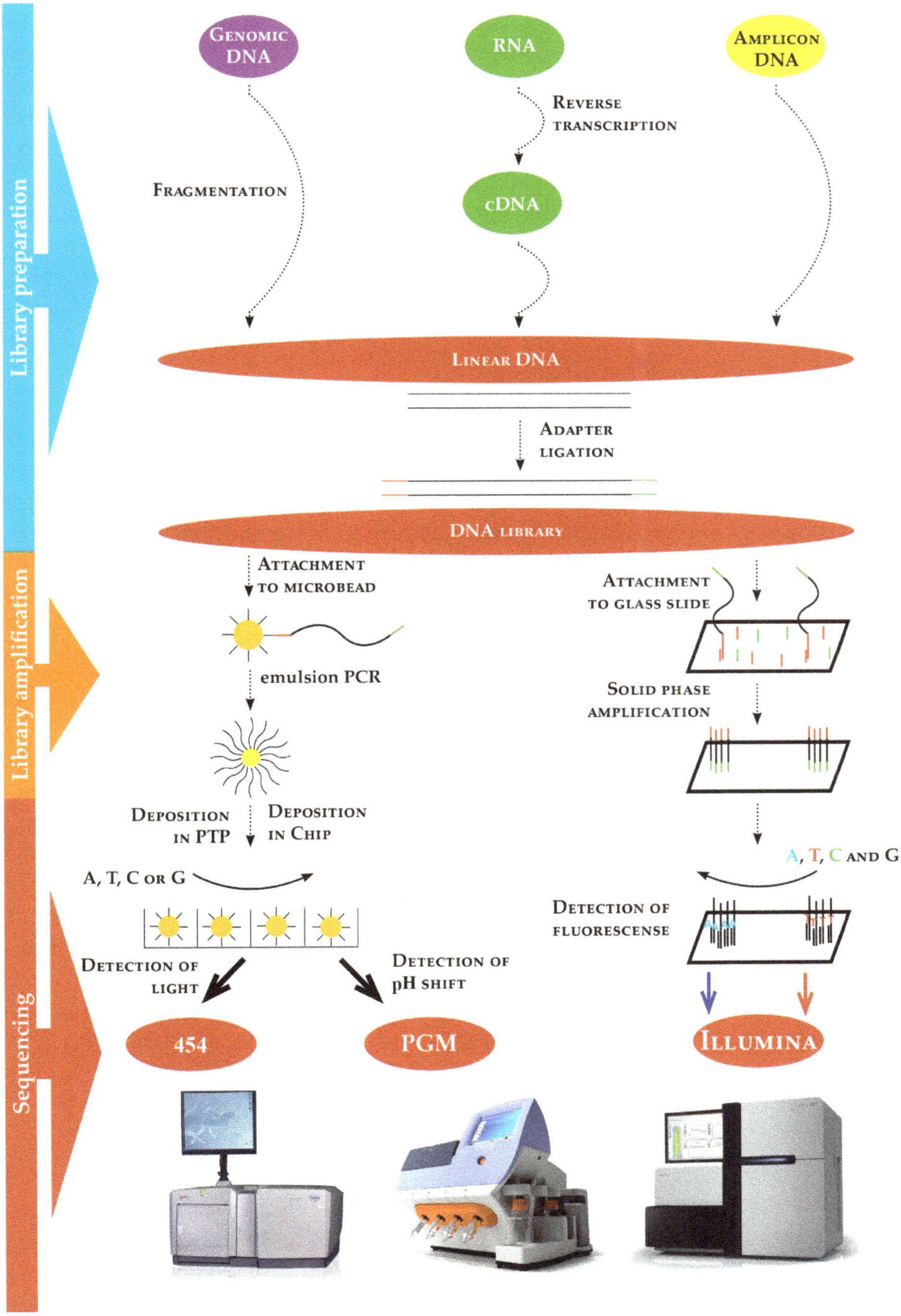

Figure 1.1 Schematic presentation of the next generation sequencing technology: the library preparation, amplification and sequencing process.

Barcoding of libraries: barcode is a unique index sequence added to each DNA fragment during library preparation. It is an optional step that finally allows large numbers of libraries to be pooled and sequenced simultaneously, during a single sequencing run. Sample multiplexing is a useful method when working with small genomes. It is a fast and cost-effective approach that improves productivity by reducing time and use of the reagents.

Library amplification by bridge PCR or emulsion PCR (EMPCR): once constructed, libraries are clonally amplified to gain signal intensity for the sequencing process. Depending on the platform used, the amplification methods vary.

Illumina instruments apply bridge amplification method to form template clusters on a flow cell. In short, library is loaded into the flow cell, fragments hybridize to the oligos attached to the flow cell surface, and then are amplified into a clonal cluster by bridge amplification.

Ion Torrent PGM platform utilizes emPCR on the OneTouch system to amplify an individual library molecule to millions of identical copies bonded to the sphere (bead), to allow ultimate detection of the signal. Finally, in the enrichment step, spheres containing amplified DNA (loaded spheres) are separated from empty spheres.

Roche 454 system, also utilizes the emPCR on unique single-stranded library fragment attached to Capture Bead, in order to create millions of clonal copies of each library.

The sequencing process: different sequencing platform vendors have devised different strategies to generate the sequence reads.

The Ion Torrent PGM, Roche 454 and the Illumina systems use sequencing by synthesis to access nucleic acid sequence from the amplified libraries. Briefly speaking, the amplified library fragments serve as a template for cDNA synthesis. The sequencing occurs in the repeated cycles of flooding the fragments with known nucleotides (differently labelled – Illumina or not labelled – 454 and Ion Proton), nucleotide addition by a DNA polymerase, detection of incorporated nucleotides and washing.

The main differences between sequencing platforms are the methods of nucleotide sequence detection.

The Illumina relies on detection of fluorescence generated by incorporation of fluorescently labelled nucleotides into the growing strand of DNA.

The PGM performs a semiconductor-based sequencing. It is based on the release of a hydrogen ion upon incorporation of a nucleotide into a growing strand of cDNA. Hydrogen ion release is directly measured as a pH change by the semiconductor chip of the Ion instrument. Because there is no detectable difference for H^+ released from the individual dNTPs, they are added in a predefined flow order. If a change in pH appears, it means that base has been incorporated into the template.

The 454 technology is based on the release of pyrophosphate – a product of the enzymatic nucleotide incorporation reaction. The pyrophosphate signal is further converted into a light signal, generated by a luciferase-mediated conversion of luciferin to oxyluciferin, upon primer extension. Generation of a light signal has led to coining of the term 'pyrosequencing' for this technology.

1.2.1 Bacterial whole genome sequencing for microbial genomic research

Shotgun metagenomic DNA sequencing enables characterization of microbial community biodiversity and function (Highlander, 2012). In this procedure, DNA is extracted from all cells in a community and sheared into fragments that are used for library preparation. Using brute force sequencing, all DNA sequences present in the sample are being read.

The limiting factor for using this powerful approach is the analysis of metagenomic sequences caused by structural complexity of the obtained data. The specific remedy to this problem, however, is use of targeted sequencing focused on conserved sequences, like rRNA coding genes. Instead of smaller complexity of the data obtained, this approach has other advantage – more reads can be assigned to a specific organism or a group of related organisms.

1.2.2 Targeted sequencing for metagenomics analysis

Profiling of the members of microbial communities and comparing the composition of the different microbial samples rely on targeted sequencing of metagenomic DNA. Most of this type of studies is based on sequencing of the hypervariable regions of the16S rRNA gene and the Large Subunit (LSU) rRNA gene as bacterial markers (see also Chapter 3). Primers for PCR amplification of hypervariable regions are designed in well-conserved, universal regions to obtain a sequence long enough to distinguish between organisms (Mayo *et al.*, 2014). Adapters for library amplification are added during the PCR using fusion primer constructs. To reduce error rates related to the PCR amplification, use of high-fidelity polymerases is recommended.

The copy numbers of rRNA genes' operons, which vary across different taxa, may falsify final analyses. Alternatively, single-copy target genes, such as *rpoB*, *recA*, *radA*, *rpoA*, *gyrB* and others are used (Volokhov *et al.*, 2012). Additionally, other genes with taxonomic signals, such as *nirS* (denitrification) and *nifH* (nitrogen fixation), have also been used in microbial profiling (Tan *et al.*, 2015).

Until now, the majority of metagenomic analyses have been performed using the 454 technology. However, currently also other platforms designed specific kits for metagenomic studies, encompassing all the steps from sample isolation, library preparation to sequencing, and including software for analysis and interpretation of results.

1.2.3 Bacterial RNA-seq for a community-wide gene expression analysis

Transcriptome is the complete set of RNAs encoded by an organism's genome. Likewise, metatranscriptome covers all the RNAs encoded by all organisms in a complex sample. Therefore, it is possible to use RNA sequencing to get an overview of 'what's happening' in the sequenced population.

RNA is an input material for metatranscriptome sequencing (see also 'Sampling, mRNA extraction, and enrichment' for more detailed information). To enrich RNA sample with mRNA, rRNAs depletion step is a prerequisite during library preparation. RNA has to be converted into cDNA which is used to prepare sequencing library sequencing (see also 'Metatranscriptomic background information on pollutant degradation' for more detailed information). Direct RNA sequencing is now available, but not in the context of mixed microbial communities.

RNA-seq has a lot of advantages over the other methods dedicated to transcriptome analysis, which include: expressed sequence tag (EST) sequencing, serial analysis of gene expression (SAGE), and microarray hybridizations (Wang *et al.*, 2009).

RNA-seq enables for:

- unbiased strand-specific identification of common and novel transcripts;
- detection of low expression genes (limited by the depth of sequencing);
- identification of exon junctions and splicing sites;
- quantitative comparison of transcript data across many experiments;
- analysis of transcripts of not yet sequenced organisms;
- identification of novel, non-coding RNAs, including small RNAs.

Metatranscriptomics reflects only potential genome functions, since it cannot account for the post-transcriptional regulation.

1.3 Metagenomic and genomic analysis of pollutant degradation

Metagenomics is a powerful tool for investigation of the composition of microbial communities of contaminated environments and for determination of the genes responsible for degradation of particular toxic compounds. To elucidate how microbial communities change under the influence of pollution, nowadays, NGS (next-generation sequencing methods) are applied. In general, a typical project should answer the following questions (Meiring *et al.*, 2011):

- What is the number and variety of species inhabiting the environment?
- Which genes or degradation pathways are present?
- Which species is responsible for a particular process?
- How is the community changing over time and under different environmental conditions?

Metagenomic studies changed our understanding of the biodegradation processes in the environment. It appears that degradation of a specific compound is commonly carried out by several or even a dozen species of microorganisms (Fang *et al.*, 2014). This approach also allows researchers to track microbiota changes that are caused by environmental conditions, e.g. temperature, pH or salinity, but also during biodegradation processes. During utilization of the contaminants, a decrease in the concentration of toxic compounds and emergence of derivatives (sometimes more toxic than the substrate) are observed; thus, the conditions in the environment change rapidly. Metagenomic research allows for exploration of degradation routes, identification of the most important genes related to degradation of the contaminants, and for tracking the whole process (see also Chapter 3).

Unfortunately, metagenomics cannot answer the question whether the identified genes are expressed. Therefore, two approaches are commonly applied. One of them involves the use of metatranscriptomics, and the other testing whether the clones carrying the cosmids or plasmids with metagenomic libraries have the ability to degrade xenobiotics. An example of this approach is the screening of a metagenomic library for phenol and catechol degradation activity (Suenaga *et al.*, 2007; Silva *et al.*, 2012).

To determine whether fosmid-containing *Escherichia coli* possesses degradation ability, quite often, fast screening tests are used. One of them is the 'catechol spray method', which involves protein extraction from plasmid- or fosmid-carrying cells and detection of enzymatic activity by monitoring the yellow colour of the product (2-hydroxymuconate semialdehyde) (Suenaga *et al.*, 2007).

Because the primary degradation pathways and the genes encoding enzymes are well known, it is quite easy to compare the results obtained by the metagenomic approach with the data deposited in databases such as KEGG (Kyoto Encyclopedia of Genes and Genomes) or MG-RAST (Metagenomic Rapid Annotations using Subsystems Technology). After the sequences are annotated, not only functional classification of the genes but also determination of the species carrying the genes become possible.

Quite often, we have to deal with the following situation: despite the presence of significant homology in genes encoding enzymes important for biodegradation, the species carrying the genes cannot be identified. One of the examples is catechol 2,3-dioxygenase coded by *xylE* catalysing ring fission in the *meta*-cleavage pathway of catechol. Silva *et al.* (2012) by applying a neighbour-joining phylogenetic approach to pyrosequencing data showed in sludge samples the presence of genes coding for this enzyme. Some of them possess strong homology to *xylE* genes from *Pseudomonas* sp. or *Thermomicrobium* sp., but others show weak relatedness to known sequences and potentially represent new sequences not previously deposited in databases.

Metagenomic analysis not only provides detailed information on the taxonomic abundance and abundance of catabolic genes in the biomass but also allows for pathway mapping and provides insights into the actual degradative capacity for specific pathways (Yadav *et al.*, 2015).

The knowledge of pathways of degradation of aromatic hydrocarbons allows investigators to find some target genes and to determine how often these genes are present in populations. Yadav *et al.* (2015) conducted relative quantification (by real-time PCR) of expression of four genes responsible for ring opening: catechol 2,3-dioxygenase from the xylene degradation pathway (*xylE*) and from phenol degradation pathway (*dmpB*), catechol 1,2-dioxygenase, *tcbC* (coding for chlorocatechol 1,2-dioxygenase), and aromatic-ring hydroxylating dioxygenases. The comparative metagenomics approach can also be a useful tool for detection of seasonal variations in metabolic pathways of aromatic hydrocarbons.

Bacterial biodegradation genes are known to form operons, which promote efficient and complete xenobiotic degradation. Metagenomics is also useful for detection and description of organization of the catabolic gene clusters. Suenaga *et al.* (2007) described organization of gene clusters for degradation of naphthalene, phenol, and aniline; they analysed a microbial community and compared the data with known gene clusters. Some differences in organization with previously described genes were observed for naphthalene and aniline. Moreover, some of the fosmids were carrying genes belonging to a novel catechol 2,3-dioxygenase subfamily.

As mentioned above, metagenomics serves as an excellent tool for determination of microbial degradation capacity in the environment. This approach enables researchers to strictly identify the responsible species for particular degradation steps. Quite often, we deal with the situation where for one of the degradation steps, the bacterium carrying the responsible gene is unculturable (Fang *et al.*, 2014).

A novel, interesting molecular approach to determination of single-cell degradation capabilities in hydrocarbon-contaminated seawater samples was proposed by Mason *et al.* (2012). Individual cells from seawater samples were sorted onto microtitre plates using cell sorters. Then, the cells were lysed, and DNA was amplified by MDA (master mix of multiple displacement amplification).

Subsequently, the single-cell MDA products were screened by means of Sanger sequencing of 16S rRNA gene amplicons, and finally, Illumina sequencing of single-cell genomes was performed. These stand-alone cells were closely related to *Oceanospirillales* and were bearing genes encoding the cyclohexane degradation pathway.

In spite of the fact that we are observing xenobiotic degradation in the environment, sometimes, it is not possible to detect all of the genes encoding the enzymes of the pathway. Then, we can only presume how the whole process occurs in the environment. Recently, databases were compiled to explore the biodegradation pathways of several pesticides. These databases facilitate identification of the degradation genes and the related bacteria by means of the BLAST algorithm. However, the databases are still limited by the availability of reference sequences and the cut-off criteria applied (Fang *et al.*, 2014).

1.4 Taxonomic analysis of the metagenomic data on microbial communities exposed to toxic organic compounds

Metagenomics, as a cultivation-independent approach, allows for analysis of microbial-community structure in contaminated environments. This way, we can detect new organisms that can play a crucial role at certain stages of degradation of xenobiotics. For taxonomic and phylogenetic analysis, 16S rRNA gene studies and NGS are commonly used, but with this approach, the resulting data are limited to *Bacteria* and *Archaea*.

Taxonomic analysis of the metagenomic data can be performed using MEGAN (MEtaGenome Analyser) or MG-RAST, which after collecting reads from a shotgun platform, compare them with sequence databases using the BLAST algorithm. After the analysis, software generates a phylogenetic tree in which the size of each circular node is proportional to the number of assignments at the taxonomic level (Huson *et al.*, 2007; Silva *et al.*, 2012). This approach greatly simplifies the data analysis and phylogenetic assignment of the species in a microbiome.

Recently, a new approach to prediction of functional composition of a microbial community's metagenome from its 16S profile was proposed (Langille *et al.*, 2013). PICRUSt (phylogenetic investigation of communities by reconstruction of unobserved states) is a two-step process that starts from the 'gene content inference' step. At the first level, gene content is precomputed for each organism in a reference phylogenetic tree, and a table of predicted abundance of gene families for each microorganism in the 16S-based phylogeny is reconstructed. This step is independent for each particular microbial community sample, and the table data are calculated only once and predict what genes are present in organisms that have not yet been sequenced judging by the genes detected in their sequenced evolutionary relatives. The next level, named 'metagenome inference', combines the resulting gene content predictions for all microbial taxa with the relative abundance of 16S rRNA genes in one or more microbial community samples, corrected for the expected 16S rRNA gene copy number, to generate the expected abundance data on gene families in the entire community.

The taxonomic analysis of the metagenomic data seems to be very important. By this method, researchers can elucidate not only the microbial community composition in a polluted environment but also changes in the communities during biodegradation of a xenobiotic or accompanying the changes in contamination (Godheja *et al.*, 2014).

Application of metagenomic methods to research on biodegradation and bioremediation has significantly facilitated characterization or monitoring of pollutant-biodegrading microbial populations (Table 1.1) and identification of novel biodegradation pathways (Desai *et al.*, 2010).

1.5 Metatranscriptomic background information on pollutant degradation

Metatranscriptomics is an alternative to functional metagenomics or proteomics and is based on the analysis of mRNA instead of DNA or proteins. The expressed genes are analysed to obtain information about the activity of all genes present in a contaminated environment or to profile the active metabolic processes while the biodegradation proceeds. In 2005, Poretsky *et al.* constructed the first mRNA-derived cDNA clone libraries from a marine ecosystem, and after that, metatranscriptomics was successfully applied to analysis of different

Table 1.1 Identification of the most common genera participating in degradation of xenobiotics

Sample	Contamination	Most common genera/ phyla	References
Marine sediment	pesticides (DDT, HCH, atrazine)	*Proteobacteria* (49.3%) *Actinobacteria* (21.7%)	Fang *et al.* (2014)
Coastal salt marshes	Oil spills	*Firmicutes* *Tenericutes* *Bacteroidetes* *Proteobacteria*	Beazley *et al.* (2012)
Petroleum refinery wastewater	Phenol	*Proteobacteria* (33%) *Firmicutes* (19%) *Actinobacteria* (18.5%) *Planctomycetes* (8%)	Silva *et al.* (2013)
Activated sludge from aeration tank	Aromatic compounds	*Proteobacteria* (41.5%) *Actinobacteria* (28.1%) *Bacteroidetes* (16.9%) *Unclassified* (13.5%)	Suenaga *et al.* (2007)

environments. Unfortunately, the metatranscriptomic approaches have not been widely used to profile degradation of complex pollutants or to analyse complex biotechnological processes, probably because of the instability of mRNAs and difficulties with their extraction (de Menezes *et al.*, 2012).

A typical metatranscriptomic project consists of several basic steps shown in Fig. 1.2.

1.5.1 Sampling, mRNA extraction, and enrichment

The most crucial step for metatranscriptomic analysis is sampling because of the short half-life (in the range of seconds to minutes) of RNA. To reduce the changes in transcriptional profiles which usually occur as a consequence of poor sampling, deep freezing in liquid nitrogen (–70°C) or immediate transfer to an RNA preservation solution are required (Carvalhais *et al.*, 2012). RNA isolation should be done immediately after sampling, but if it is not possible, the storage in liquid nitrogen is recommended.

Another important issue is isolation of pure and good-quality mRNA. Efficient extraction of mRNA is a significant technical challenge, especially from environmental samples, where transcripts are scarce in comparison to rRNA or DNA. The extraction step is very important because its efficiency determines which transcripts will be regarded as essential for biodegradation. Moreover, various compounds in the environmental samples may inhibit the downstream sample processing and should be removed (Kuske *et al.*, 2015). Currently, there are several commercially available extraction kits that provide efficient mRNA isolation from contaminated environments (Carvalhais *et al.*, 2012).

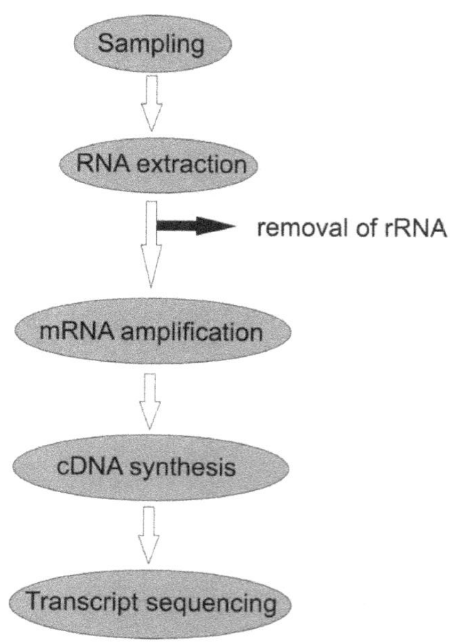

Figure 1.2 The main steps of a metatranscriptomic protocol.

mRNA samples from some of the contaminated areas, such as soil, should be enriched before amplification. It is estimated that total mRNA content in environmental samples is very low, approximately 1–5% of the entire pool of RNA (He *et al.*, 2010). Therefore, enrichment is an important step that must be considered before cDNA synthesis. To date, several methods for mRNA recovery from environmental samples have been used, such as

- duplex specific nuclease (DSN) treatment
- subtractive hybridization Microbe express
- Ribo-Zero rRNA Removal Kit
- exonuclease treatment (mRNA-ONLY)
- size separation by gel electrophoresis.

The presence of both eukaryotic and bacterial mRNAs in environmental samples seems to be a problem. Bacterial and archaeal mRNAs are not polyadenylated in contrast to eukaryotic mRNAs; therefore, methods for specific capture of eukaryotic cDNAs are not applicable.

Sometimes, mRNA amounts are very low, and at the next step, in vitro transcription-mediated linear amplification is performed to increase the number of all mRNA molecules.

1.5.2 cDNA synthesis and sequencing of transcripts

The RNA yield from environmental samples is commonly low. To obtain sufficient starting material for sequencing, sometimes additional amplification steps are required (Carvalhais and Schenk, 2013):

- polyadenylation of bacterial RNA (eukaryotic RNA is already polyadenylated);
- reverse transcription to generate double-stranded cDNA;
- removal of polyA tails by enzymatic digestion.

However, reproducibility of cDNA templates synthesized by PCR is poor; this situation results in the selection and therefore amplification of the most abundant transcripts (Subkhankulova *et al.*, 2006; Gadkar *et al.*, 2014). To overcome these problems, whole-transcriptome amplification (WTA) can be used to amplify a limited amount of experimental RNA sufficient for its use in downstream applications. In general, WTA protocols based on application of Phi29 DNA polymerase are technically relatively easy to implement (Gadkar *et al.*, 2014).

In first metatranscriptomic studies, the traditional cDNA plasmid libraries were used as templates for sequencing, but nowadays, high-throughput parallel sequencing is commonly employed (Carvalhais *et al.*, 2012; Carvalhais and Schenk, 2013). The NGS methods and background information on preparation of libraries are described in detail in part 2.2.

1.5.3 Application of metatranscriptomic studies to analysis of biodegradation of pollutants

The main challenge of 'omics techniques is to identify interactions and relations between different microorganisms during biodegradation processes. Although there is ample knowledge about metabolism of individual xenobiotics in individual microbial species in pure culture, the role of microbial consortia in xenobiotic degradation processes in complex environments remains poorly understood (de Menezes *et al.*, 2012). Only a few metatranscriptomic studies have been focused on microbial communities with biodegradation abilities, but there are also some data concerning contaminated soil or activated sludge from wastewater.

de Menezes *et al.* (2012) evaluated the response of a soil microbial community to phenanthrene. After categorization of genes into functional groups, a significant increase was observed in the numbers of transcripts associated with fungal and bacterial metabolism of aromatic compounds in polycyclic aromatic hydrocarbon (PAH)-amended soil. Moreover, transcripts related to protein metabolism, the respiration stress response, membrane transport, and nitrogen metabolism were overexpressed. Among all transcripts associated with degradation of PAHs, up-regulation of dioxygenases was observed (they catalyse the main steps of xenobiotic degradations, acting at the initial stages of phenanthrene metabolism).

In other studies, metatranscriptomics was used to identify initial steps of anaerobic benzene catabolism (Luo *et al.*, 2014). Analysis of non-ribosomal sequences allowed the researchers to identify several genes transcribed in the presence of this aromatic compound and to assign them to particular taxa of bacteria. Anaerobic benzoate degradation genes (*ubiD-like, bzlA, bamC,* and *bamD*) from strict anaerobes (*Clostridiales*) were transcribed only when the culture was contaminated with benzene. When benzoate (a degradation product) appeared in the cultures, relative anaerobes (*Azoarcus* species) were most abundant, and an increase in transcription of relevant genes (*bzdA, bzdN,* and *bzdO*) was observed. These findings indicate that mineralization of benzene starts with its activation by a strict anaerobe belonging to *Peptococcaceae,* involving a carboxylation step to form benzoate.

A completely different approach to the metatranscriptomic research on xenobiotic biodegradation is single-cell metatranscriptome sequencing, which is intended to determine key functions of individual microorganisms. Embree *et al.* (2014) demonstrated the existence of transcripts possibly involved in hexadecane degradation by *Smithella* sp. Metatranscriptomic data revealed the involvement of hypothetical proteins (with no known homology) in the activation of alkanes. Extended studies on other species coexisting in this community revealed not only the presence of other prevalent organisms but also their interactions with *Smithella* sp.

At present, it is difficult to compare the metatranscriptomic data used in biodegradation studies. The main problem is a limited number of data and a small number of environments and processes that have been examined. Moreover, there is also a need to compare the sizes of metatranscriptomic libraries and to use similar sequencing platforms during the experiments. Besides, the samples are often sequenced at different depth or intensity, and different software packages are used for analysis (de Menezes *et al.*, 2012).

1.6 Conclusion

NGS technologies have already been applied to many biological studies including those dealing with microbial degradation or transformation of various anthropogenic pollutants. Metagenomic and metatranscriptomic studies changed our understanding of biodegradation processes in the environment and provide significant and remarkable insights into the molecular biology of the organisms living in contaminated environments. These methods not only enabled identification of novel species involved in decontamination of the environment but also allow for exploration of degradation routes, identification of the most important genes related to degradation of contaminants, and for tracking of the whole processes. So far, many data concerning bacteria have been collected; however, little is known about eukaryotic organisms (e.g. fungi), which also take part in the decontamination processes. This is because isolation of mRNA from environmental samples is still a technical challenge. In conclusion, the goal for researchers in the coming decade is to combine and improve 'meta' techniques to facilitate the development of effective clean-up strategies.

Acknowledgements

The authors would like to thank professor Adam Jaworski (University of Łódź) for valuable suggestions and helpful comments on this chapter.

References

Beazley, M.J., Martinez, R.J., Rajan, S., Powell, J., Piceno, Y.M., Tom, L.M., Andersen, G.L., Hazen, T.C., Van Nostrand, J.D., Zhou, J., Mortazavi, B., and Sobecky, P.A. (2012). Microbial community analysis of a coastal salt marsh affected by the deepwater horizon oil spill. PLoS ONE *7*, e41305.

Cardoso, A.M., Vieira, R.P., Paranhos, R., Clementino, M.M., Albano, R.M., and Martins, O.B. (2011). Hunting for extremophiles in Rio de Janeiro. Front. Microbiol. *2*, 2–100.

Carvalhais, L.C., Dennis, P.G., Tyson, G.W., and Schenk, P.M. (2012). Application of metatranscriptomics to soil environments. J Microbiol. Methods *91*, 246–251.

Carvalhais, L.C., and Schenk, P.M. (2013). Sample processing and cDNA preparation for microbial metatranscriptomics in complex soil communities. Methods Enzymol. *531*, 251–267.

Cowan, D.A., Ramond, J.B., Makhalanyane, T.P., and De Maayer, P. (2015). Metagenomics of extreme environments. Curr. Opin. Microbiol. *25*, 97–102.

de Menezes, A., Clipson, N., and Doyle, E. (2012). Comparative metatranscriptomics reveals widespread community responses during phenanthrene degradation in soil. Environ. Microbiol. *14*, 2577–2588.

Desai, C., Pathak, H., and Madamwar, D. (2010). Advances in molecular and '-omics' technologies to gauge microbial communities and bioremediation at xenobiotic/anthropogen contaminated sites. Bioresour. Technol. *101*, 1558–1569.

Embree, M., Nagarajan, H., Movahedi, N., Chitsaz, H., and Zengler, K. (2014). Single-cell genome and metatranscriptome sequencing reveal metabolic interactions of an alkane-degrading methanogenic community. ISME J. *8*, 757–767.

Fang, H., Cai, L., Yu, Y., and Zhang, T. (2013). Metagenomic analysis reveals the prevalence of biodegradation genes for organic pollutants in activated sludge. Bioresour. Technol. *129*, 209–218.

Fang, H., Cai, L., Yang, Y., Ju, F., Li, X., Yu, Y., and Zhang, T. (2014). Metagenomic analysis reveals potential biodegradation pathways of persistent pesticides in freshwater and marine sediments. Sci. Total Environ. *470–471*, 983–992.

Gadkar, V.J., Arseneault, T., and Filion, M. (2014). Fidelity and representativeness of two isothermal multiple displacement amplification systems to preamplify limiting amounts of total RNA. Mol. Biotechnol. *56*, 377–385.

Glenn, T.C. (2011). Field guide to next-generation DNA sequencers. Mol. Ecol. Resour. *11*, 759–769.

Godheja, J., Shekhar, S.K., and Modi, D.R. (2014). Advances in molecular biology approaches to gauge microbial communities and bioremediation at contaminated sites. Int. J. Environ. Bioremed. Biodegrad. *2*, 167–177.

He, S.M., Wurtzel, O., Singh, K., Froula, J.L., Yilmaz, S., Tringe, S.G., Wang, Z., Chen, F., Lindquist, E.A., Sorek, R., and Hugenholtz, P. (2010). Validation of two ribosomal RNA removal methods for microbial metatranscriptomics. Nat. Methods *7*, 807–812.

Highlander, S.K. (2012). High throughput sequencing methods for microbiome profiling: application to food animal systems. Anim. Health Res. Rev. *13*, 40–53.

Huson, D.H., Auch, A.F., Qi, J., and Schuster, S.C. (2007). MEGAN analysis of metagenomic data. Genome Res. *17*, 377–386.

Knief, C. (2014). Analysis of plant microbe interactions in the era of next generation sequencing technologies. Front. Plant Sci. *5*, Article 216.

Kuske, Ch.., Hesse, C.N., Challacombe, J.F., Cullen, D., Herr, J.R., Mueller, R.C., Tsang, A., and Vilgalys, R. (2015). Prospects and challenges for fungal metatranscriptomics of complex communities. Fungal Ecol. *14*, 133–137.

Langille, M.G., Zaneveld, J., Caporaso, J.G., McDonald, D., Knights, D., Reyes, J.A., Clemente, J.C., Burkepile, D.E., Vega Thurber, R.L., Knight, R., Beiko, R.G., and Huttenhower, C. (2013). Predictive functional profiling of microbial communities using 16S rRNA marker gene sequences. Nat. Biotechnol. *31*, 814–821.

Liu, L., Li, Y., Li, S., Hu, N., He, Y., Pong, R., Lin, D., Lu, L., and Law, M. (2012). Comparison of next-generation sequencing systems. J. Biomed. Biotechnol. *2012*, 1–11.

Luo, F., Gitiafroz, R., Devine, C.E., Gong, Y., Hug, L.A., Raskin, L., and Edwards, E.A. (2014). Metatranscriptome of an anaerobic benzene-degrading, nitrate-reducing enrichment culture reveals involvement of carboxylation in benzene ring activation. Appl. Environ. Microbiol. *80*, 4095–4107.

Mardis, E.R. (2013). Next-generation sequencing platforms. Annu. Rev. Anal. Chem. *6*, 287–303.

Mason, O.U., Hazen, T.C., Borglin, S., Chain, P.S., Dubinsky, E.A., Fortney, J.L., Han, J., Holman, H.Y., Hultman, J., Lamendella, R., et al. (2012). Metagenome, metatranscriptome and single-cell sequencing reveal microbial response to Deepwater Horizon oil spill. ISME J. *6*, 1715–1727.

Mayo, B., Rachid, C.T., Alegría, A., Leite, A.M., Peixoto, R.S., and Delgado, S. (2014). Impact of next generation sequencing techniques in food microbiology. Curr. Genomics. *15*, 293–309.

Meiring, T., Bauer, R., Scheepers, I., Ohlhoff, C., Tuffin, M., and Cowan, D. (2011). Metagenomics and beyond: current approaches and integration with complementary technologies. Ch. In Metagenomics: Current approaches and integration with complementary technologies. Marco, D., ed. Caister Academic Press, pp. 1–26.

Metzker, M.L. (2010). Sequencing technologies – the next generation. Nat. Rev. Genet. *11*, 31–46.

Morey, M., Fernández-Marmiesse, A., Castiñeiras, D., Fraga, J.M., Couce, M.L., and Cocho, J.A. (2013). A glimpse into past, present, and future DNA sequencing. Mol. Genet. Metab. *110*, 3–24.

Pareek, C.S., Smoczynski, R., and Tretyn, A. (2011). Sequencing technologies and genome sequencing. J. Appl. Genet. *52*, 13–35.

Poretsky, R.S., Bano, N., Buchan, A., LeCleir, G., Kleikemper, J., Pickering, M., Pate, W.M., Moran, M.A., and Hollibaugh, J.T. (2005). Analysis of microbial gene transcripts in environmental samples. Appl. Environ. Microbiol. *71*, 4121–4126.

Shokralla, S., Spall, J.L., Gibson, J.F., and Hajibabaei, M. (2012). Next-generation sequencing technologies for environmental DNA research. Mol. Ecol. *21*, 1794–1805.

Silva, C.C., Hayden, H., Sawbridge, T., Mele, P., Kruger, R.H., Rodrigues, M.V., Costa, G.G., Vidal, R.O., Sousa, M.P., Torres, A.P., Santiago, V.M., and Oliveira, V.M. (2012). Phylogenetic and functional diversity of metagenomic libraries of phenol degrading sludge from petroleum refinery wastewater treatment system. AMB Express. 2, 18.

Silva, C.C., Hayden, H., Sawbridge, T., Mele, P., De Paula, S.O., Silva, L.C., Vidigal, P.M., Vicentini, R., Sousa, M.P., Torres, A.P., Santiago, V.M., and Oliveira, V.M. (2013). Identification of genes and pathways related to phenol degradation in metagenomic libraries from petroleum refinery wastewater. PLoS ONE *18*, e61811.

Subkhankulova, T., and Livesey, F.J. (2006). Comparative evaluation of linear and exponential amplification techniques for expression profiling at the single-cell level. Genome Biol. *7*, R18.

Suenaga, H., Ohnuki, T., and Miyazaki, K. (2007). Functional screening of a metagenomic library for genes involved in microbial degradation of aromatic compounds. Environ. Microbiol. *9*, 2289–2297.

Tan, B., Ng, C., Nshimyimana, J.P., Loh, L.L., Gin, K.Y., and Thompson, J.R. (2015). Next-generation sequencing (NGS) for assessment of microbial water quality: current progress, challenges, and future opportunities. Front. Microbiol. *6*, Article 1027.

Voelkerding, K.V., Dames, S.A., and Durtschi, J.D. (2009). Next-generation sequencing: from basic research to diagnostics. Clin. Chem. *55*, 641–658.

Volokhov, D.V., Simonyan, V., Davidson, M.K., and Chizhikov, V.E. (2012). RNA polymerase beta subunit (rpoB) gene and the 16S-23S rRNA intergenic transcribed spacer region (ITS) as complementary

molecular markers in addition to the 16S rRNA gene for phylogenetic analysis and identification of the species of the family Mycoplasmataceae. Mol. Phylogenet. Evol. *62*, 515–528.
Wang, Z., Gerstein, M., and Snyder, M. (2009). RNA-Seq: a revolutionary tool for transcriptomics. Nat. Rev. *10*, 57–63.
Yadav, T.C., Pal, R.R., Shastri, S., Jadeja, N.B., and Kapley, A. (2015). Comparative metagenomics demonstrating different degradative capacity of activated biomass treating hydrocarbon contaminated wastewater. Bioresour. Technol. *188*, 24–32.
Zhang, J., Chiodini, R., Badr, A., and Zhang, G. (2011). The impact of next-generation sequencing on genomics. J. Genet. Genom. *38*, 95–109.

Heavy Metals Resistance, Metabolism and Transformation – Genomic, Metagenomic and Metatranscriptomic Studies

2

Łukasz Dziewit and Łukasz Drewniak

Contents

Abstract

Heavy metal-contaminated regions are common across the planet. The treatment of such regions (especially those co-contaminated with other toxic substances), remains one of the most costly environmental challenges currently faced by many countries. It appears that the employment of microorganisms in biotransformation of heavy metals is a good solution to this problem, however a key step in the waste management technologies is the understanding of the mechanism of heavy metals resistance, metabolism and transformation. Recent advances in genomics, metagenomics and metatranscriptomics increased their role in identifying traits that may maximize the benefits of heavy metals bioremediation technologies.

2.1 Introduction

Microbial communities are often the main driving force of various processes in many ecosystems, as they include primary producers, consumers, and decomposers (Guerrero *et al.*, 2002). Most of bacteria and fungi found in the environment primarily use easily assimilable nutrients, but under extreme conditions, many of them are capable of utilizing various toxic compounds, including heavy metals or metalloids. Microorganisms can transform heavy metals (e.g. by oxidation, reduction, methylation, or complexation) and use them, e.g. as sources of energy, terminal electron acceptors and structural elements of enzymes, or they can simply remove metal ions from their cells, using various detoxification mechanisms. The knowledge about the physiology and ecology of microorganisms involved in heavy metal transformations sheds light on their role in the ecosystems (including the phenomenon of so-called self-purification) and biogeochemical cycles of metals and metalloids. The recent genomic, metagenomic and metatranscriptomic studies provided data on the microbial mechanisms of transformation of heavy metals, thus enabling practical application of such processes in bioremediation.

In this chapter, the physiological and genetic background of the mechanisms of microbial heavy metal resistance, metabolism and transformations is described. The results of several genomic, metagenomic and metatranscriptomic studies of heavy metal transforming microbes, which found application in bioremediation technologies, are also presented.

2.2 Mechanisms of microbial resistance to heavy metals

Heavy metals are toxic to microbial cells even at low concentrations; however, some microorganisms developed sufficient resistance mechanisms. These heavy metal resistant microorganisms attract a great interest of scientists because of their potential biotechnological application in biomining of expensive and precious heavy metals, and bioremediation of heavy metal-contaminated sites. Therefore, a detailed analysis of the resistance mechanisms is important for understanding the biology of heavy metal resistant strains. To date, three main types of heavy metal resistance mechanisms have been described, i.e. (i) efflux of toxic ions, (ii) enzymatic transformations (oxidation, reduction, methylation and demethylation) and (iii) incorporation or precipitation of heavy metals as complexes by metal-binding proteins (Silver and Phung le, 2005).

Efflux systems constitute the largest group of metal resistance systems. Active transport mechanisms are used by the microorganisms to rapidly export toxic metals from the cytoplasm, and the process can be non-ATPase or ATPase-linked. There are three main types of proteins involved in the export of heavy metals from the cell, i.e. (i) resistance-nodulation-cell division proteins, belonging to RND superfamily, (ii) cation diffusion facilitators (CDF family), and (iii) P-type ATPases (Nies, 2003; Silver and Phung le, 2005).

The RND protein superfamily members in *Bacteria* and *Archaea* are involved in the transport of heavy metals, hydrophobic compounds, amphiphiles, nodulation factors, and proteins. In the analysed bacterial genomes, the gene encoding bacterial RND protein is usually accompanied by two other genes that encode a member of the membrane fusion protein family MFP (periplasmic adapter protein family) and a third protein that belongs to the family of outer membrane factors (OMFs). MFPs and OMFs enhance the function of RND proteins. Together they form an efficient efflux system capable of exporting the appropriate substrate from the cytoplasm (the cytoplasmic membrane or the periplasm) directly outside the cell (Nies, 2003). Examples of such efflux complexes are the following systems: CzcCBA – mediating resistance to Co(II), Zn(II) and Cd(II) (Fig. 2.1); CnrCBA – mediating resistance to Co(II) and Ni(II); NccCBA – mediating resistance to Cd(II), Co(II) and Ni(II); CzrCBA – mediating resistance to Cd(II) and Zn(II); CusCFBA – mediating resistance to Cu(I) and Ag(I), SilCFBA – mediating resistance to Ag(I) (Nies, 2003; Silver and Phung le, 2005).

The cation diffusion facilitator family (CDF) groups the single polypeptide chemiosmotic efflux systems, whose primary substrates are Zn(II), Co(II), Ni(II), Cd(II), and Fe(II). CDF-mediated transport is driven by a concentration gradient, a chemiosmotic gradient, ΔpH or a potassium gradient. The identified CDF-like efflux systems include: CzcD – mediating small degree resistance to Co(II), Cd(II) and Zn(II) (Fig. 2.1); ZitB and ZntA – mediating resistance to Zn(II); and FieF – a ferrous iron efflux system (Nies, 2003).

The third type of proteins involved in the heavy metals export, P-type ATPases (P-ATPases), is in fact a basic defence mechanism. P-type ATPases are mostly single polypeptides forming a phosphorylated intermediate state during the ion transport cycle (Silver and Phung le, 2005). The prominent members of the P-type ATPases involved in heavy metal export are: CadA – Cd(II) resistance-mediating protein, ZntA – conferring resistance to Zn(II) and Pb(II) (Fig. 2.1), CopA – conferring resistance to Cu(I), PbrA – Pb(II) resistance-mediating protein.

Heavy metal resistance may also involve enzymatic transformations of toxic ions. A model example of an enzymatic detoxification system in microorganisms is mercury resistance mechanism. Bacteria resistant to Hg(II) usually contain the *mer*

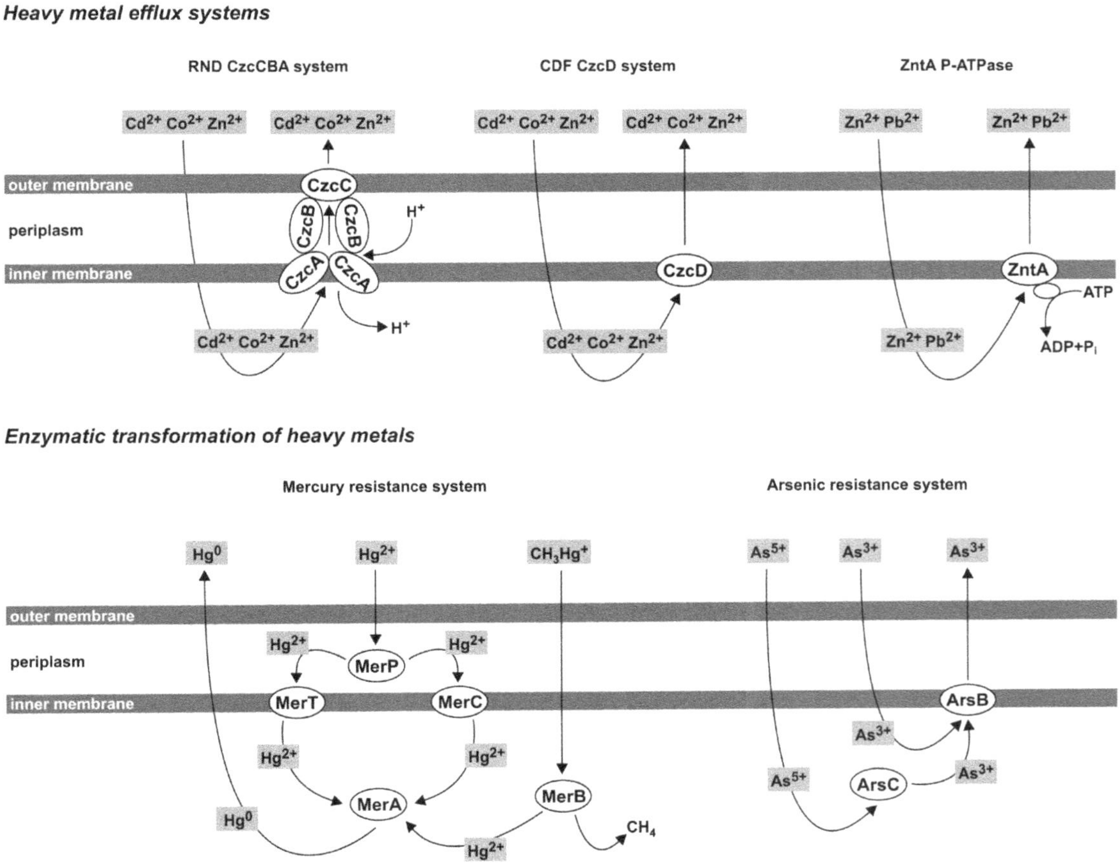

Figure 2.1 Schematic representation of various heavy metals resistance mechanisms.

operon in their genome, which may be composed of the following genes: (i) *merA* coding for mercuric reductase, (ii) *merB* encoding organomercurial lyase, (iii-v) *merC, merP* and *merT* encoding transport proteins and (vi-vii) *merR* and *merD* coding for regulatory proteins. In the process of Hg(II) detoxification, the periplasmic binding proteins (MerC, MerP and MerT) transport the toxic ions into the cytoplasm, where they are delivered to the NADPH-dependent mercuric reductase, which catalyses the reduction of Hg(II) to volatile, less-toxic Hg^0. Additionally, in some cases (e.g. a broad-spectrum mercury resistance) MerB lyase may confer resistance to organomercurials (e.g. methylmercury and phenylmercury) by catalysing the hydrolysis of the C–Hg bond before Hg(II) reduction (Nascimento and Chartone-Souza, 2003) (Fig. 2.1).

Another well-described enzymatic detoxification system mediates As(V) resistance. A crucial element of this system is ArsC – arsenate reductase involved in intracellular reduction of As(V) to the As(III) form. The produced As(III) is then removed from the cell using ArsB efflux pump encoded within the *ars* operon (Fig. 2.1). It was shown that the *ars* operons are common within the bacterial genomes (Bruins *et al.*, 2000).

The last of the described heavy metal resistance mechanism involves incorporation or precipitation of heavy metals as complexes by metal-binding proteins. An intracellular sequestration is in fact a form of protection of the cellular components against the toxic effect of heavy metal ions [usually Cd(II), Cu(II) and Zn(II)] by their accumulation in complexated form within the cytoplasm. The metal-binding proteins involved in this process – metallothioneins – are small, cysteine-rich proteins, whose best described representative is SmtA of *Synechococcus* sp. PCC 7942 that binds to Cd(II) and Zn(II) (Blindauer, 2011).

The progress in genomics and metagenomics resulted in identification of thousands of the heavy metal resistance genes. However, information on these genes, including their sequences and molecular functions, was rather scattered. Therefore, BacMet (http://bacmet.biomedicine.gu.se) – a manually curated database of antibacterial biocide- and metal-resistance genes was created. It contains the experimentally verified resistance genes, as well as putative resistance genes, which are organized according to their molecular function and the resistance phenotypes they encode (Pal *et al.*, 2014). The database became a valuable source of information for comparative genomic and metagenomic analyses.

2.3 Mechanisms of dissemination of heavy metals resistance and metabolism genes

Most of the contaminated environments undergo rapid biogeochemical changes and therefore bacteria inhabiting such ecological niches had to develop mechanisms enabling their adaptation to the changing environmental conditions. A major factor contributing to bacterial diversification and adaptation is the horizontal gene transfer (HGT) phenomenon, in which mobile genetic elements (MGEs) play a key role (Heuer and Smalla, 2012). There are two main groups of mobile genetic elements contributing to the dissemination of heavy metals resistance and metabolism genes in all kinds of environments, i.e. plasmids and transposable elements.

Plasmids are self-replicating, extra chromosomal MGEs, which are key vectors of DNA exchange among bacteria. They have a modular structure, and their core region (plasmid backbone) is composed of a set of conserved modules coding for replication, stability and conjugal transfer functions, which are crucial for plasmid maintenance and spread. Many plasmids also carry accessory genes determining various phenotypic traits (Siefert, 2009; Thomas, 2000). Plasmids are not crucial for the viability of the host, but as they carry a variable assortment of accessory genes (which make them an important adaptive force), they often contribute to their hosts' phenotypic diversity.

The analysis of plasmids brings valuable information concerning adaptation of bacteria to the changing environmental conditions. Several types of phenotypic modules conferring various phenotypes were identified within plasmids, including: (i) antibiotics and heavy metals resistance – resistance plasmids (e.g. Rahube *et al.*, 2014), (ii) utilization of various (including toxic) compounds – catabolic plasmids (e.g. Trefault *et al.*, 2004), (iii) virulence – virulence plasmids (e.g. Uchiya *et al.*, 2015), (iv) bacteriocins and antibiotics production (e.g. Kirby *et al.*, 1975) and (v) interactions with plants (e.g. Brom *et al.*, 2002).

Among the phenotypic modules most frequently identified within plasmids are those responsible for heavy metals resistance and metabolism. It was shown that within the bacterial plasmids genes conferring resistance to most (if not all) heavy metals can be found, basically irrespective of the analysed environment. This indicates that such genes, though still responsible for the selective pressure in the contaminated habitats, are now also a common and important component of the global mobilome in all the analysed ecological niches. In that form, as plasmid-carried genetic modules, they can be considered as a reservoir of adaptive features, which can be easily gained and, when necessary, used by the bacterial host to overcome the toxic conditions. Examples of plasmids carrying heavy metals resistance and metabolism genes are presented in Table 2.1.

It is also worth mentioning that in the process of dissemination of heavy metals resistance and metabolism genes in a given environment an essential role is played by broad host range and/or conjugal plasmids, which can be transferred from one host to another via conjugation. As their replication systems are functional in a wide range of bacteria they can easily cross the taxonomic boundaries, e.g. the plasmid pSinA of *Sinorhizobium* sp. M14 carrying genes that confer resistance to arsenic, cobalt, zinc, cadmium and mercury, as well as genetic modules responsible for arsenite oxidation and transport of phosphates and (probably) arsenates (Drewniak *et al.*, 2013).

The second group of the MGEs, significantly contributing in dissemination of heavy metals resistance and metabolism genes are transposable elements (TEs), mainly transposons. Transposons

Table 2.1 Examples of plasmids carrying heavy metals resistance and metabolism genes

Plasmid name	Plasmid host	Place of the bacterium isolation	Heavy metal resistance and metabolism genes identified within plasmid	Reference
pBSC2–1	*Methylocystis* sp. SC2	Highly polluted aquifer, Germany	*mer* – mercury resistance *czc* – cadmium, zinc and cobalt resistance copper transport genes	Dam *et al.* (2012)
pGLAAG01	*Glaciecola* sp. 4H-3–7+YE-5	Suruga Bay sediments, Japan	*mer* – mercury resistance *cnr* – nickel and cobalt resistance *czc* – cadmium, zinc and cobalt resistance *cop* – copper resistance *znt* – copper resistance	Klippel *et al.* (2011)
pR55	*Klebsiella pneumoniae*	hospital, France	*mer* – mercury resistance	Doublet *et al.* (2012)
pSinA	*Sinorhizobium* sp. M14	arsenic-contaminated gold mine, Poland	*aio* – arsenite oxidation *ars* – arsenic resistance *mer* – mercury resistance *czc* – cadmium, zinc and cobalt resistance	Drewniak *et al.* (2013)
pUR2940	*Staphylococcus aureus*	hospital, Spain	*cad* – cadmium resistance *cop* – copper resistance	Gomez-Sanz *et al.* (2013)

are mobilized to change their location in DNA by the activity of a transposase, an enzyme which catalyses transposition, i.e. recombination that does not require sequence homology between the TE and the target site (Mahillon and Chandler, 1998).

Two groups of transposons, composite and non-composite, can be distinguished. Composite transposons consist of random segments of genomic DNA (core region) flanked by a pair of simple, cryptic TEs – insertion sequences or other transposons. Within the central region various phenotypic modules can be found, e.g. genes conferring resistance to antibiotics and heavy metals. The mechanism of composite transposons formation plays a very important role in HGT, as this is in fact a natural way of mobilizing random fragments of DNA to transposition, which significantly enriches the bacterial mobilome and shapes bacterial genomes (Inui *et al.*, 2005; Merlin *et al.*, 2000; Stibitz and Davies, 1987; van der Meer *et al.*, 1991). Naturally, composite transposons are involved in dissemination of heavy metals resistance and metabolism genes, e.g. 65-kb Tn*5385* carrying genes for resistance to mercury, and additionally to several antibiotics, including penicillin, erythromycin, gentamycin, streptomycin and tetracycline (Bonafede *et al.*, 1997).

The second group of transposons, non-composite transposons, encode their own transposases, which interact with highly conserved inverted sequences flanking the transposon. These transposons often encode functions beneficial for their hosts in a particular ecological niche, e.g. Tn*AO22a* conferring resistance to mercury, present within a broad host range plasmid pLM16A1 of *Achromobacter* sp. LM16 isolated, from a copper mine in Poland (Dziewit *et al.*, 2015).

An interesting, recently observed and thoroughly investigated, phenomenon is co-localization of antibiotic and heavy metals resistance genes within the same mobile genetic element, e.g. the aforementioned transposon Tn*5385* and plasmid pUR2940. This has a strong impact on the overall dissemination of the resistance traits in an environment, since it drives the co-selection process, i.e. selection for one phenotype inadvertently selects for its unintended companion (Gaze *et al.*, 2013).

The molecular and functional characterization of the heavy metals resistance and metabolism genes, as well as genomic analysis of the genetic vectors (especially plasmids) enabling their spread may become a starting point for a development of novel bioremediation technologies. Such strategies may involve the bioaugmentation of the indigenous microflora in the contaminated sites with strains carrying natural and/or genetically modified plasmids, which carry appropriate genes conferring particular metal transformation, e.g. pSinA plasmid and pAIO1/pARS1 system providing their bacterial hosts arsenic resistance and arsenite oxidation abilities (Drewniak *et al.*, 2015; Drewniak *et al.*, 2013).

2.4 Genomics of metallotolerant bacteria

Over the past 20 years the progress in bacterial genomics has changed our understanding of microbial species and our approach to all fields of microbiology. Since the first report on the complete genome sequence of *Haemophilus influenzae* in 1995 (Fleischmann *et al.*, 1995), over 4000 complete bacterial genome sequences have been obtained and another 40,000 genome projects are in progress (however most of them will probably never reach the complete genome status and they will remain in the form of draft genomes). As sequence data began to accumulate, it quickly became clear that their detailed analysis and interpretation is impossible. For this reason genomics shortly became a fast screening technique for identification and basic *in silico* analysis of microorganisms, mainly pathogenic strains or those with a significant application potential, for example in bioremediation.

Analysis of the microbial genomes revealed that almost all microorganisms carry genes involved in heavy metal resistance and/or metabolism, which is understandable as several heavy metal ions, e.g. Cu^{2+} or Zn^{2+} are co-factors of important enzymes. However, there are some microorganisms whose genomes are particularly rich in such genes. One of the best examples of heavy metals hypertolerant bacteria is *Cupriavidus metallidurans* CH34, which is considered a model organism in this field of environmental microbiology. This strain is resistant to 20 different heavy metal and metalloid ions [Ag(I), As(III), As(V), Au(I), Au(III), Bi(III), Cd(II), Co(II), Cr(VI), Cs(I), Cu(I), Cu(II), Hg(II), Ni(II), Pb(II), Se(IV), Se(VI), Sr(II), Tl(I), and Zn(II)] and its remarkable phenotype is determined by genes forming at least 25 clusters. Many of these genes are located within two plasmids, pMOL28 and pMOL30 (Janssen *et al.*, 2010). A unique genomic content and interesting physiological properties made the strain CH34, and its genetically modified derivatives, useful in several bioremediation technologies, e.g. a broad-spectrum mercury-resistant strain MSR33, capable of removing mercury and other heavy metals from polluted waters, was created by introducing the plasmid pTP6 into *C. metallidurans* CH34 (Rojas *et al.*, 2011).

The applicability of genomics is well visible in various heavy metal-removal biotechnologies, especially at the designing stage, when the appropriate microorganisms are selected. For example, several genomes of 'arsenic' bacteria have recently been sequenced, including: *Thiomonas* sp. 3As, *Achromobacter arsenitoxydans* SY8, *Halomonas* sp. HAL1 and *Acidovorax* sp. NO1, *Ochrobactrum tritici* SCII24^T and the analyses of their genomes were helpful in the exploration of the metal resistance and metabolism mechanisms. Moreover, some of them, after being subjected to targeted genomic manipulation, gained new features useful in bioremediation, e.g. *O. tritici* SCII24^T mutants became hyper-accumulators of arsenic (Sousa *et al.*, 2015). Such bioengineered strains show a great potential for application in various novel bioremediation technologies.

Although the progress in genomics enabled the overall genome-scale metabolic analyses, also in respect to heavy metals metabolism, the use of 'omics' data to generate predictive models is still in its infancy and is limited by many factors. On the other hand, this is a good starting point for new discoveries in the area of systems biology, which may become a milestone in modern bioremediation strategies using microorganisms.

2.5 Heavy metals transformation in the light of an overall microbial metabolism and potential application in bioremediation

Microbial interactions with heavy metals are determined by various environmental factors (such as concentration, speciation and mobility of metals, pH, Eh, temperature, organic content, community structure, etc.) and are dependent on the physiology of cells and bacterial needs (Gadd, 2010). Many metals and metalloids that exist in numerous oxidation states (such as arsenic, iron, manganese) may act as electron carriers during redox reactions of electron transport chains to generate chemical energy for bacterial cells. Some heavy metals (e.g. cobalt, copper, iron or manganese) may serve as micronutrients, which act as redox centres for metalloproteins (such as cytochromes, blue copper proteins, and iron–sulfur proteins) or function as cofactors of various proteins (Grey and Ellis Jr., 2006). Apart from direct involvement of heavy metals in microbial metabolism, there are also physiological processes, which indirectly affect metal transformations. Examples are processes leading to the precipitation of heavy metals by increasing the pH (e.g. nitrification, ammonium production) and/or production of sulfides, phosphates or other precipitation agents. All of the aforementioned transformations play an important role in biogeochemical cycles of most heavy metals and may also find application in bioremediation of contaminated environments, as they may lead to mobilization or immobilization of metals (Fig. 2.2).

2.5.1 Oxidation of metals and metalloids

Metals and metalloids can be used as sources of energy in electron donating reactions by numerous chemolithoautotrophic bacteria. Oxidation of the reduced form of an inorganic compound is coupled with the reduction of oxygen and carbon dioxide fixation. Energy for the assimilation of CO_2 and cell growth may also be obtained from oxidation of heavy metals under anoxic conditions, where an oxyanion such as nitrate or chlorate is used as the final electron acceptor (Rhine *et al.*, 2006; Sun *et al.*, 2009).

Ferrous iron is an example of metal that is oxidized in many acidic environments by microbial communities dominated by the representatives of the genera *Acidithiobacillus* and *Leptospirillum* (Johnson and Hallberg, 2003). The activity of iron-oxidizing microorganisms leads to the production of ferric iron, a strong oxidizing agent that contributes to the mobilization (leaching) of other heavy metals co-occurring with Fe-S minerals. For this reason, autotrophic iron oxidation is one of the key process in bioremediation of metal-contaminated solid materials, including soil (Zagury *et al.*, 1994).

Another group of chemolithoautotrophic microorganisms, which play an important role in environmental pollution and bioremediation are arsenite-oxidizing bacteria. They promote the dissolution of arsenic-bearing minerals and release arsenic into the aqueous phase (Drewniak and Sklodowska, 2013). In turn, microbial oxidation of soluble As(III), coupled with the respiratory electron transport chain, leads to the formation of less toxic form, As(V), which can be removed from the environment by standard physical and chemical methods (e.g. sorption, coagulation or reverse osmosis).

Microbial oxidation process that leads to a direct removal of heavy metal ions from aqueous environments is carried out by manganese oxidizing bacteria. Several studies have reported that

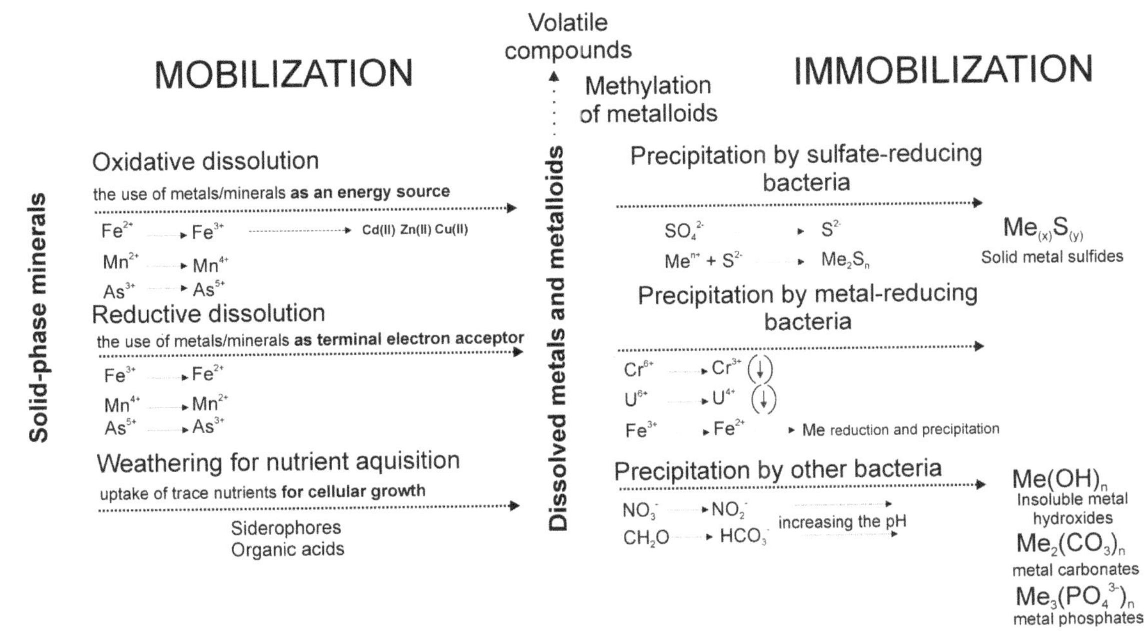

Figure 2.2 Microbiological transformations of metals in the light of the overall metabolism and potential application in bioremediation.

bacterial oxidation of Mn(II) leads to the formation of Mn(III) and Mn(IV) oxides, which are then deposited on the surface of cells (Tebo *et al.*, 2005). Interestingly, this process can be carried out non-enzymatically, as a result of non–specific interactions with cellular or extracellular products, or may be catalysed by multicopper oxidases (MCOs) (Tebo *et al.*, 2005). Bacterial MCOs have a wide range of substrate specificity, including oxidation of Cu(I) to Cu(II) (Singh *et al.*, 2004) or Fe(II) to Fe(III) (Huston *et al.*, 2002).

2.5.2 Reduction of metals and metalloids

Under reducing conditions, many microorganisms can utilize metals (e.g. iron, manganese) or metalloids (e.g. arsenite, selenite) as terminal electron acceptors in respiration processes. In such processes, the reduction of metals/metalloids is accompanied by oxidation of organic or inorganic substrates and the synthesis of ATP. Reduction of metals may also proceed indirectly, e.g. in processes catalysed by biogenic Fe(II) or with the use of an enzyme operating under aerobic conditions [e.g. NAD(P)H-dependent reductases]. Regardless of the mechanisms, the reduction process may occur in the aqueous phase or may be carried on the surface of minerals, and, depending on the compound, can lead to the mobilization or immobilization of heavy metals. (Gadd, 2010).

Dissimilatory iron reduction is one of the most common microbial respiration process in soil, sediments and subsurface environments (Lloyd, 2003). Various representatives of *Archaea* [e.g. *Pyrobaculum aerophilum*, *Archaeoglobus fulgidus* (Vargas et al., 1998)] and *Bacteria* [e.g. *Geobacter metallireducens*, *Shewanella oneidensis* (Lovley, 2002), *Wolinella succinogenes* (Lovley and Phillips, 1988)] are capable of reducing insoluble Fe(III) oxides and mobilize Fe(II). The reductive dissolution of iron oxides may also result in the release of other heavy metals that are bound to the Fe(III) minerals (Lloyd, 2003). Interestingly, under favourable conditions, iron-reducing bacteria can also be involved in the formation of reduced minerals such as magnetite (Fe_3O_4) or siderite ($FeCO_3$) (Zavarzina, 2004; Chistyakova *et al.*, 2008). Regardless of the final product, the mechanism of Fe(III) reduction may rely on direct contact with the surface of the cell or may be carried out via an extracellular electron shuttle (Lloyd, 2003).

Microbial reduction of Mn(IV) proceeds according to a similar mechanism as Fe(III) respiration, and is very often carried out by the same strains, as in the case of *Shewanella oneidensis* (Lovley *et al.*, 1989) or *Geobacter metallireducens* (Lovley and Phillips, 1988). Furthermore, as in the case of dissimilatory iron reduction, manganese respiration may contribute to the mobilization of potentially toxic levels of Mn(II) and co-occurring trace metals or may lead to its immobilization by the formation of secondary minerals such as rhodochrosite ($MnCO_3$) (Lee *et al.*, 2011).

Transformation of Cr(VI) into Cr(III) is carried out by a wide range of bacteria, in particular by the representatives of *Pseudomonas*, *Aeromonas* and *Shewanella* species (Wang, 2000). Many of these organisms are obligate anaerobes and catalyse the reduction of Cr(VI) via electron transport systems containing cytochromes (Tebo and Obraztsova, 1998). In turn, NADH and endogenous cell reserves are used as electron donors for Cr(VI) reduction by bacteria that prefer aerobic conditions (Park *et al.*, 2000). Reduction of Cr(VI) may also be catalysed indirectly by biogenic Fe(II), resulting from the activity of iron-reducing bacteria (Bishop *et al.*, 2014) or by non-specific reactions catalysed by intermediate organic compounds, such as amino acids or organic acids (Myers *et al.*, 2000; Robins *et al.*, 2013).

The enzymatic reduction of U(VI) to U(IV) mainly results from the activity of iron reducing bacteria (e.g. *Shewanella* spp. and *Geobacter* spp.), which can use uranium as a terminal electron acceptor in anaerobic respiration (Lovley *et al.*, 1991). However, it is worth mentioning that U(VI) can be reduced enzymatically by other (not iron-reducing) bacteria (e.g. *Clostridium* spp. and *Desulfovibrio desulfuricans*) that are unable to conserve energy for growth via this transformation (Lovley and Phillips, 1992; Francis *et al.*, 1994).

Not only metals may be utilized in the dissimilatory reduction processes, but also metalloid oxyanions such as arsenate and selenate, which are alternative electron acceptors for microbial respiration in many anoxic environments. Anaerobic transformation of Se(VI) and As(V) by processes that support growth was described for numerous bacterial strains isolated from various contaminated environments (Paez-Espino *et al.*, 2009; Slyemi and Bonnefoy 2012; Nancharaiah and Lens 2015). The reduction of SeO_4^{2-} results in the formation of a red precipitate of elemental selenium (Se^0), while the reduction of AsO_4^{3-} leads to the production of more toxic and mobile AsO_3^{3-} oxyanions (Stolz and Oremland, 1999; Oremland and Stolz, 2003).

2.5.3 Biologically induced mineralization

Microbial metabolic processes may contribute to mineral precipitation, and thus to metal immobilization. In most cases, precipitation is not directly controlled by the organism, but occurs as a response to the interactions between metal (or metalloid) and metabolic exudates from the organism. The prime example is sulfate-reducing bacteria. Under reducing conditions, these microorganisms produce sulfides, which can react with aqueous metals (e.g. Cd or Zn) and precipitate secondary sulfide minerals (e.g. CdS or ZnS). In contrast, iron-reducing bacteria promote metal precipitation by the production of (bio)genic Fe(II), which can catalyse reduction of U(VI) or Cr(VI).

Dissimilatory sulfate reduction and other anaerobic respiration processes (e.g. iron reduction, nitrate reduction) may also promote carbonate production and its precipitation with metals. The reason for the precipitation of metals (mostly in the form of hydroxides) is also the increase of the pH caused by iron reducing bacteria, sulfate reducing bacteria, denitrifying and nitrifying bacteria (Warren *et al.*, 2001; Perez-Gonzalez *et al.*, 2010).

2.6 Metagenomics and metatranscriptomics of heavy metal-contaminated environments

Microorganisms are treated as key agents in heavy metal transformation in various environments, since they may be responsible for the release and dissemination of many metals and metalloids or may contribute to the (self)purification of contaminated areas. The paths and mechanisms of microbial transformation of heavy metals and the consequences of this interaction (mobilization/immobilization of metals) are closely linked to the microbial response to the environmental conditions. In order to gain an understanding of microbe–metal interactions, numerous investigations of environmental isolates and microbial communities were performed. Physiological and genomic analyses of pure cultures of microorganisms allowed for the determination of the relationship between microorganisms and heavy metals at the cellular level. On the other hand, studies on the physiology and (meta)genomics of the microbial communities allowed for the determination of the nature and extent of the interactions at the ecosystem level.

The first trail demonstrating the effects of the influence of the presence of heavy metals on a microbial ecosystem is the biodiversity of a given community. Currently, the most appropriate way to explore and follow dynamics of the structure of microbial communities is to use of culture-independent metagenomic approach. The use of sequencing of phylogenetic markers such as 16S rRNA, *rpoB* or *recA* genes allows to identify most of the microorganisms that are engaged in biogeochemical cycles and play a crucial role in the overall metabolic fluxes (see also Chapter 3). In the context of microbial communities structure in heavy metal-contaminated environments the most frequently studied are acid mine drainages (AMDs), which are characterized by low pH and high concentration of dissolved metals (Sanchez-Andrea *et al.*, 2011; Hogsden and Harding, 2012; Streten-Joyce *et al.*, 2013; Mendez-Garcia *et al.*, 2014). Biodiversity analyses based on 16S rRNA gene sequencing showed that AMD environments are inhabited mostly by planktonic microorganisms dominated by the following: (i) chemolithotrophic bacteria of the genera *Acidithiobacillus* and *Leptospirillum*; (ii) heterotrophic genus *Acidiphilium*; and (iii) acidophilic, moderately thermophilic (*Thermoplasmatales*) to extremely thermophilic (*Sulfolobales*) *Archaea* (Baker and Banfield, 2003; Johnson and Hallberg, 2003). Metagenomics investigation of a microbial community found in the extreme acidic environment in Rio Tinto (SW Spain) revealed that AMD communities might also be dominated by bacteria representing non-acidophilic genera, such as *Aeromonas* and *Acinetobacter* (Garcia-Moyano *et al.*, 2007). Furthermore, metagenomic sequencing of 16S rRNA genes demonstrated that a significant part of AMD ecosystems are also formed by biofilms, microbial mats or macroscopic microbial growth forms (Jones *et al.*, 2012). In turn, the analyses of 16S rRNA genes showed that environments with a moderate pH exhibit much greater biodiversity compared to locations with an extremely low pH (Baker and Banfield, 2003; Johnson and Hallberg, 2003; Lin *et al.*, 2006).

Clone libraries of 454 pyrosequencing of 16S rRNA gene sequences obtained from environmental metagenomes may also enable tracking of spatial and temporal changes during bioremediation of various heavy metal contaminated sites. An example is the tracking of changes of bacterial community structure during *in situ* uranium bioremediation of contaminated groundwater (Hwang *et al.*, 2009). The studies revealed that the levels of nitrate, uranium, sulfide, sulfate and ethanol were highly correlated with particular bacterial populations' changes.

Prokaryotic community analyses performed by 16S rRNA gene amplicon sequencing may also allow for an indirect estimation of metabolic potential of the community. However, much better for drawing conclusions about physiological capabilities and metal–microorganisms interactions within contaminated ecosystems is the use of functional (PCR-based) target gene sequencing or shotgun metagenome sequencing. Both approaches provide detailed information about the abundance and the diversity of genes that encode specific groups of enzymes involved in adaptive evolution of microorganisms, and enable the assignment of specific gene clusters directly linked to a particular ecological function, even at low abundances within a metagenome.

As mentioned in the previous section ('Heavy metals transformation in the light of an overall microbial metabolism and potential application in bioremediation'), the key microbiological processes affecting the biogeochemical cycles are those that contribute to mobilization and/or immobilization of metals and metalloids. Thus, among the molecular markers that may be used for analyses of microbial communities inhabiting heavy metal contaminated areas are genes encoding proteins involved in broadly defined metal homeostasis that leads to mobilization/immobilization processes. Heavy metal resistance genes are prime targets of metagenomics based on sequence-driven screening, as they inform about the general diversity and physiological condition of the entire community. More precise knowledge about the structure of the microbial community and its interaction with the contaminated environment is delivered via metabolic and/or structural genes. Biofilm formation-related genes provide information about possible heavy metal biosorption, as it is commonly known that extracellular polymeric substances are effective barriers for trapping metals (Cao *et al.*, 2011; Tourney and Ngwenya, 2014). Genes encoding cytochromes and other proteins involved in respiration processes, in which metals [e.g. Fe(III), Mn(IV), Cr(VI)] or metalloids [e.g. As(V) or Se(VI)] are reduced in electron accepting reactions indicate that various mobilization or immobilization processes (depending on the reduced elements) may occur. The presence of genes encoding dissimilatory sulfate reductase indicates that microbial communities may contribute to metal precipitation in the form of sulfides. In turn, the presence of nitrate and nitrite reductases, as well as nitrogenases genes may indicate a potential increase of the pH through ammonia production, which in turn triggers the precipitation of metal hydroxides. Metabolic potential for mineral dissolution in oxidative conditions may be detected by identification of genes involved in CO_2 fixation (e.g. RuBisCO) and electron donating reactions with the use of metals [e.g. Fe(II)] or metalloids [e.g. As(III)]. Metal mobilization processes may also be recognized by the identification of genes coding for proteins involved in heavy metals complexation by organic acids (e.g. citrate synthase, aconitate hydratase) or organic ligands (e.g. siderophores).

Not all of these functions can be analysed by PCR-based sequence driven screening, since this approach requires designing DNA probes and primers derived from the conserved regions of known gene or protein families. Many of the aforementioned functions are catalysed by enzymes with a broad spectrum of activity or by proteins, for which it is hard to find suitable conserved regions. Cytochrome C is a representative of a heterogeneous family of proteins, which are known for their ability to use oxidized species of Fe(III), Cr(VI), and Mn(IV) as terminal electron acceptors, but which are also involved in respiration processes utilizing an enormous number of organic compounds. For the analysis of such ambiguous functions, shotgun metagenome sequencing is a much more practical and suitable approach. It enables the analysis of the entire genetic and functional diversity of communities by a multidimensional bioinformatic approach, combined with environmental data.

High throughput shotgun sequencing also enables the reconstruction of community metabolism, as well as the reconstruction of the entire genomes of dominant species within microbial communities. Reconstruction of near-complete genomes of *Leptospirillum* group II and *Ferroplasma* type II, and partial recovery of three other genomes with the use of random shotgun sequencing of DNA was described by Tyson *et al.* (2004). This was possible because the natural acidophilic biofilm, was dominated by a small number of species and, additionally, the metagenome was characterized by a relatively low frequency of genomic rearrangements. The example of metagenome-based metabolic reconstruction is the shotgun sequencing analysis of the microbial community of uranium-contaminated groundwater (Hemme *et al.*, 2010). High sequence coverage of the metagenome allows for partial reconstruction of the major metabolic pathways, including carbon metabolism, respiration using inorganic compounds and heavy metal detoxification.

Targeted metagenomics provides general information about the genetic diversity and metabolic potential of microbial communities, but it cannot answer the question whether the identified genes are expressed or not. To reveal the overall activity of microbial communities and gene expression profiles of natural microbial assemblages, metatranscriptomic approach should be used. Such techniques involve random sequencing of RNA expressed by the microbial community (Abram, 2015). One of the few examples of coupled analyses of both DNA and RNA sequences from heavy metal contaminated sites is the metagenomic and metatranscriptomic analyses of microbial communities from acid mine drainage (AMD) (Chen *et al.*, 2015). This study is of a comparative character and demonstrates how microbial communities respond and adapt to the extreme physicochemical conditions that occur in AMD environments. The majority of the transcripts were associated with functions that were closely related to the physicochemical characteristics of the AMD systems. One example may be the high level of expression of genes encoding key proteins for ferrous iron oxidation, including rusticyanin of *Acidithiobacillus* spp. and cytochromes of *Leptospirillum* spp., which suggest that Fe(II) oxidation is one of the major pathways for energy generation within the AMD communities.

Metatranscriptomic analyses also have many limitations, such as insufficient transcriptional information for the less active populations. This problem is mainly associated with the ineffective removal of rRNA sequences from the total RNA and inadequately deep sequencing, which at the current development of sequencing technologies is becoming less important.

2.7 Conclusion

Genomic, metagenomic and metatranscriptomic analyses of microbes already enabled a deep insight into mechanisms of heavy metals resistance, metabolism and transformation, which were recognized and thoroughly described. Physiological, and genomic analyses of pure cultures of metallotolerant microorganisms allowed understanding and modelling of metabolic traits linked to overall heavy metals biotransformation and the interaction between microorganisms and heavy metals at the cellular level. On the other hand, metagenomic and metatranscriptomic approaches applied to analyse the various microbial communities from contaminated regions allowed recognizing the nature and extent of the interactions at the ecosystem level. The aim of the future studies is to properly understand and apply the results of studies with various 'omics' technologies in a biotreatment of heavy metal-contaminated sites.

References

Abram, F. (2015). Systems-based approaches to unravel multi-species microbial community functioning. Comput. Struct. Biotechnol. J. *13*, 24–32.

Baker, B.J., and Banfield, J.F. (2003). Microbial communities in acid mine drainage. FEMS Microbiol. Ecol. *44*, 139–152.

Bishop, M.E., Glasser, P., Dong, H.L., Arey, B., and Kovarik, L. (2014). Reduction and immobilization of hexavalent chromium by microbially reduced Fe-bearing clay minerals. Geochim. Cosmochim. Acta *133*, 186–203.

Blindauer, C.A. (2011). Bacterial metallothioneins: past, present, and questions for the future. J. Biol. Inorg. Chem. *16*, 1011–1024.

Bonafede, M.E., Carias, L.L., and Rice, L.B. (1997). Enterococcal transposon Tn*5384*: evolution of a composite transposon through cointegration of enterococcal and staphylococcal plasmids. Antimicrob. Agents Chemother. *41*, 1854–1858.

Brom, S., Girard, L., Garcia-de los Santos, A., Sanjuan-Pinilla, J.M., Olivares, J., and Sanjuan, J. (2002). Conservation of plasmid-encoded traits among bean-nodulating *Rhizobium* species. Appl. Environ. Microbiol. *68*, 2555–2561.

Bruins, M.R., Kapil, S., and Oehme, F.W. (2000). Microbial resistance to metals in the environment. Ecotoxicol. Environ. Saf. *45*, 198–207.

Cao, Y.Y., Wei, X., Cai, P., Huang, Q.Y., Rong, X.M., and Liang, W. (2011). Preferential adsorption of extracellular polymeric substances from bacteria on clay minerals and iron oxide. Colloid Surface B *83*, 122–127.

Chen, L.X., Hu, M., Huang, L.N., Hua, Z.S., Kuang, J.L., Li, S.J., and Shu, W.S. (2015). Comparative metagenomic and metatranscriptomic analyses of microbial communities in acid mine drainage. ISME J. *9*, 1579–1592.

Chistyakova, N.I., Rusakov, V.S., Nazarova, K.A., Koksharov, Y.A., Zavarzina, D.G., and Greneche, J.M. (2008). Iron minerals formed by dissimilatory iron-and sulfur reducing bacteria studied by Mossbauer spectrometry. Hyperfine Interact. *182*, 55–63.

Dam, B., Kube, M., Dam, S., Reinhardt, R., and Liesack, W. (2012). Complete sequence analysis of two methanotroph-specific *repABC*-containing plasmids from *Methylocystis* sp. strain SC2. Appl. Environ. Microbiol. *78*, 4373–4379.

Doublet, B., Boyd, D., Douard, G., Praud, K., Cloeckaert, A., and Mulvey, M.R. (2012). Complete nucleotide sequence of the multidrug resistance IncA/C plasmid pR55 from *Klebsiella pneumoniae* isolated in 1969. J. Antimicrob. Chemother. *67*, 2354–2360.

Drewniak, L., Dziewit, L., Ciezkowska, M., Gawor, J., Gromadka, R., and Sklodowska, A. (2013). Structural and functional genomics of plasmid pSinA of *Sinorhizobium* sp. M14 encoding genes for the arsenite oxidation and arsenic resistance. J. Biotechnol. *164*, 479–488.

Drewniak, L., and Sklodowska, A. (2013). Arsenic-transforming microbes and their role in biomining processes. Environ. Sci. Pollut. Res. *20*, 7728–7739.

Drewniak, L., Ciezkowska, M., Radlinska, M., and Sklodowska, A. (2015). Construction of the recombinant broad-host-range plasmids providing their bacterial hosts arsenic resistance and arsenite oxidation ability. J. Biotechnol. *196–197*, 42–51.

Dziewit, L., Pyzik, A., Szuplewska, M., Matlakowska, R., Mielnicki, S., Wibberg, D., Schlüter, A., Pühler, A., and Bartosik, D. (2015). Diversity and role of plasmids in adaptation of bacteria inhabiting the Lubin copper mine in Poland, an environment rich in heavy metals. Front. Microbiol. *6*, 152.

Fleischmann, R.D., Adams, M.D., White, O., Clayton, R.A., Kirkness, E.F., Kerlavage, A.R., Bult, C.J., Tomb, J.F., Dougherty, B.A., Merrick, J.M., *et al.* (1995). Whole-genome random sequencing and assembly of *Haemophilus influenzae* Rd. Science *269*, 496–512.

Francis, A.J., Dodge, C.J., Lu, F.L., Halada, G.P., and Clayton, C.R. (1994). Xps and Xanes studies of uranium reduction by *Clostridium* sp. Environ. Sci. Technol. *28*, 636–639.

Gadd, G.M. (2010). Metals, minerals and microbes: geomicrobiology and bioremediation. Microbiology *156*, 609–643.

Garcia-Moyano, A., Gonzalez-Toril, E., Aguilera, A., and Amils, R. (2007). Prokaryotic community composition and ecology of floating macroscopic filaments from an extreme acidic environment, Rio Tinto (SW, Spain). Syst. Appl. Microbiol. *30*, 601–614.

Gaze, W.H., Krone, S.M., Larsson, D.G., Li, X.Z., Robinson, J.A., Simonet, P., Smalla, K., Timinouni, M., Topp, E., Wellington, E.M., *et al.* (2013). Influence of humans on evolution and mobilization of environmental antibiotic resistome. Emerg. Infect. Dis. *19*, e120871.

Gomez-Sanz, E., Kadlec, K., Fessler, A.T., Zarazaga, M., Torres, C., and Schwarz, S. (2013). Novel *erm*(T)-carrying multiresistance plasmids from porcine and human isolates of methicillin-resistant *Staphylococcus aureus* ST398 that also harbor cadmium and copper resistance determinants. Antimicrob. Agents Chemother. *57*, 3275–3282.

Gray, H.B., and Ellis Jr., W.R. (2006). Electron transfer. In Bioinorganic Chemistry, I. Bertini, H.B. Gray, S.J. Lippard, and J.S. Valentine, eds. (Mill Valley, CA: University Science Books), pp. 315–364.

Hemme, C.L., Deng, Y., Gentry, T.J., Fields, M.W., Wu, L.Y., Barua, S., Barry, K., Tringe, S.G., Watson, D.B., He, Z.L., *et al.* (2010). Metagenomic insights into evolution of a heavy metal-contaminated groundwater microbial community. ISME J. *4*, 660–672.

Heuer, H., and Smalla, K. (2012). Plasmids foster diversification and adaptation of bacterial populations in soil. FEMS Microbiol. Rev. *36*, 1083–1104.

Hogsden, K.L., and Harding, J.S. (2012). Consequences of acid mine drainage for the structure and function of benthic stream communities: a review. Freshw. Sci. *31*, 108–120.

Huston, W.M., Jennings, M.P., and McEwan, A.G. (2002). The multicopper oxidase of *Pseudomonas aeruginosa* is a ferroxidase with a central role in iron acquisition. Mol. Microbiol. *45*, 1741–1750.

Hwang, C.C., Wu, W.M., Gentry, T.J., Carley, J., Corbin, G.A., Carroll, S.L., Watson, D.B., Jardine, P.M., Zhou, J.Z., Criddle, C.S., *et al.* (2009). Bacterial community succession during in situ uranium bioremediation: spatial similarities along controlled flow paths. ISME J. *3*, 137–137.

Inui, M., Tsuge, Y., Suzuki, N., Vertes, A.A., and Yukawa, H. (2005). Isolation and characterization of a native composite transposon, Tn*14751*, carrying 17.4 kilobases of *Corynebacterium glutamicum* chromosomal DNA. Appl. Environ. Microbiol. *71*, 407–416.

Janssen, P.J., Van Houdt, R., Moors, H., Monsieurs, P., Morin, N., Michaux, A., Benotmane, M.A., Leys, N., Vallaeys, T., Lapidus, A., *et al.* (2010). The complete genome sequence of *Cupriavidus metallidurans* strain CH34, a master survivalist in harsh and anthropogenic environments. PLoS ONE *5*, e10433.

Johnson, D.B., and Hallberg, K.B. (2003). The microbiology of acidic mine waters. Res. Microbiol. *154*, 466–473.

Jones, D.S., Albrecht, H.L., Dawson, K.S., Schaperdoth, I., Freeman, K.H., Pi, Y.D., Pearson, A., and Macalady, J.L. (2012). Community genomic analysis of an extremely acidophilic sulfur-oxidizing biofilm. ISME J. *6*, 158–170.

Kirby, R., Wright, L.F., and Hopwood, D.A. (1975). Plasmid-determined antibiotic synthesis and resistance in *Streptomyces coelicolor*. Nature *254*, 265–267.

Klippel, B., Lochner, A., Bruce, D.C., Davenport, K.W., Detter, C., Goodwin, L.A., Han, J., Han, S., Land, M.L., Mikhailova, N., *et al.* (2011). Complete genome sequence of the marine cellulose- and xylan-degrading bacterium *Glaciecola* sp. strain 4H-3-7+YE-5. J. Bacteriol. *193*, 4547–4548.

Lee, J.H., Kennedy, D.W., Dohnalkova, A., Moore, D.A., Nachimuthu, P., Reed, S.B., and Fredrickson, J.K. (2011). Manganese sulfide formation via concomitant microbial manganese oxide and thiosulfate reduction. Environ. Microbiol. *13*, 3275–3288.

Lin, L.H., Wang, P.L., Rumble, D., Lippmann-Pipke, J., Boice, E., Pratt, L.M., Sherwood Lollar, B., Brodie, E.L., Hazen, T.C., Andersen, G.L., *et al.* (2006). Long-term sustainability of a high-energy, low-diversity crustal biome. Science *314*, 479–482.

Lloyd, J.R. (2003). Microbial reduction of metals and radionuclides. FEMS Microbiol. Rev. *27*, 411–425.

Lovley, D.R., and Phillips, E.J.P. (1988). Novel mode of microbial energy-metabolism – organic-carbon oxidation coupled to dissimilatory reduction of iron or manganese. Appl. Environ. Microbiol. 54, 1472–1480.

Lovley, D.R., Phillips, E.J.P., and Lonergan, D.J. (1989). Hydrogen and formate oxidation coupled to dissimilatory reduction of iron or manganese by *Alteromonas putrefaciens*. Appl. Environ. Microbiol. *55*, 700–706.

Lovley, D.R., Phillips, E.J.P., Gorby, Y.A., and Landa, E.R. (1991). Microbial reduction of uranium. Abstr. Pap. Am. Chem. Soc. *202*, 8-Geoc.

Lovley, D.R., and Phillips, E.J.P. (1992). Reduction of uranium by *Desulfovibrio desulfuricans*. Appl. Environ. Microbiol. *58*, 850–856.

Lovley, D.R. (2002). Dissimilatory metal reduction: from early life to bioremediation. ASM News *68*, 231–237.

Mahillon, J., and Chandler, M. (1998). Insertion sequences. Microbiol. Mol. Biol. Rev. *62*, 725–774.

Mendez-Garcia, C., Mesa, V., Sprenger, R.R., Richter, M., Diez, M.S., Solano, J., Bargiela, R., Golyshina, O.V., Manteca, A., Ramos, J.L., *et al.* (2014). Microbial stratification in low pH oxic and suboxic macroscopic growths along an acid mine drainage. ISME J. *8*, 1259–1274.

Merlin, C., Mahillon, J., Nesvera, J., and Toussaint, A. (2000). Gene recruiters and transporters: the modular structure of bacterial mobile elements. In The horizontal gene pool, C.M. Thomas, ed. (Harwood Academic Publishers, Amsterdam, the Netherlands), pp. 363–409.

Myers, C.R., Carstens, B.P., Antholine, W.E., and Myers, J.M. (2000). Chromium(VI) reductase activity is associated with the cytoplasmic membrane of anaerobically grown Shewanella putrefaciens MR-1. J. Appl. Microbiol. *88*, 98–106.

Nancharaiah, Y.V., and Lens, P.N.L. (2015). Ecology and biotechnology of selenium-respiring bacteria. Microbiol. Mol. Biol. Rev. *79*, 61–80.

Nascimento, A.M., and Chartone-Souza, E. (2003). Operon *mer*: bacterial resistance to mercury and potential for bioremediation of contaminated environments. Genet. Mol. Res. *2*, 92–101.

Nies, D.H. (2003). Efflux-mediated heavy metal resistance in prokaryotes. FEMS Microbiol. Rev. *27*, 313–339.

Oremland, R.S., and Stolz, J.F. (2003). The ecology of arsenic. Science *300*, 939–944.

Paez-Espino, D., Tamames, J., de Lorenzo, V., and Canovas, D. (2009). Microbial responses to environmental arsenic. Biometals *22*, 117–130.

Pal, C., Bengtsson-Palme, J., Rensing, C., Kristiansson, E., and Larsson, D.G. (2014). BacMet: antibacterial biocide and metal resistance genes database. Nucleic Acids Res. *42*, D737–743.

Park, C.H., Keyhan, M., Wielinga, B., Fendorf, S., and Matin, A. (2000). Purification to homogeneity and characterization of a novel *Pseudomonas putida* chromate reductase. Appl. Environ. Microbiol. *66*, 1788–1795.

Perez-Gonzalez, T., Jimenez-Lopez, C., Neal, A.L., Rull-Perez, F., Rodriguez-Navarro, A., Fernandez-Vivas, A., and Ianez-Pareja, E. (2010). Magnetite biomineralization induced by Shewanella oneidensis. Geochim. Cosmochim. Acta *74*, 967–979.

Rahube, T.O., Viana, L.S., Koraimann, G., and Yost, C.K. (2014). Characterization and comparative analysis of antibiotic resistance plasmids isolated from a wastewater treatment plant. Front. Microbiol. *5*, 558.

Rhine, E.D., Phelps, C.D., and Young, L.Y. (2006). Anaerobic arsenite oxidation by novel denitrifying isolates. Environ. Microbiol. *8*, 899–908.

Robins, K.J., Hooks, D.O., Rehm, B.H.A., and Ackerley, D.F. (2013). *Escherichia coli* NemA is an efficient chromate reductase that can be biologically immobilized to provide a cell free system for remediation of hexavalent chromium. PLoS ONE *8*, e59200.

Rojas, L.A., Yanez, C., Gonzalez, M., Lobos, S., Smalla, K., and Seeger, M. (2011). Characterization of the metabolically modified heavy metal-resistant *Cupriavidus metallidurans* strain MSR33 generated for mercury bioremediation. PLoS ONE *6*, e17555.

Sanchez-Andrea, I., Rodriguez, N., Amils, R., and Sanz, J.L. (2011). Microbial diversity in anaerobic sediments at Rio Tinto, a naturally acidic environment with a high heavy metal content. Appl. Environ. Microbiol. *77*, 6085–6093.

Siefert, J.L. (2009). Defining the mobilome. Methods Mol. Biol. *532*, 13–27.

Silver, S., and Phung le, T. (2005). A bacterial view of the periodic table: genes and proteins for toxic inorganic ions. J. Ind. Microbiol. Biotechnol. 32, 587–605.

Singh, S.K., Grass, G., Rensing, C., and Montfort, W.R. (2004). Cuprous oxidase activity of CueO from *Escherichia coli*. J. Bacteriol. *186*, 7815–7817.

Slyemi, D., and Bonnefoy, V. (2012). How prokaryotes deal with arsenic. Environ. Microbiol. Rep. *4*, 571–586.

Sousa, T., Branco, R., Piedade, A.P., and Morais, P.V. (2015). Hyper accumulation of arsenic in mutants of *Ochrobactrum tritici* silenced for arsenite efflux pumps. PLoS ONE *10*, e0131317.

Stibitz, S., and Davies, J.E. (1987). Tn*602*: a naturally occurring relative of Tn*903* with direct repeats. Plasmid *17*, 202–209.

Stolz, J.F., and Oremland, R.S. (1999). Bacterial respiration of arsenic and selenium. FEMS Microbiol. Rev. 23, 615–627.

Streten-Joyce, C., Manning, J., Gibb, K.S., Neilan, B.A., and Parry, D.L. (2013). The chemical composition and bacteria communities in acid and metalliferous drainage from the wet-dry tropics are dependent on season. Sci. Total Environ. *443*, 65–79.

Sun, W., Sierra-Alvarez, R., Fernandez, N., Sanz, J.L., Amils, R., Legatzki, A., Maier, R.M., and Field, J.A. (2009). Molecular characterization and in situ quantification of anoxic arsenite-oxidizing denitrifying enrichment cultures. FEMS Microbiol. Ecol. *68*, 72–85.

Tebo, B.M., and Obraztsova, A.Y. (1998). Sulfate-reducing bacterium grows with Cr(VI), U(VI), Mn(IV), and Fe(III) as electron acceptors. FEMS Microbiol. Lett. *162*, 193–198.

Tebo, B.M., Johnson, H.A., McCarthy, J.K., and Templeton, A.S. (2005). Geomicrobiology of manganese(II) oxidation. Trends Microbiol. *13*, 421–428.

Thomas, C.M. (2000). Paradigms of plasmid organization. Mol. Microbiol. *37*, 485–491.

Tourney, J., and Ngwenya, B.T. (2014). The role of bacterial extracellular polymeric substances in geomicrobiology. Chem. Geol. *386*, 115–132.

Trefault, N., De la Iglesia, R., Molina, A.M., Manzano, M., Ledger, T., Perez-Pantoja, D., Sanchez, M.A., Stuardo, M., and Gonzalez, B. (2004). Genetic organization of the catabolic plasmid pJP4 from *Ralstonia eutropha* JMP134 (pJP4) reveals mechanisms of adaptation to chloroaromatic pollutants and evolution of specialized chloroaromatic degradation pathways. Environ. Microbiol. *6*, 655–668.

Tyson, G.W., Chapman, J., Hugenholtz, P., Allen, E.E., Ram, R.J., Richardson, P.M., Solovyev, V.V., Rubin, E.M., Rokhsar, D.S., and Banfield, J.F. (2004). Community structure and metabolism through reconstruction of microbial genomes from the environment. Nature *428*, 37–43.

Uchiya, K., Takahashi, H., Nakagawa, T., Yagi, T., Moriyama, M., Inagaki, T., Ichikawa, K., Nikai, T., and Ogawa, K. (2015). Characterization of a novel plasmid, pMAH135, from *Mycobacterium avium* subsp. hominissuis. PLoS ONE *10*, e0117797.

van der Meer, J.R., Zehnder, A.J., and de Vos, W.M. (1991). Identification of a novel composite transposable element, Tn*5280*, carrying chlorobenzene dioxygenase genes of *Pseudomonas* sp. strain P51. J. Bacteriol. *173*, 7077–7083.

Vargas, M., Kashefi, K., Blunt-Harris, E.L., and Lovley, D.R. (1998). Microbiological evidence for Fe(III) reduction on early Earth. Nature *395*, 65–67.

Wang, Y.T. (2000). Microbial Reduction of Chromate. In Environmental Microbe-Metal Interactions, D.R. Lovley, ed. (ASM Press, Washington, DC, USA).

Warren, L.A., Maurice, P.A., Parmar, N., and Ferris, F.G. (2001). Microbially mediated calcium carbonate precipitation: Implications for interpreting calcite precipitation and for solid-phase capture of inorganic contaminants. Geomicrobiol. J. *18*, 93–115.

Zagury, G.J., Narasiah, K.S., and Tyagi, R.D. (1994). Adaptation of indigenous iron-oxidizing bacteria for bioleaching of heavy-metals in contaminated soils. Environ. Technol. *15*, 517–530.

Zavarzina, D.G. (2004). Formation of magnetite and siderite by thermophilic Fe(III)-reducing bacteria. Paleontol. J. *38*, 585–589.

Molecular Markers in Biodegradation Processes

3

Aleksandra Ziembińska-Buczyńska

Contents

Abstract

Biological decomposition of various substrates is performed by a large number of microorganisms. Microbial performance in biodegradation processes has been observed and analysed for years only for microbiological cultivation methods. For the last 30 years, molecular biology tools that are useful for describing the microbial world have been discovered, improved and introduced for routine microbial studies. These molecular methods are performed on a wide variety of molecular markers – housekeeping genes that are always present in the microbial genome as well as functional genes that are responsible for the production of a particular enzyme. Biodegradation processes are performed in the environment by microbial communities; thus, it is important to study the microbiological part of these biodegradation processes utilizing a combination of molecular tools and markers. This chapter provides brief characteristics of the housekeeping genes used in microbial identification together with a description of the functional genes that are commonly used as molecular markers in the biodegradation of cellulose, hemicellulose, lignin, chlorinated compounds, plastics and polycyclic aromatic hydrocarbons.

3.1 Introduction

A molecular marker is a particular fragment of genetic material (genetic *locus*) that is easily tracked and quantified using molecular tools. This part of the DNA may be linked with a particular gene or a feature of interest (Hayward *et al.*, 2015). In the case of microorganisms, a wide variety of molecular markers, present among both coding and non-coding DNA fragments, are used to identify the bacteria and fungi and assess their biochemical potential. Furthermore, the decision regarding the type of molecular marker to use for an analysis depends on its aim. For identification and phylogenetic analysis, the most popular microbial molecular markers belong to the housekeeping genes. They are typically large molecules constantly present in the microbial cell and in relatively large number. They also possess both conserved and polymorphic fragments. This type of molecule is useful for such analysis because they are shared by all microorganisms. However, microorganisms are an extremely diverse group, presenting a flexible metabolism and ability to decompose myriad substrates. This diversity is why functional genes are much more useful in the description of microbial biochemical potential, including the biodegradation of different compounds. Both the housekeeping and functional genes used for microbial research need to present several features to be considered good molecular markers. These marker genes must be highly polymorphic in their sequences to be distinguishable in various microbial species, and they should also have conserved regions to allow appropriate PCR primer design (Liu *et al.*, 2012).

Molecules such as the 16S and 23S rRNA coding genes for bacteria or the 18S rRNA coding gene for fungi are regarded as the gold standard for microbial identification and phylogenetic research. However, there are several other housekeeping genes useful in microbial identification and research. Table 3.1 presents multiple examples of molecular markers commonly used in microbial identification.

Molecular markers are used as a standard for microbial research that is based on molecular methods. Since the discovery of PCR over 30 years ago, these analyses have become increasingly complex. PCR and its modifications are still the most frequently used methods in microbial identification, but for the characterization of complex microbial communities, next generation sequencing (NGS) has become the technique most often used. Recently, NGS has also been used with functional genes, enabling functional genomics of the investigated biocoenosis (see also: Chapter 1). In the case of microbial research, especially for characterization of the biodegradation processes, a combination of molecular tools and markers is required.

3.2 Functional genes as the molecular markers for biodegradation processes

3.2.1 Molecular markers in cellulose, hemicellulose and lignin biodegradation

The main component of biomass is lignocellulose, which consists of three types of biopolymers: cellulose, hemicellulose and lignin. To decompose these substances, enzymes from the hydrolases and oxidases groups are required. Since the 1990s, molecular research regarding these enzymes has been developed; thus, several well-known molecular markers can be used for the biodegradation research of these biopolymers (Pérez *et al.*, 2002).

Lignins are known to be very difficult materials to biodegrade. Because they are large and insoluble molecules, the enzymes that decompose lignin are often unspecific, which has also enabled them to decompose a larger number of lignin-related compounds (nitro-substituted and chlorinated molecules, polycyclic aromatic hydrocarbons, etc.) (Mester and Tien, 2000). The enzymatic system for lignin degradation is typically described based on the white rot fungi. The enzymes responsible for this process are: laccase (phenol oxidase), manganese dependent peroxidases (MnPs) and lignin peroxidases (LiPs) (de la Rubia *et al.*, 2002).

Laccase is an extracellular enzyme has been extensively described in numerous fungi since the 1990s (Saloheimo *et al.*, 1991). One example of the fungal laccases are the blue copper oxidases that catalyse the one electron oxidation of phenolic compounds, aromatic amines or other electron-rich substances together with the reduction of H_2 to H_2O (Cullen and Kersten, 2004). These glycoproteins typically contain four relatively conserved regions that directly bind copper ions in their active site (Luis *et al.*, 2004), which is why they have potential as a molecular marker to monitor lignin decomposition. However, laccase is not exclusive to the fungal world; there are several reports of laccase-like proteins present in bacterial cells, where these molecules are intercellular or periplasmic enzymes (Baldrian, 2005). There are many genes encoding this enzyme and therefore, several isoenzymes can be produced by microbial cells (Chen *et al.*, 2003). The number of isoenzymes produced by fungal cells depends on their taxonomical classification. Most of the filamentous fungi produce several laccase isoenzymes because they possess between four and eight laccase coding genes (Janusz *et al.*, 2013). For *Plerotus ostreatus*, there are eleven or twelve laccase coding genes present in its cells, and this fungi can produce at least eight different laccase isoenzymes, among which only a few have been isolated and described. These genes have been named *pcx1*, *poxc*, *poxa1b*, *poxa3*, *pox3*, *pox4* and *pox5*. *Pleurotus sajor-caju* possesses the *lac1*, *lac2*, *lac3*, *lac4*, *lacB* and *lacC* genes (Soden and Dobson, 2003), *Pleorotus eryngii* has *pel* (Rodriguez *et al.*, 2008) and *Trametes pubescens* has *lap2* (Galhaup *et al.*, 2002). Laccase isoenzyme diversity may be explained not only by gene number but also by posttranslational modifications such as glycosylation, proteolytic processing, etc. (Janusz *et al.*, 2013).

Table 3.1 Examples of molecular markers used in microbial identification

Molecular marker type	Molecular marker	Characteristics/usage	References
Bacterial, coding	16S rRNA gene	ca. 1.5 kb; housekeeping gene with both conserved and polymorphic regions. Most often used molecular marker; known to be a gold standard in bacterial identification; Higher degree of conservation than functional genes, thus more resistant to mutations allows for precise identification; the gene sequence database has been enlarged recently, Thus it is difficult to identify bacteria using a short fragment of the 16S rRNA sequence. There are some species that are better identified with the 23S rRNA gene or the ISR region	Bojesen *et al.* (2007), Liu *et al.* (2012)
Bacterial, coding	23S rRNA gene	Useful for medically important strain identification and bacteraemia diagnostics, Applied to a DNA microarray-based approach	Liu *et al.* (2012)
bacterial, non-coding	16S-23S rDNA ISR	Variable size and sequence among bacteria; useful for microbial classification Used with RFLP method to characterize total bacterial communities as well as bacteria at the species level	Ruiz *et al.* (2000); Liu *et al.* (2012)
Bacterial, coding	*gyrB* gene	Encodes β-subunit of DNA gyrase Used as a marker for bacterial phylogeny In some identification cases such as *Pseudomonas putida* used instead of classical 16 S rRNA approach due to the higher substitution frequency of gyrB than that of the 16S rRNA gene	Yamamoto and Harayama (1995); Liu *et al.* (2012)
bacterial, Coding	*rpoB* gene	Encodes the β subunit of DNA-dependent RNA polymerase Possesses conserved regions particularly useful in bacterial classification Often used for bacterial genotyping in combination with other markers (in case of *Legionella* identification, for example, used with the 16S rRNA gene and the *mip* gene that is responsible for surface exposed protein synthesis)	Liu *et al.* (2012)
Bacterial, coding	*dnaK* gene	Encodes heat shock protein Ubiquitous and sequence is highly conserved Used as identification marker instead of the 16S rRNA gene Research performed on *Lactobacillus casei* revealed higher polymorphism; the bootstrap value at its phylogenetic tree is higher than that for the 16S rRNA gene tree and it creates more clear clads	Huang and Lee (2011); Liu *et al.* (2012)
Bacterial, coding	*recA* gene	Important gene in DNA recombination and DNA metabolism Two groups of gene libraries were constructed: broad-specific primers (BUR1 and BUR2) and those linked with the *Burkholderia* genus (BUR3 and BUR4)	Vermis *et al.* (2002); Liu *et al.* (2012)
bacterial, coding	*dsrAB* gene	Code α and β subunits of the enzyme reducing sulfite to sulfide; useful for sulfate-reducing bacteria (SRB) analysis Also observed in *Archaea* Its phylogeny is congruent with the 16S rRNA coding gene	Liu *et al.* (2012)
Fungal, coding	18S rRNA gene	ca. 400 bp; housekeeping gene for eukaryotic cells Highly homologous, repeated in genome	Embong *et al.* (2008)
fungal, noncoding	ITS1 and ITS2	ITS1 is localized between the 18S and 5.8 rRNA genes ITS2 is localized between the 5.8, and 28S rRNA genes Highly polymorphic	Embong *et al.* (2008)

The laccase genes can also be analysed by sequencing, as performed by Kim *et al.* (2015), to compare SNPs (Single Nucleotide Polymorphisms). This research has been performed on shiitake fungi (*Lentinula edodes*) to propose an effective system for the identification of different types of shiitake and to explore the potential influence of the observed SNP polymorphisms on fungi phenotype.

The primers designed for laccase gene detection in microbial materials are based on the conserved DNA sequences located around the two pairs of histidine in the N-terminal copper-binding regions of the basidiomycete enzyme. The product obtained from laccase gene amplification using degenerate primers should be approximately 200 bp, but in some cases the product could be smaller due to lack of the intron inside the amplified region (D'Souza *et al.*, 1996).

In his research, Chen *et al.* (2003) used three pairs of PCR primers (*lac1*, *lac2*, and *lac3*) to screen a broad range of *Basidiomycetes* fungi for the presence of laccase-like coding genes. In this study, one degenerate primer used for PCR amplification (*lac3*) presented better screening properties for laccases than *lac1* and *lac2*. Nonetheless, the same *lac* genes that enable laccase gene detection in a large number of fungi fail in some cases. This failure is most likely due to such a variety of laccase coding genes, and this analysis should be supported with the use of another molecular marker linked to lignin degradation, such as manganese-dependent peroxidase.

The main function of manganese-dependent peroxidase (MnP) is to oxidize Mn^{2+} to Mn^{3+} using H_2O_2 as an oxidant (Cullen and Kersten, 2004). Manganese peroxidase, as well as laccase, is only able to oxidize the phenolic residues of lignins. The mediators, however, are able to extend the range of these enzymes substrates to non-phenolic structures (Giardina *et al.*, 2000; Cullen and Kersten 2004). The cooperation of the other proteins with laccase and manganese peroxidase has widened the role of these enzymes in lignin decomposition (Hilden *et al.*, 2000). Most fungi secrete two different types of manganese peroxidase: MnP1 and MnP2 (de la Rubia *et al.*, 2002). This enzyme is a glycosylate haem protein that is encoded by the *mnp* genes, and it also occurs in several isoforms coded for by four *mnp* gene families (Hilden *et al.*, 2005; Janusz *et al.*, 2013, after Hakala *et al.*, 2006). The enzyme consists of 330–370 amino acids with short sequences including N-terminal MAF and AAP regions conserved from the *mnp* genes (Li *et al.*, 1999; Janusz *et al.*, 2013 after Hilden *et al.*, 2000). The *mnp* genes are widely used as a molecular marker for the investigation of microorganisms that are able to biodegrade lignin, which have garnered increasing interest for use in biotechnological applications (such as the improvement of feedstock quality for the production of second-generation biofuels) (Morgenstern *et al.*, 2010). For example, the research performed by Chen *et al.* (2001) screened the *Phanerochaete chrysosporium* fungus for *mnp* genes using PCR to increase its biodegradation potential. In a study by Pointing *et al.* (2005), this marker was used to screen for lignin degradation ability among 38 strains of *Basidiomycetes* and *Xylariaceae* fungi. Moreover, Morgenstern *et al.* (2010) used the *mnp* genes for Real-time PCR gene expression monitoring in the white rot fungi grape vine pathogen *Fomitiporia mediterranea* (*Agaricomycotina, Hymenochaetales*). This research suggested that lignin degradation by white rot taxa relies largely on ligninolytic peroxidases and is not efficiently achieved by laccases alone.

The third important enzyme in the lignin degradation system is lignin peroxidase, which catalyses the biodegradation of different phenolic and non-phenolic compounds based on their high redox potential caused by the reduction of H_2O_2. This enzyme is not as widely known as the other two enzymes of the lignin degradation system. The family of this gene is large and in some species, such as *Trametes versicolor*, up to 16 isoenzymes have been identified (Johansson *et al.*, 1993; Janusz *et al.*, 2013). The variability of isoenzymes is present because of the large number of *lip* genes coding these enzymes. Ten *lip* genes (from *lipA* to *lipJ*) were described in *Phanerochaete chrysosporium.* This molecular marker can be used to study the presence and abundance of microorganisms in the environment that possess lignin peroxidase. Furthermore, the type of marker gene present in the biological material depends on the environmental factors. Northern blot analysis revealed that the *lipA* gene was more abundant in nitrogen-limited media, while *lipD* dominated in carbon-starved cultures (Holzbaur and Tien, 1988). Nuclease protection assays enabled *lipE* gene identification in both carbon- and nitrogen-starved cultures (Reiser *et al.*, 1993). Stewart *et al.* (1992) performed quantitative reverse transcriptase-PCR (qRT-PCR) analysis, which confirmed up-regulation of the *lipC* and *lipJ* genes under nitrogen starvation. Interestingly, all

LiP genes but *lipF* were detected in anthracene-contaminated soil cultures (Bogan *et al.*, 1996).

Cellulose-degrading microorganisms belong to both bacteria and fungi, and they possess a wide variety of cellulolytic enzymes. These enzyme-coding genes are used as molecular markers for cellulose biodegradation monitoring.

Cellulose biodegradation can be performed by both aerobic and anaerobic microorganisms. The microorganisms capable of degrading cellulose produce a battery of enzymes with different specificities, all working together. Cellulases hydrolyse the β-1,4-glycosidic linkages of cellulose. Traditionally, they are divided into two classes: the endoglucanases hydrolyse internal bonds, releasing new termini, while the cellobiohydrolases decompose the existing or endoglucanase-generated chain ends (Pérez *et al.*, 2002). Genes coding these two classes of enzymes are widely used as molecular markers for cellulose biodegradation monitoring. The most common model bacteria used for cellulose degradation studies is *Clostridium thermocellum*. This bacterium possesses a complex multienzymatic exocellular system called the cellulosome, which consists of numerous enzymes organized around a scaffolding protein termed CipA (Guglielmi and Béguin, 1998). According to previous research, this complex contains ca. 30 cellulosome-related genes (Bayer *et al.*, 2000; Schwarz 2001; Zverlov *et al.*, 2005). Among these genes, *CelO, CbhA* and *CelK* are the markers for cellobiohydrolase performance, while *CelB, CelG, CelA, CelD, CleN, CelR, CelQ, CelF, CelT* and *CelS* are the markers for endoglucanase presence and functionality. The *CelH* and *CelE* genes are recognized as multifunctional components encoding endoglucanases (Zverlov *et al.*, 2005).

Hemicellulose content in the lignocellulosic plant biomass is estimated at 20–30%. The main constituent of hemicellulose is xylan, which is a promising biodegradation substrate for the production of useful end products. It can be degraded by numerous microorganisms producing xylanases. Microbial biodegradation is an interesting alternative for xylan decomposition to enable further product usage; thus, microorganisms able to synthetize xylanases in large amounts are being investigated. These enzymes also belong to the cellulosome complex. Biodegradation research is usually performed with the *Xyn* genes. For *C. thermocellum*, the cellulosome contains *XynA, XynC, XynD* and *XynU* (Zverlov *et al.*, 2005).

3.2.2 Molecular markers in chlorinated compound biodegradation

Owing to their large metabolic potential, microorganisms are able to biodegrade xenobiotics, and their role in bioremediation of polluted environments has already been clearly recognized. In the case of biodegradation monitoring during the bioremediation processes, it is wise to use a combination of molecular analyses. The best option is to use assays that search for the known microbial degrader using housekeeping genes (the 16S or 23S rRNA genes for bacteria or the 18S rRNA gene for fungi) together with functional genes to monitor the feasibility of a treatment and to follow up on an ongoing bioremediation treatment (see also Chapter 1 and Power *et al.*, 1998; Marzorati *et al.*, 2010).

The chlorinated compounds are a group of recalcitrant pollutants that are present in groundwater and biodegraded by several bacterial species. Research performed by Marzorati *et al.* (2010) using 1,2-dichloroethane (1,2-DCA) as a reference compound, the major pollutant from this group, to investigate the effectiveness of the metabolic process of biodegradation was performed using both 16S rRNA and functional genes as molecular markers. The key enzymes in anaerobic dehalogenation processes are the reductive dehalogenases (RDs). They are cobalamin-dependent oxygen-sensitive enzymes. The marker gene used for these enzymes is the *dca* gene. Marzorati *et al.* (2010) used the primers PceAFor1 and PcaBRev to amplify 2000 bp of the reductive dehalogenase *dcaABCT* gene cluster to prepare a clone library. Using a 16S rRNA gene-based approach, they also identified the dominant microbial genera in the environment as the members of *Firmicutes*. The use of functional markers in biodegradation reveals the capacity of metabolic functionalities to be shared within the bacterial community to cooperatively perform a particular process. This approach is particularly useful in functionality monitoring of the biocoenosis under stress conditions (Marzorati *et al.*, 2010).

For aerobic dehalogenation, four groups of enzymes are able to perform dehalogenation. They are the hydrolases, reductases, monooxygenases and dioxygenases (Kumar and Bae, 2014). The enzymes most often analysed for aerobic dehalogenation are the hydrolytic dehalogenases (FcbB). According to research by Chae *et al.* (2008), 4-chlorobenzoyl-CoA (4CBA-CoA) dehalogenase was chosen as the enzyme responsible for dehalogenation in the consortium. Their analysis was performed with the *FcbB* gene using degenerate primers, which produced a 158-bp amplicon (5′-SCGMMGDGCRATGGAG-3′, forward; 5′-CYTGSAGRTGBGTSGG-3′, reverse).

One of the most important monooxygenase type dehalogenases is pentachlorophenol 4-monooxygenase. The molecular marker used for the effectiveness of this enzyme in halogenated compound biodegradation is the gene encoding this enzyme (*pcpB*), which is relatively common among a variety of bacteria (Kumar and Bae, 2014). Moreover, *pcpB* is most often used as the molecular marker for pentachlorophenol biodegradation monitoring because of its performance in the beginning of the process. In their study, Cai and Xun (2002) analysed the organization of the *pcp* genes in *Sphingobium chlorophenolicum* ATCC 39723. They discovered 5 *pcp* genes involved in PCP degradation: the *pcpB*, *pcpC*, *pcpA*, and *pcpE* genes were catabolic, while *pcpR* was regulatory. The *pcpD* gene is uncharacterized (Orser and Lange, 1994) and *pcpM* is a hypothetical regulatory gene. These genes are scattered on two fragments. The *pcpB*, *pcpD*, and *pcpR* are clustered and possess the same orientation; *pcpB* and *pcpD* are most likely cotranscribed. However, *pcpR* is likely not cotranscribed with *pcpBD* because a rho-independent terminator sequence is located 38 bp downstream of *pcpD*. The *pcpC*, *pcpA*, *pcpM*, and *pcpE* genes are localized on another fragment, but they are apparently arranged in discrete locations and transcribed from different promoters (Cai and Xun, 2002).

One example of a dehalogenating dioxygenase is tetrachlorobenzene dioxygenase (TecA). This enzyme was studied in *Burkholderia* sp. strain PS12 in which it dechlorinates 1,2,4,5-tetrachlorobenzene to 3,4,6-trichlorocatechol (Donelly *et al.*, 1993; Beil *et al.*, 1999; Kumar and Bae, 2002). In case *Burkholderia* sp. strain PS12, the enzyme is encoded by plasmid-located *tecA*, but according to Pollmann *et al.* (2001), this enzyme can perform deoxygenation, dechlorination and monooxygenation because it could also be coded for by the *tecB* gene that was discovered in *Ralstonia* sp. strain PS12 and is located downstream of *tecA* (Pollmann *et al.*, 2001). The research using these markers was based on cloning and sequencing (Pollmann *et al.*, 2001).

3.2.3 Molecular markers in plastic biodegradation

Plastics have become an important type of long chain polymer. This group of compounds became a part of everyday life more than 50 years ago and their importance is still increasing. The plastics used most often now belong to the synthetic group of polymers. These types of molecules are difficult for microorganisms to biologically decompose. This situation is a result of their relatively short time present in nature. Microorganisms have not had enough time to design new enzymatic structures for synthetic polymer degradation (Mueller, 2006; Shah *et al.*, 2008). Some synthetic plastics, such as polyester, polyurethane or polyethylene with starch blend, can be biodegraded by microorganisms, but most of the compounds from this group are non-biodegradable. That is why there is a growing concern about biodegradable plastics that possess similar properties as the synthetic ones but are biocompatible, more readily biodegradable and produced from renewable resources.

Compounds such as polyhydroxybutyrate (PHB), polyhydroxyvalerate (PHBV), polylactic acid (PLA) and polycaprolactone (PCL) belong to the biodegradable plastics. These compounds were introduced to the market over 15 years ago, but until now there has been no clear evidence of their biodegradation in the landfill environment (Shah *et al.*, 2008). This uncertainty is why they are still only a small part of the plastics market (Shah *et al.*, 2008), despite the fact that their effective biodegradation has been explored extensively.

The natural and synthetic plastics can be biodegraded by a large group of bacteria and fungi (Gu *et al.*, 2000). According to Shah *et al.* (2008), at least two groups of microbial enzymes perform during plastics biodegradation: extracellular and intracellular depolymerases. The genes responsible for plastics degradation can be used as molecular markers for these microorganisms' detection and activity measurements. Fig. 3.1 presents the mechanism of the plastic degradation under aerobic conditions.

One of the most often used groups of bioplastics in which biodegradation can be studied are the polyhydroxyalcanoaetes (PHAs) such as

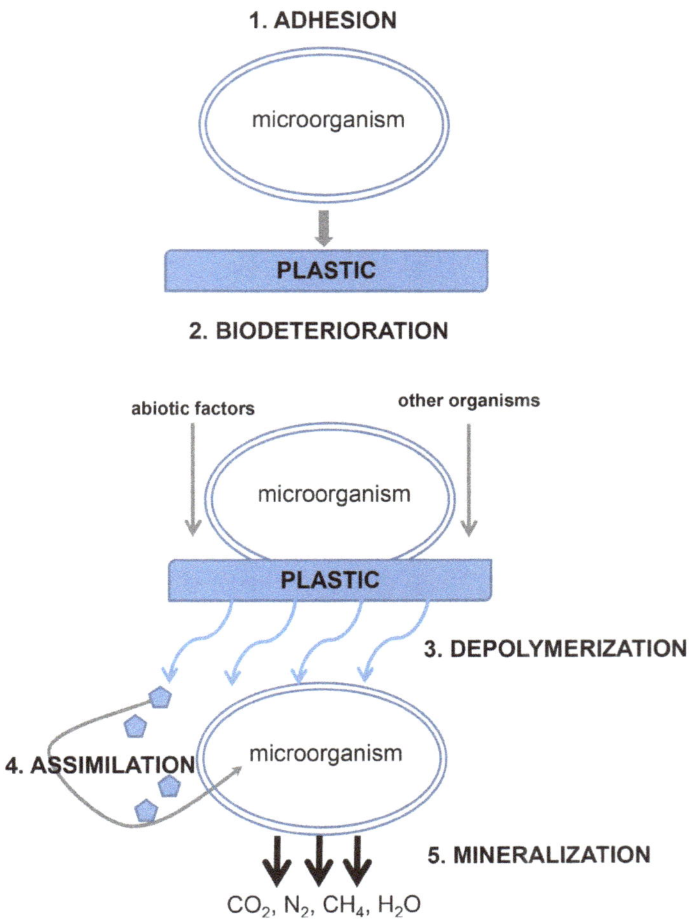

Figure 3.1 Mechanism of the plastic degradation under aerobic conditions.

polyhydroxybutyrate (PHB) and polyhydroxyvalerate (PHBV). *Pseudomonas lemoignei* is recognized as the bacteria most often used to study PHA degradation (Schöber *et al.*, 2000). *P. lemoignei* is unique among the PHA-degrading bacteria because it is able to synthesize at least five different extracellular PHA depolymerases. The PHA depolymerase structural genes (*phaZ1* to *phaZ5*) are commonly used as molecular markers mainly for identification of PHB bacteria and monitoring its activity. The molecular cloning studies performed by Briese *et al.* (1994), Jendrossek *et al.* (1993), Jendrossek *et al.* (1995) and Schöber *et al.* (2000) enabled the discovery of six genes, *phaZ1*, *phaZ2*, *phaZ3*, *phaZ4*, *phaZ5* and *phaZ6*, encoding PHB depolymerases C, B, D, E, A, and PHBV depolymerase, respectively. According to the results obtained by Schöber *et al.* (2000), the PHBV depolymerase encoded by the *phaZ6* gene is responsible for the most effective PHBV degradation; however, PHB depolymerase C, encoded by the *phaZ1* gene, is active in PHBV degradation, but the effectiveness of this process is lower.

In 2001, Sei *et al.* proposed a wide set of PCR primers and Southern blot hybridization probes for PHB-degrading bacteria possessing the fibronectin type III linker type-PHB depolymerases useful for their detection and monitoring. Their research was based on 14 PHB depolymerase gene sequences obtained from GeneBank (Keller *et al.*, 1984). The experimental set of bacterial strains and their PHB depolymerase sequences was composed of *Comamonas* sp. (gene *phaZ*; GenBank accession No. U16275), *Ralstonia pickettii* T1 (gene *phaZ*, GenBank accession No. J04223), *Streptomyces exfoliates* K10 (gene *phaZ*; GenBank accession No. U58990) and *Pseudomonas lemoignei* (gene *phaZ4*; GenBank accession No. U12976). Tests of the PCR primers and hybridization probes were performed on 57 bacterial strains freshly isolated from soil.

The research of Birgham *et al.* (2012) was performed on *Ralstonia eutropha*, which is regarded as the model for PHA production. A large number of microorganisms producing PHA are also able to biodegrade this compound because they needed to immobilize PHA inside the cell. Thus, the effects of PHA depolymerase (PhaZ) expression on PHA homeostasis in *R. eutropha* was examined. The studies were performed with RT-qPCR using primers for the *phaZ1* to *phaZ7* genes. The authors suspect that the performance of the *phaZ* genes has not only an impact on PHA degradation but also on the 3-dimensional granule architecture.

Polylactic acid (PLA) is also used relatively often as a biodegradable polymer. It is a linear, aliphatic polyester of lactic acid produced naturally via fermentation processes. Its ester linkages are sensitive to both chemical hydrolysis and enzymatic degradation (Shah *et al.*, 2008). To increase its biodegradability, starch is often used during PLA-based plastic production. Several fungi and bacteria have been reported as PLA-based bioplastic biodegraders (Shah *et al.*, 2008). In 1996, Torres *et al.* used *Fusarium moniliforme* and *Penicillum roqueforti* to degrade PLA oligomers. Pranamuda *et al.* (1997) and Pranamuda and Tokiwa (1999) reported PLA biodegradation by *Amycolatopsis* sp. Furthermore, the *Bacillus brevis* bacterium was mentioned as a PLA biodegrader by Tomita *et al.* (1999). Low molecular weight PLA can also be biodegraded by *Rhizopus delemar* via a lipase-based metabolic pathway (Fukuzaki *et al.*, 1989).

There have been several reports of PLA biodegradation in soil or compost (Ghorpade *et al.*, 2001), thus the most effective biodegradation is most likely performed by communities of microorganisms, not by single strains. Nonetheless, the research to establish a molecular marker useful for PLA biodegradation monitoring must be performed on pure cultures to allow gene discovery and analysis. In 2003, Akutsu-Shigeno *et al.* published molecular cloning and sequencing results of the poly(DL-lactic acid) depolymerase gene from *Paenibacillus amylolyticus* Strain B-13, which was previously described as plastic degrading (Teeraphatpornchai *et al.*, 2003). They also expressed a functional version of the enzyme in *Escherichia coli.* The gene responsible for PLA degradation was named *plaA* and encoded a lipase. Its expression product presented a wide range of substrate specificities for biodegradable plastics with ester bonds (Akutsu-Shigeno *et al.*, 2003). However, as described (Li *et al.*, 2002), the PLA polymers are biodegraded mainly by proteinase-like enzymes. Matsuda *et al.* (2005) characterized an extracellular poly(L-lactic acid) depolymerase encoded by the *pld* gene isolated from *Amycolatopsis* sp. strain K104-1 at the molecular level. The authors cloned the gene into *Streptomyces lividans* 1326 carrying the pUPLD plasmid for molecular characterization of the marker.

Synthetic plastics are difficult to degrade via biological pathways due to the relatively short time they have been present in the environment. Their biodegradation is very slow and it needs to be preceded by physiochemical decomposition caused by environmental factors (Shah *et al.*, 2008). Examples of synthetic plastics that are biodegradable by microorganisms are nylon and polyester polyurethane (PUR) (Nomura *et al.*, 2013; Shimao, 2001).

Despite the fact that the first generation of plastics contains a bond that is relatively common in natural and biodegradable polymers, the amide bond (-CONH-), nylon is rather difficult to decompose by microorganisms (Nomura *et al.*, 2013). In over 20 years of research, only two bacterial hydrolases able to degrade nylon oligomers have been described and were isolated from *Achromobacter guttatus, Flavobacterium* and *Pseudomonas* (Kinoshita *et al.*, 1972; Negoro *et al.*, 1994). Deguchi *et al.* published articles in 1997 and 1998 containing research in which lignin-degrading fungi, *Phanerochaete chrysosporium* (ATCC 34541), *Trametes versicolor* (1F07043) and IZU-154, were responsible for nylon decomposition. Their results revealed that nylon-degrading enzymes are identical to manganese-dependent peroxidases but their requirements for performance are different. Their research suggests that molecules such as H_2O_2 and the superoxide anion radical are involved in nylon degradation. Interestingly, this enzyme is coded by the same molecular marker – the *mnp* gene. This enzyme was also revealed in the fifty-eight fungi studied as nylon degrading by Freidrich *et al.* (2007). Therefore, it is not surprising that the same molecular marker, the *mnp* gene, would be useful in nylon-degrading microorganism research. However, such discoveries underline the necessity of the use of a combination of markers instead of single marker based research.

Polyester polyurethanes (PURs) are one of two groups of the polymers derived from the condensation of polyisocyanates and polyols, creating intramolecular urethane bonds (the carbamate ester bond, NHCOO). In contrast to the ether polyurethanes, this group of polymers can be degraded by microorganisms (Nomura *et al.*, 2013), as certain bacteria and fungi have been able to biodegrade polyester polyurethanes. Among these bacteria, *Pseudomonas fluorescens* (Howard and Blake, 1998) and *Comamonas acidovorans* (Nakajima-Kambe et al., 1995) are capable of PUR biodegradation, while *P. chloroaphis* can not only degrade but can also grow on polyester polyurethane as its sole carbon and energy source (Howard *et al.*, 1999). The fungal species *Fusarium solani, Aureobasidium pullulans, Cladosporium* sp. and *Curvularia senegalensis* are also capable of biodegrading this compound (Crabbe *et al.*, 1994). According to many sources (Nakajima-Kambe *et al.*, 1995; Akutsy *et al.*, 1998; Howard and Blake, 1998; Stern and Howard, 2000), the enzymes that are most likely responsible for PUR biodegradation belong predominantly to the esterases. Bacterial polyurethanases can be used as a molecular marker for PUR biodegradation research. Such genes have been already cloned and characterized in *Pseudomonas fluorescens* (gene *pulA*; Vega *et al.*, 1999) and *Comamonas acidovorans* strain TB-35 (structural gene *pudA*; Nomura *et al.*, 1998). Two polyester polyurethanase genes (*pueA* and *pueB*) from *Pseudomonas chlororaphis* encoding a lipase responsible for PUR biodegradation have also been described (Stern and Howard, 2000; Howard *et al.*, 2007) and can be used as PUR biodegradation molecular markers.

3.2.4 Molecular markers in polycyclic aromatic hydrocarbon biodegradation

Polycyclic aromatic hydrocarbons (PAHs) are hydrophobic organic molecules composed of two or more benzene rings. They can appear in the environment naturally or can be manmade. PAHs are a very large group of environmental contaminants and most are persistent in the ecosystem (Peng *et al.*, 2008). They contaminate soil and sediment and are considered to be dangerous to life because of their toxic, mutagenic and in some cases carcinogenic properties (Patnaik, 1992). Because they have been recognized as an environmental risk, there are numerous studies on possible strategies for their removal from the environment (see also Chapter 9). PAHs can be adsorbed, volatilized, photolysed, or degraded chemically, but the most important way is microbial decomposition – biodegradation (Haritash and Kaushik, 2009). They can be biodegraded by a wide group of microorganisms from both the bacterial and fungal worlds. Among bacteria, *Pseudomonas aeruginosa, Pseudomons fluoresens, Mycobacterium* spp., *Haemophilus* spp., *Rhodococcus* spp., and *Paenibacillus* spp. have been indicated and among the fungi, lignolytic fungi *Phanerochaete chrysosporium, Bjerkandera adusta* and *Pleurotus ostreatus* are known to be PAH-degraders (Haritash and Kaushik, 2009). Due to such a large diversity among the PAHs and their common presence in different parts of the environment, both aerobic and anaerobic biodegradation pathways are known, but under anaerobic conditions, the biodegradation is much slower (Coates *et al.*, 1996).

The enzymes involved in the PAH biodegradation processes belong to the oxygenase and dehydrogenase groups of enzymes. Additionally, the lignolytic enzymes previously described in section 3.2.1, laccase, manganese dependent peroxidase and lignin peroxidase, take part in PAH biodegradation, mainly in fungi.

For the aerobic biodegradation, the most extensively studied bacterial biodegradation systems are naphthalene (Resnick *et al.*, 1996; Annweiler *et al.*, 2000), phenanthrene (Menn *et al.*, 1993; Kiyohara *et al.*, 1994), anthracene and acenaphthene (Dean-Ross *et al.*, 2001; Pinyakong *et al.*, 2004). The mechanism of this biodegradation process is based on the presence of molecular oxygen to initiate the enzymatic decomposition of the PAH ring, and the first enzyme, crucial for the biodegradation process, is dioxygenase (Peng *et al.*, 2008). The genes encoding the dioxygenases are common molecular markers for PAH biodegradation monitoring. The last 20 years of PCR-based research has enabled monitoring of decomposition for several PAH compounds.

For example, for naphthalene dioxygenase, the *nahAc* gene from *Pseudomonas putida* NCBI 9816–4 (Kauppi *et al.*, 1998), the *phnAc* gene described for the first time in 1999 in *Burkholderia* sp. strain RP007 (Laurie and Lloyd-Jones, 1999; Wilson *et al.*, 2003), and the *nidA* and *nidB* genes from *Mycobacterium vanbaalenii* strain PYR-1 (Khan *et al.*, 2001; Brezna *et al.*, 2003) are all used as molecular markers for naphthalene biodegradation monitoring.

The *nahAc* gene is the molecular marker most often used for naphthalene degradation monitoring among bacteria, most likely because it was the first molecular marker discovered for naphthalene degradation monitoring. It was also used for cultivable PAH-degrading bacteria because early studies of PAH biodegradation were performed on the lab scale and often on pure bacterial and fungal cultures. However, recent research performed on environmental samples has indicated that the *nahAc* gene is less commonly present than the *phnAc* gene among the PAH-degrading communities, most likely because it is linked with cultivable bacteria that represent the minority of bacterial communities in the environment. The *phnAc* gene was obtained in PCR-based studies performed on New Zealand soil bacteria that were able to biodegrade naphthalene; however, in the same sample only the cultivable part of this community possessed the *nahAc* gene (Lloyd-Jones *et al.*, 1999). Moreover, real-time PCR analysis showed that *phnAc* was present at the same or greater concentration than *nahAc* in contaminated New Zealand soils and that adding naphthalene to a pristine soil increased the quantity of *phnAc* but not *nahAc* (Laure and Lloyd-Jones, 2000). Research by Wilson *et al.* (2003) performed with PCR, hybridization, cloning and sequencing revealed that the *nahAc* gene was distributed only among bacteria from highly contaminated sediments, while the *phnAc* gene was distributed only among taxonomically diverse isolates from the less-contaminated hillside soil. Brezna *et al.* (2003) used *nidA* and *nidB* marker genes in PCR and PFGE with Southern hybridization for the molecular characterization of *Mycobacterium* species.

PAHs are a ubiquitous recalcitrant in many environments and their biodegradation is also performed in anaerobic environments, but is slower and its mechanism is not entirely understood; thus, it is still extensively researched. The anaerobic metabolism of PAHs is linked with benzoyl-CoA prior to dearomatization and ring cleavage (Kuntze *et al.*, 2008). According to Fuchs *et al.* (2011), there are two benzoyl-CoA reductases responsible for anaerobic biodegradation of the aromatic ring: the ATP-dependent and ATP-independent reductases. The first enzyme consists of four different subunits encoded by the *bcrA* to *bcrD* genes (Breese *et al.*, 1998; Boll, 2005; Sun *et al.*, 2014). It is associated generally with only denitrifying bacteria (Boll and Fuchs, 1995; Song and Ward, 2005). The second one, ring opening enzyme 6-oxocyclohex-1-ene-1-carbonyl-CoA hydrolase, appears to be nearly the same in almost all known facultative and obligate PAH-degrading anaerobic microorganisms (Kuntze *et al.*, 2008; Li *et al.*, 2011) The *bamA* gene coding this enzyme is a perfect choice as a molecular marker for monitoring this process. This molecule is detected under denitrifying, iron-reducing, sulfate-reducing, and fermentative conditions (Porter and Young, 2013). Recently, the *bssA* gene encoding the α-subunit of benzylsuccinate synthase has been shown to be useful for anaerobic aromatic biodegradation (Li *et al.*, 2012).

The *bamA* gene has already been used in the research performed by Kuntze *et al.* (2008) to detect anaerobic PAH-degrading microorganisms in environmental samples. The primers designed in this study were used in other research by Kuntze *et al.* (2011) in which the *bamA* gene was used to monitor microorganisms that were able to biodegrade benzene in field studies. Li *et al.* (2012) analysed the diversity of the bacteria performing the anaerobic degradation of toluene and benzoate under denitrifying conditions. The *bamA* and *bssA* genes were used as two functional molecular markers by Sun *et al.* (2014) with qPCR for the quantitative analysis of toluene-degrading bacteria in soil, granular, activated and digester sludge. Staats *et al.* (2011) also used both molecular markers in qPCR to investigate the diversity, distribution and abundance of PAHs from anaerobic monoaromatic degradation bacteria in field studies of the aquifer polluted by the Banisveld landfill.

In 2007, Winderl *et al.* designed primers for *bssA* gene detection for use in their studies of anaerobic toluene and hydrocarbon degraders in contaminated aquifers. The authors suggested that this gene could be used as a 'universal' marker for BTEX (xylene) degrader monitoring. They also emphasized that the *bssA* gene could be used for rapid *bssA* gene fingerprinting (e.g. by DGGE or

T-RFLP analyses) not only in PCR detection but also in qPCR quantitative analysis.

3.3 Conclusion

The use of molecular markers in biodegradation process studies increases knowledge in the field of microbial metabolism and biochemical properties. Each year provides more information about new genes and their proven functionality in the microbial world. A great deal of molecular tools, both methods and markers, have been used in biodegradation monitoring to enable the discovery of the complexity of the microbial genome and metabolome. As presented in this review, each enzyme participating in biodegradation can be encoded by several functional genes, but there is most likely much more work necessary to discover all possible genes that enable the production of these enzymes.

References

Akutsu-Shigeno, Y., Teeraphatpornchai, T., Teamtisong, K., Nomura, O., Uchiyama, H., Nakahara, T., and Nakajima-Kambe, T. (2003). Cloning and sequencing of a poly(DL-lactic acid) depolymerase gene from *Paenibacillus amylolyticus* strain TB-13 and its functional expression in *Escherichia coli*. Appl. Environ. Microbiol. *69*, 2498–2504.

Akutsy, Y., Nakajima-Kambe, T., Nomura, N., and Nakahara, T. (1998). Purification and properties of a polyester polyurethane degrading enzyme from *Comamonas acidovorans* TB-35. Appl. Environ. Microbiol. *64*, 62–67.

Annweiler, E., Richnow, H.H., Antranikian, G., Hebenbrock, S., Garms, C., Franke, S., Francke, W., and Michaelis, W. (2000). Naphthalene degradation and incorporation of naphthalene-derived carbon into biomass by the thermophile *Bacillus thermoleovorans*. Appl. Environ. Microbiol. *66*, 518–523.

Arora, P.K., and Bae, H. (2014). Role of Dehalogenases in Aerobic Bacterial Degradation of Chlorinated Aromatic Compounds., J Chem 2014, Article ID 157974, 10 pages.

Baldrian, P. (2006). Fungal laccases – occurrence and properties, FEMS Microbiol Rev *30*, 215–242.

Bayer, E.A., Shoham, Y., and Lamed, R. (2000). Cellulose-decomposing prokaryotes and their enzyme systems In The Prokaryotes: An Evolving Electronic Resource for the Microbiological Community, M. Dworkin, S. Falkow, E. Rosenberg, K.-H. Schleifer, and E. Stackebrandt, eds. (Springer-Verlag, New York) pp. 1–62.

Beil, S., Timmis, K.N., and Pieper, D.H.(1999). 'Genetic and biochemical analyses of the tec operon suggest a route for evolution of chlorobenzene degradation genes. J. Bacteriol. *181*, 341–346.

Bogan, B., Schoenike, B., Lamar, R., and Cullen, D. (1996). Expression of lip genes during growth in soil and oxidation of anthracene by *Phanerochaete chrysosporium*. Appl. Environ. Microbiol. *62*, 3697–3703.

Bojesen, A.M., Vazquez, M.E., Robles, F., Gonzalez, C., Soriano, E.V., Olsen, J.E., and Christensen, H. (2007). Specific identification of *Gallibacterium* by a PCR using primers targeting the 16S rRNA and 23S rRNA genes. Vet Med, *123*, 262–268.

Boll, M., and Fuchs, G. (1995). Benzoyl-coenzyme A reductase (dearomatizing), a key enzyme of anaerobic aromatic metabolism – ATP dependence of the reaction, purification and some properties of the enzyme from *Thauera aromatica* strain K172. Eur. J. Biochem. *234*, 921–933.

Boll, M. (2005). Dearomatizing benzene ring reductases. J. Mol. Microbiol. Biotechnol. *10*, 132–142.

Breese, K., Boll, M., Alt-Morbe, J., Schagger, H., and Fuchs, G. (1998). Genes coding for the benzoyl-CoA pathway of anaerobic aromatic metabolism in the bacterium *Thauera aromatica*. Eur. J. Biochem. *256*, 148–154.

Brezna, B., Khan, A.A., and Cerniglia, C.E. (2003). Molecular characterization of dioxygenases from polycyclic aromatic hydrocarbon-degrading *Mycobacterium* spp. FEMS Microbiol. Lett. *223*, 177–183.

Briese, B.-H., Schmidt, B., and Jendrossek, D. (1994). *Pseudomonas lemoignei* has five poly(hydroxyalkanoic acid) (PHA) depolymerase genes: a comparative study of bacterial and eukaryotic depolymerases. J. Environ. Polym. Degrad. *2*, 75–87.

Brigham, C.J., Reimer, E.N., Rha, C., and Sinskey, A.J. (2010). Examination of PHB depolymerases in *Ralstonia eutropha:* further elucidation of the roles of enzymes in PHB homeostasis. AMB Express 2, 26.

Cai, M., and Xun, L. (2002). Organization and regulation of pentachlorophenol-degrading genes in *Sphingobium chlorophenolicum* ATCC 39723. J. Bacteriol. *184*, 4672–4680.

Chae, J.-C., Song, B., and Zylstra, G.J. (2008). Identification of genes coding for hydrolytic dehalogenation in the metagenome derived from a denitrifying 4-chlorobenzoate degrading consortium. FEMS Microbiol. Lett. *281*, 203–209.

Chen, D.M., Taylor, A.F.S., Burke, R.M., and Cairney, J.W.G. (2001). Identification of genes for lignin peroxidases and manganese peroxidases in ectomycorrhizal fungi. New Phytol. *152*, 151–158.

Chen, D.M., Bastias, B.A., Taylor, A.F.S., and Cairney, J.W.G. (2003). Identification of laccase-like genes in ectomycorrhizal basidiomycetes and transcriptional regulation by nitrogen in *Piloderma byssinum*. New Phytol. *157*, 547–554.

Coates, J.D., Anderson, R.T., and Lovley, D.R. (1996). Oxidation of polycyclic aromatic hydrocarbons under sulfate-reducing conditions, Appl. Environ. Microbiol. *62*, 1099–1101.

Crabbe, J.R., Campbell, J.R., Thompson, L., Walz, S.L., and Schwultz, W.W. (1994). Biodegradation of a colloidal ester-based polyurethane by soil fungi. Int. Biodeterior. Biodegrad. 33, 103–113.

Cullen, D.C., and Kersten, P.J.K. (2004). Enzymology and molecular biology of lignin degradation. In Biochemistry and Molecular Biology, R. Brambl and G.A. Marzluf, eds. (Springer-Verlag:Berlin), pp. 249–273.

de la Rubia, T., Linares, A., Mester, J., Munoz-Dorado, J., Romera, J., and Martinez, J. (2002). Characterization of manganese-dependent peroxidase isoenzymes from the ligninolytic fungus *Phanerochaete flavido-alba*. Res. Microbiol. *153*, 547–554.

Dean-Ross, D., Moody, J.D., Freeman, J.P., Doerge, D.R., and Cerniglia, C.E. (2001). Metabolism of anthracene by a *Rhodococcus* species. FEMS Microbiol. Lett. *204*, 205–211.

Deguchi, T., Kakezawa, M., and Nishida, T. (1997). Nylon biodegradation by lignin degrading fungi. Appl. Environ. Microbiol. *63*, 329–331.

Deguchi, T., Kitaoka, Y., Kakezawa, M., and Nishida, T. (1998). Purification and characterization of a nylon-degrading enzyme. Appl. Environ. Microbiol. *64*, 1366–1371.

Donnelly, P.K., Entry, J.A., and Crawford, D.L. (1993). Degradation of atrazine and 2,4-dichlorophenoxyacetic acid by mycorrhizal fungi at three nitrogen concentrations in vitro. Appl. Environ. Microbiol. *59*, 2642–2647.

D'Souza, T.M., Boominathan, K., and Reddy, C.A. (1996). Isolation of laccase gene-specific sequences from white rot and brown rot fungi by PCR. Appl Environ Microbiol. *62*, 3739–3744.

Embong, Z., Hitam, W.H.W., Rashid, C., Kamarudin, B., Abidin, S., Osman, S., Zainuddin, Z.F., and Ravichandran, M. (2008). Specific detection of fungal pathogens by 18S rRNA gene PCR in microbial keratitis. BMC Ophthalmol. *8*, 7.

Friedrich, J., Zalar, P., Mohorcic, M., Klun, U., and Krzan, A. (2007). Ability of fungi to degrade synthetic polymer nylon-6. Chemosphere *67*, 2089–2095.

Fuchs, G., Boll, M., and Heider, J. (2011). Microbial degradation of aromatic compounds – from one strategy to four. Nat Rev Microbiol *9*, 803–816.

Fukuzaki, H., Yoshida, M., Asano, M., and Kumakura, M. (1989). Synthesis of copoly (D, L-lactic acid) with relative low molecular weight and in vitro degradation. Eur. Polym. J. *25*, 1019–1026.

Galhaup, C., Goller, S., Peterbauer, C.K., Strauss, J., and Haltrich, D. (2002). Characterization of the major laccase isoenzyme from *Trametes pubescens* and regulation of its synthesis by metal ions. Microbiology *148*, 2159–2169.

Ghorpade, V.M., Gennadios, A., and Hanna, M.A. (2001). Laboratory composting of extruded poly(lactic acid) sheets. Bioresour. Technol. *76*, 57–61.

Giardina, P., Palmieri, G., Fontanella, B., Rivieccio, V., and Sannia, G. (2000). Manganese peroxidase isoenzymes produced by *Pleurotus ostreatus* grown on wood sawdust. Arch. Biochem. Biophys. *376*, 171–179.

Gu, J.D., Ford, T.E., Mitton, D.B., and Mitchell, R. (2000). Microbial corrosion of metals. In The Uhlig Corrosion Handbook. W. Revie, ed. (John Wiley: New York), pp. 915–927.

Guglielmi, G., and Béguin, P. (1998). Cellulase and hemicellulase genes of *Clostridium thermocellum* from five independent collections contain few overlaps and are widely scattered across the chromosome. FEMS Microbiol. Lett. *161*, 209–215.

Hakala, T.K., Hilden, K., Maijala, P., Olsson, C., and Hatakka, A. Differential regulation of manganese peroxidases and characterization of two variable MnP encoding genes in the white-rot fungus *Physisporinus rivulosus*. Appl. Microbiol. Biotechnol. 73, 839–849.

Haritash, A.K., and Kaushik, C.P. (2009). Biodegradation aspects of polycyclic aromatic hydrocarbons (PAHs): A review. J. Hazard. Mater. *169*, 1–15.

Hayward, A.C., Tollenaere, R., Dalton-Morgan, J., and Batley, J. (2015). Molecular marker applications in plants. Methods Mol. Biol. *2015*, 13–27.

Hilden, L., Johansson, G., Pettersson, G., Li, J.B., Ljungquist, P., and Henriksson, G. (2000). Do the extracellular enzymes cellobiose dehydrogenase and manganese peroxidase form a pathway in lignin biodegradation? FEBS Lett. *477*, 79–83.

Hilden, K., Martinez, A.T., Hatakka, A., and Lundell, T. (2005). The two manganese peroxidases Pr-MnP2 and Pr-MnP3 of *Phlebia radiata*, a lignin-degrading basidiomycete, are phylogenetically and structurally divergent. Fungal Genet. Biol. *42*, 403–419.

Holzbaur, E., and Tien, M. (1988). Structure and regulation of a lignin peoxidase gene from *Phanerochaete chrysosporium*. Biochem. Biophys. Res. Commun. *155*, 626–633.

Howard, G.T., and Blake, R.C. (1998). Growth of *Pseudomonas fluorescens* on a polyester-polyurethane and the purification and characterization of a polyurethanase-protease enzyme. Int. Biodeterior Biodegrad *42*, 213–220.

Howard, G.T., Ruiz, C., and Hilliard, N.P. (1999). Growth of *Pseudomonas chlororaphis* on a polyester-polyurethane and the purification and characterization of a polyurethanase-esterase enzyme. Int. Biodeterior. Biodegrad. *43*, 7–12.

Howard, G.T., Mackie, R.I., Cann, I.K.O., Ohene-Adjei, S., Aboudehen, K.S., Duos, B.G., and Childers, G.W. (2007). Effect of insertional mutations in the pueA and pueB genes encoding two polyurethanases in *Pseudomonas chlororaphis* contained within a gene cluster. J. Appl. Microbiol. *103*, 2074–2083.

Huang, C.-H., and Lee, F.-L. (2011). The dnaK gene as a molecular marker for the classification and discrimination of the Lactobacillus casei group. Antoine van Leeuwenhoek 99: 319–327.

Janusz, G., Kucharzyk, K.H., Pawlik, A., Staszczak, M., and Paszczynski, A.J. (2013). Fungal laccase, manganese peroxidase and lignin peroxidase: gene expression and regulation. Enzyme Microb. Technol. *10*, 1–12.

Jendrossek, D., Müller, B., and Schlegel, H.G. (1993). Cloning and characterization of the poly(hydroxyalkanoic acid)-depolymerase gene locus, *phaZ1*, of *Pseudomonas lemoignei* and its gene product. Eur J Biochem *218*, 701–710.

Jendrossek, D., Frisse, A., Behrends, A., Andermann, M., Kratzin, H.D., Stanislawski, T., and Schlegel, H.G. (1995). Biochemical and molecular characterization of the *Pseudomonas lemoignei* polyhydroxyalkanoate depolymerase system. J. Bacteriol. *177*, 596–607.

Johansson, T., Welinder, K.G., and Nyman, P.O. (1993). Isozymes of lignin peroxidase and manganese (II) peroxidase from the white-rot basidiomycete *Trametes versicolor*. Partial sequences, peptide maps, and amino acid and carbohydrate compositions. Arch Biochem Biophys *300*, 57–62.

Kauppi, B., Lee, K., Carredano, E., Parales, R., Gibson, D., Eklund, H., and Ramaswamy, S. (1998). Structure of an aromatic-ring-hydroxylating dioxygenasenaphthalene 1,2-dioxygenase. Structure *6*, 571–586.

Keller, C., Corcoran, M., and Roberts, R.J. (1984). Computer programs handling numceic acid sequences. Nucleic Acid Res. *12*, 397–386.

Khan, A.A., Wang, R.F., Cao, W.W., Doerge, D.R., Wennerstrom, D., and Cerniglia, C.E. (2001). Molecular cloning, nucleotide sequence, and expression of genes encoding a polycyclic aromatic ring dioxygenase from *Mycobacterium* sp. strain PYR-1. Appl. Environ. Microbiol. *67*, 3577–3585.

Kim, K.H., Ka, K.H., Kang, J.H., Kim, S., Lee, J.W., Jeon, B.K., Yun, J.K., Park, S.R., and Lee, J. (0105). Identification of single nucleotide polymorphism markers in the laccase gene of shiitake mushrooms (*Lentinula edodes*). Mycobiol *43*, 75–80.

Kinoshita, S., Kageyama. S., Iba, K., Yamada, Y., and Okada, H. (1972). Utilization of cyclic dimmer and linear oligomers of ε-aminocaproic acid by *Acromobacter guttatus* K172. Agric. Bio.l Chem. *39*, 1219–1223.

Kiyohara, H., Torigoe, S., Kaida, N., Asaki, T., Iida, T., Hayashi, H., and Takizawa, N. (1994). Cloning and characterization of a chromosomal gene cluster, pah, that encodes the upper pathway for phenanthrene and naphthalene utilization by *Pseudomonas putida* OUS82. J. Bacteriol. *176*, 2439–2443.

Kuntze, K., Shinoda, Y., Moutakki, H., McInerney, M.J., Vogt, C., Richnow, H.-H., and Boll, M. (2008). 6-Oxocyclohex-1- ene-1-carbonyl-coenzyme A hydrolases from obligately anaerobic bacteria: characterization and identification of its gene as a functional marker for aromatic compounds degrading anaerobes. Environ. Microbiol. *10*, 1547–1556.

Kuntze, K., Vogt, C., Richnow, H.H., and Boll, M. (2011). Combined application of PCR-based functional assays for the detection of aromatic-compound-degrading anaerobes. Appl. Environ. Microbiol. *77*, 5056–5061.

Laurie, A.D., and Lloyd-Jones, G. (1999). The *phn* genes of *Burkholderia* sp. strain RP007 constitute a divergent gene cluster for polycyclic aromatic hydrocarbon catabolism. J. Bacteriol. *181*, 531–540.

Laurie, A.D., and Lloyd-Jones, G. (2000). Quantification of *phnAc* and *nahAc* in contaminated New Zealand soils by competitive PCR. Appl. Environ. Microbiol. *66*, 1814–1817.

Li, D., Li, N., Ma, B., Mayfield, M.B., and Gold, M.H. (1999). Characterization of genes encoding two manganese peroxidases from the lignin-degrading fungus *Dichomitus squalens*. Biochim. Biophys. Acta *1434*, 356–364.

Li, S.M., Molina, I., Martinez, M.B., and Vert. M. (2002). Hydrolytic and enzymatic degradations of physically crosslinked hydrogels prepared from PLA/PEO/PLA triblock copolymers. J. Mater. Sci. Mater. Med. *13*, 81–86.

Li, Y.N., Porter, A.W., Mumford, A., Zhao, X.-H., and Young, L.Y. (2011). Bacterial community structure and bamA gene diversity in anaerobic degradation of toluene and benzoate under denitrifying conditions. J. Appl. Microbiol. *112*, 269–279.

Li, Y.N., Porter, A.W., Mumford, A., Zhao, X.H., and Young, L.V. (2012). Bacterial community structure and bamA gene diversity in anaerobic degradation of toluene and benzoate under denitrifying conditions. J. Appl. Microbiol. *112*, 269–279.

Liu, W., Li, L., Khan, M.A., and Zhu, F. (2012). Popular molecular markers in bacteria. Mol Gen Mikrobiol Virusol 3, 14–17.

Lloyd-Jones, G., Laurie, A.D., Hunter, D.W.F., and Fraser. R. (1999). Analysis of catabolic genes for naphthalene and phenanthrene degradation in contaminated New Zealand soils. FEMS Microbiol. Ecol. *29*, 69–79.

Luis, P., Walthera, G., Kellnera, H., Martin, F., and Buscot, F. (2004). Diversity of laccase genes from basidiomycetes in a forest soil, Soil Biol. Biochem. *36*, 1025–1036.

Marzorati, M., Balloi, A., de Ferra, F., Corallo, L., Carpani, G., Wittebolle, L., Verstraete, W., and Daffonchio, D. (2010). Bacterial diversity and reductive dehalogenase redundancy in a 1,2-dichloroethane-degrading bacterial consortium enriched from a contaminated acquifer. Microb. Cell Fact. *9*, 1–9.

Matsuda, E., Abe, N., Tamakawa, H., Kaneko, J., and Kamio, Y. (2005). Gene cloning and molecular characterization of an extracellular poly(L-lactic acid) depolymerase from *Amycolatopsis* sp. Strain K104–1. J Bacterol *187*, 7333–7340.

Menn, F.M., Applegate, B.M., and Sayler, G.S. (1993). NAH Plasmid mediated catabolism of anthracene and phenanthrenen to naphthoic acids. Appl. Environ. Microbiol. *59*, 1938–1942.

Mester, T., and Tien, M. (2000). Oxidation mechanism of ligninolytic enzymes involved in the degradation of environmental pollutants, Int Biodeter Biodegr *46*, 51–59.

Morgenstern, I., Robertson, D.L., and Hibbett, D.S. (2010). Characterization of three *mnp* genes of *Fomitiporia mediterranea* and report of additional class II peroxidases in the order Hymenochaetales. Appl. Environ. Microbiol. *76*, 6431–6440.

Mueller, R.J. (2006). Biological degradation of synthetic polyesters – enzymes as potential catalysts for polyester recycling. Proc Biochem *41*, 2124–2128.

Nakajima-Kambe, T., Onuma, F., Kimpara, N., and Nakahara, T. (1995). Isolation and characterization of a bacterium which utilizes polyester polyurethane as a sole carbon and nitrogen source. FEMS Microbiol. Lett. *129*, 39–42.

Negoro, S., Kato, K., Fujiyama, K., and Okada, H. (1994). Thenylon oligomer biodegradation system of *Flavobacterium* and *Pseudomonas*. Biodegradation *5*, 185–194.

Nomura, N., Shigeno-Akutsu, Y., Nakajima-Kambe, T., and Nakahara, T. (1998). Cloning and sequence analysis of a polyurethane esterase of *Comamonas acidovorans* TB-35. J Ferment Bioeng *86*, 339–334.

Nomura, N., Deguchi, T., Shigeno-Akutsu, Y., Nakajima-Kambe T., and Nakahara, T. (2001). Gene structures and catalytic mechanisms of microbial enzymes able to blodegrade the synthetic solid polymers nylon and polyester polyurethaoe. Biotechnol Genet Eng Rev *18*, 125–147.

Orser, C.S., and Lange. C.C. (1994). Molecular analysis of pentachlorophenol degradation. Biodegradation *5*, 277–288.

Patnaik, P. (1992). Health hazard, which includes toxic, corrosive, carcinogenic, and teratogenic properties, exposure limits. A Comprehensive Guide to the Hazardous Properties of Chemical Substances (Van Nostrand Reinhold: New York), pp. 425–445.

Peng, R.H., Xiong, A., Xue, Y., Fu, X.-Y., Gao, F., Zhao, W., Tian, Y.-S., and Yao, Q.-H. (2008). Microbial biodegradation of polyaromatic hydrocarbons. FEMS Microbiol. Rev. *32*, 927–955.

Pérez, J., Muñoz-Dorado, J., de la Rubia, T., and Martínez, J. (2002). Biodegradation and biological treatments of cellulose, hemicellulose and lignin: an overview. Int. Microbiol. *5*, 53–63.

Pinyakong, O., Habe, H., Kouzuma, A., Nojiri, H., Yamane, H., and Omori, T. (2004). Isolation and characterization of genes encoding polycyclic aromatic hydrocarbon dioxygenase from acenaphthene and acenaphthylene degrading *Sphingomonas* sp. strain A4. FEMS Microbiol. Lett. *238*, 297–305.

Pointing, S.B., Pelling, A.L., Smith, G.J.D., Hyde, K.D., Reddy, and C.A. (2005). Screening of basidiomycetes and xylariaceous fungi for lignin peroxidase and laccase gene-specific sequences. Mycol. Res. *109*, 115–112.

Pollmann, K., Beil, S., and Pieper, D.H. (2001). Transformation of chlorinated benzenes and toluenes by *Ralstonia* sp. strain PS12 tecA (tetrachlorobenzene dioxygenase) and tecB (chlorobenzene dihydrodiol dehydrogenase) dene products. Appl. Environ. Microbiol. *67*, 4057–4063.

Poter, A.W., and Young, L.Y. (2013). The *bamA* gene for anaerobic ring fission is widely distributed in the environment. Front, Microbiol, *10*, 4:302.

Power, M., van der Meer, J.R., Tchelet, R., Egli, T., and Eggen, R. (1998). Molecular based methods can contribute to assessments of toxicological risks and bioremediation strategies. J. Microbiol. Meth. *32*, 107–119.

Pranamuda, H., Tokiwa, Y., and Tanaka, H. (1997). Polylactide degradation by an *Amycolatopsis* sp. Appl. Environ. Microbiol. *63*, 1637–1640.

Pranamuda, H., and Tokiwa, Y. (1999). Degradation of poly (l-lactide) by strains belonging to genus *Amycolatopsis*. Biotechnol. Lett. *21*, 901–905.

Reiser, J., Walther, I.S., Fraefel, C., and Fiechter, A. (1993). Methods to investigate the expression of lignin peroxidase genes by the white-rot fungus *Phanerochaete chrysosporium*. Appl. Environ. Microbiol. *59*, 2897–2903.

Resnick, S.M., Lee, K., and Gibson, D.T. (1996). Diverse reactions catalyzed by naphthalene dioxygenase from *Pseudomonas* sp. strain NCIB 9816. J. Ind. Microbiol. Biotechnol. *17*, 438–457.

Rodriguez, E., Ruiz-Duenas, F.J., Kooistra, R., Ram, A., Martinez, A.T., and Martinez, M.J. (2008). Isolation of two laccase genes from the white-rot fungus *Pleurotus eryngii* and heterologous expression of the pel3 encoded protein. J. Biotechnol. *134*, 9–19.

Ruiz, A., Poblet, M., Mas, A., and Guillamon, J.M. (2000). Identification of acetic acid bacteria by RFLP of PCR-amplified 16S rDNA and 16S-23S rDNA intergenic spacer Int. J. Syst. Evol. Microbiol. *50*, 1981–1987.

Saloheimo, M., Niku-Paavola, M.-L., and Knowles, J.K.C. (1991). Isolation and structural analysis of the laccase gene from the lignin- degrading fungus *Phlebia radiate*. J Gen Microbiol *137*, 1537–1544.

Schöber, U., Thiel, C., and Jendrossek, D. (2000). Poly(3-Hydroxyvalerate) depolymerase of *Pseudomonas lemoignei*. Appl. Environ. Microbiol. *66*, 1385–1392.

Schwarz, W.H. (2001). The cellulosome and cellulose degradation by anaerobic bacteria. Appl. Microbiol. Biotechnol. *56*, 634–649.

Sei, K., Nakao, M., Mori, K., Ike, M., Kohno, T., and Fujita, M. (2001). Design of PCR primers and a gene probe for extensive detection of poly(3-hydroxybutyrate) (PHB)-degrading bacteria possessing fibronectin type III linker type-PHB depolymerases. Appl Microbiol Biotechnol. *55*, 801–806.

Shah, A.A., Hasan, F., Hameed, A., and Ahmed, S. (2008). Biological degradation of plastics: A comprehensive review. Biotechnol Adv *26*, 246–265.

Shimao, M. (2001). Biodegradation of plastics. Curr Opin Biotechnol *12*, 242–247.

Soden, D.M., and Dobson, A.D. (2003). The use of amplified flanking region-PCR in the isolation of laccase promoter sequences from the edible fungus *Pleurotus sajor-caju*. J. Appl. Microbiol. *95*, 553–562.

Song, B., and Ward, B.B. (2005). Genetic diversity of benzoyl coenzyme A reductase genes detected in denitrifying isolates and estuarine sediment communities. Appl. Environ. Microbiol. *71*, 2036–2045.

Staats, M., Braster, M., and Roling, W.F.M. (2011). Molecular diversity and distribution of aromatic hydrocarbon-degrading anaerobes across a landfill leachate plume. Environ. Microbiol. *13*, 1216–1227.

Stewart, P., Kersten P., Vanden Wymelenberg A., Gaskell J., and Cullen D.. 1992. The lignin peroxidase gene family of *Phanerochaete chrysosporium* complex regulation by carbon and nitrogen limitation and the identification of a second dimorphic chromosome. J. Bacteriol. *174*, 5036–5042.

Sun, W., Sun, X., and Cupples, A.M. (2014).Presence, diversity and enumeration of functional genes (bssA and bamA) relating to toluene degradation across a range of redox conditions and inoculum sources. Biodegradation *25*, 189–203.

Teeraphatpornchai, T., Nakajima-Kambe, T., Shigeno-Akutsu, Y., Nakayama, M., Nomura, N., Nakahara, T., and Uchiyama, H. (2003). Isolation and characterization of a bacterium that degrades various polyester-based biodegradable plastics. Biotechnol Lett *25*, 23–28.

Tomita, K., Kuraki, Y., and Nagai, K. (1999). Isolation of thermophiles degradating poly (L-lactic acid). J Biosci Bioeng *87*, 752–755.

Torres, A., Li, S., Roussos, S., and Vert, M. (1996). Screening of microorganisms for biodegradation of poly (lactic acid) and lactic acid-containing polymers. Appl. Environ. Microbiol. *62*, 2393–1397.

Vega, R.E., Main, T., and Howard, G.T. (1999). Cloning and expression in *Escherichia coli* of a polyurethane-degrading enzyme from *Pseudomonas fuorescens*. Int. Biodeterior. Biodegrad. *43*, 49–55.

Vermis, K., Coenye, T., Mahenthiralingam, E., Nelis, H.J., and Vandamme, P. (2002). Evaluation of species-specific recA-based PCR tests for genomovar level identification within the *Burkholderia cepacia* complex. J. Med. Microbiol. *51*, 937–940.

Wilson, M.S., Herrick, J.B., Jeon, Ch.O., Hinman, D.E., and Madsen, E.L. (2003). Horizontal transfer of *phnAc* dioxygenase genes within one of two phenotypically and genotypically distinctive naphthalene-degrading guilds from adjacent soil environments. Appl. Environ. Microbiol. *69*, 2172–2181.

Winderl, Ch., Schaefer, S., and Lueders, T. (2007). Detection of anaerobic toluene and hydrocarbon degraders in contaminated aquifers using benzylsuccinate synthase (*bssA*) genes as a functional marker. Environ. Microbiol. *9*, 1035–1046.

Yamamoto, S., and Harayama, S. (1995). PCR amplification and direct sequencing of gyrB genes with universal primers and their application to the detection and taxonomic analysis of *Pseudomonas putida* strains. Appl. Environ. Microbiol. *61*, 1104–1109.

Zverlov, V.V., Kellermann, J., and Schwarz, W.H. (2005). Functional subgenomics of *Clostridium thermocellum* cellulosomal genes: Identification of the major catalytic components in the extracellular complex and detection of three new enzymes. Proteomics 5, 3646–3653.

Metabolomics and Crucial Enzymes in Microbial Degradation of Contaminants

4

Rafał Szewczyk and Konrad Kowalski

Contents

Abstract

Metabolic pathways in living organisms are series of chemical reactions and typically consist of the enzymatic activities of intra- or extracellular enzymes. This chapter focuses on the metabolomic workflow characterization, which may be briefly characterized as the initial discovery, profiling, and final quantitation of the compound or a class of compounds of interest. This approach incorporates different instrumentation techniques and bioinformatics tools described in the chapter. The metabolomics of biodegradation processes is presented here as a direct result of enzymatic activity of the microbes, plants, and higher organisms towards xenobiotics, and includes examples of the typical degradation pathways catalysed by enzymes involved in certain steps of toxic substance removal. In the last section, the examples of targeted and untargeted biodegradation process characterizations at the cellular metabolism level are described.

4.1 Introduction

Metabolomics is one of the largest branches of system biology, focused on the small molecules, metabolites (metabolome), present in a certain biological system (cell, tissue, organ, biological fluid, or organism). The rapidly emerging field of metabolomics combines strategies to identify and quantify cellular metabolites using analytical technologies with the application of statistical and multi-variant methods for information extraction and data interpretation. All omics approaches, including genomics, transcriptomics, proteomics, and metabolomics, are considered important tools to be applied and utilized in order to understand the biology of an organism and its response to environmental stimuli or genetic perturbation (Roessner and Bowne, 2009).

4.1.1 Definitions

Metabolomics and metabonomics are terms that were used interchangeably at the time of their creation. Recently, some attempts have been made to distinguish between the specific meanings of both terms. According to this distinction, metabolomics mainly focuses on quantitative determination of certain metabolites in biological material, as their molecular fingerprint, while metabonomics takes into account both quantitative and qualitative analyses of the metabolic response, which arises from the interactions between the analysed system and its environment (Nicholson and Lindon, 2008). These changes in metabolic profiles described by metabonomics can be a result of the interactions with other biological systems or represent a response to stress factors, such as biocides. Differences in these approaches can be simply visualized as differences in colony morphology during microorganism cultivation on solid media, where metabolomics is used to examine colony characteristics on universal medium, while the cultivation on differentiation media is used to examine cellular responses to medium components.

The composition of the metabolome of a particular biological system is the sum of both physiological and biochemical naturally occurring processes and their response to different external factors. The analyses of primary and secondary metabolites allow the determination of the biochemical phenotype of the organism as a whole, or specific tissues or cell types, and their comparison with each other in certain environmental conditions.

4.2 From discovery to quantitation

4.2.1 Biomarker identification through sample analysis

Low molecular weight primary and secondary metabolites present in biological materials belong to the different classes of natural products formed during various enzymatic activities. These compounds vary enormously in different physicochemical properties, such as molecular weight range, polarity, vapour pressure, volatility, or thermal stability. These differences create the biggest technical problem in metabolomics – the lack of universal sample preparation procedure. Each experimental procedure needs to be developed for a certain sample type (matrix), type of detection, and more or less defined group of analytes. Depending on their intrinsic properties, the compounds can be separated using gas chromatography, liquid chromatography, or capillary electrophoresis. Various methods can be used for the detection of their respective metabolites as well. Currently, the instrumental methods of metabolite profiling are changing, both the separation of complex mixtures of substances (chromatographic systems) and the identification of the individual components of these mixtures, using different methods for the detection of sample components: nuclear magnetic resonance spectroscopy (NMR), mass spectrometry (MS), Fourier transform infrared spectrophotometry (FT IR), spectrophotometry in ultraviolet (UV), and Raman spectroscopy. A large number of different experimental approaches have been developed based on this wide variety of ways of performing metabolomic experiments. Different methodological approaches used for the analyses of primary and secondary metabolites are as follows:

Unknown metabolite screening – simultaneous qualitative and quantitative analyses of the highest possible number of metabolites (Martin *et al.*, 2015).

Targeted metabolomics analysis – the analysis of metabolites produced in a specific metabolic pathway or specific target enzymes.

Metabolite profiling – quantitative examination of defined metabolite groups. This approach is commonly used in medical research (clinical and pharmacological).

Metabolic fingerprinting – an approach to sample classification based on the comparison of rates of intracellular metabolites. Recorded signals are obtained using instruments that provide a high-speed analysis, without the need for separation and identification of individual components present in the test samples (MALDI-TOF, NMR, IR, DIMS) (Howlett *et al.*, 2014; Vázquez-Fresno *et al.*, 2014);.

Metabolic footprinting – a method for the sample classification based on the comparison of rates of extracellular metabolites (exometabolome). Here, the same analytical techniques as for the metabolic fingerprinting are used (Ryffel *et al.*, 2015; Sue *et al.*, 2011).

Metabolite flux analysis (fluxomics) – an approach focused on the analysis of radiolabelled metabolites (permanent isotopes ^{13}C and/or ^{15}N). Labelled substrates are introduced into the environment as the indicator molecules for particular metabolic pathways (Guerrasio *et al.*, 2013; Hanke *et al.*, 2013).

4.2.2 Commonly used statistical methods

The number of data obtained in metabolomics studies is extremely large, and advanced statistical methods and bioinformatics tools are necessary in order to interpret qualitative and quantitative results. Each sample needs to be prepared and tested in an appropriate number of repetitions, because of the diversity of biological samples, and the possibility of making technical mistakes during the preparation of samples and instrumental analysis. The main problem during metabolomic analysis is the determination of sets of potentially interesting metabolites in the analysed samples. Multivariate comparison between groups of samples is performed by advanced statistical methods, such as principal component analysis (PCA), partial least squares-discriminant analysis (PLS-DA), or independent constituent analysis (ICA) (Scholz and Selbig, 2007). These methods provide graphical visualization of sample differences based on differences between the analysed variables. Multivariate statistical analysis allows comparisons of very complex data, but it is necessary to normalize the obtained values. The normalization process leads to the unification of statistical analysis results. This is done by excluding all systematic errors or other non-biological factors that may affect the result (e.g. injection volume and accurate weight of sample). Normalization can be performed either as an integral part of the experiment (by using an internal standard) or after the experiment, using the supported statistical software tools, such as MetTailor (http://mettailor.sourceforge.net) (Sysi-Aho *et al.*, 2007; Chen *et al.*, 2015). Comparative analysis can be performed on the collected data following the adequate data normalization. Due to a large amount of signals obtained, particularly by the high resolution LC-MS/MS analysis, comparative analysis should only include information concerning the chemical diversity of samples. Preliminary statistical visualization of the groups present in the analysed sample populations, done by PCA, PLS-DA, or ICA, needs to be additionally confirmed by the determination of significant differences between groups, which can be determined using the classical statistical analysis methods, such as Student's t-test or analysis of variance (ANOVA) (Chu *et al.*, 2015). Recently, most metabolomics instrumentation (e.g. for GC-MS, LC-MS, ICP-MS, CE-MS) manufacturers have started providing their customers with appropriate statistical tools: Mass Profiler (Agilent), MarkerView (Sciex), or Profile Analysis (Bruker). These tools and workflows lead to the identification of a small group of unknown biomarker compounds that are further analysed, and subjected to a detailed qualitative metabolite identification.

4.2.3 Qualitative approaches for compound identification

A crucial step in every metabolomic screening study is a proper compound identification. In mass spectrometry analysis, this identification is based on both isotopic patterns of pseudo-molecular ions and fragmentation patterns.

Monoisotopic mass, isotopic pattern, and potentially occurring adducts in pseudo-molecular ion spectra give accurate information about the molecular formula of an unknown compound. Most mass spectrometer manufacturers, who provide qualitative software, include formula-finder scripts that can suggest a potential molecular formula, based on the defined limit values of each chemical element in an analysed compound. *In silico* predictions of the chemical structure based on MS/MS spectra of unknown compounds are far more complicated. Depending on the obtained fragmentation spectra, two approaches to structure confirmation are possible. First, searching for the spectral similarity in commercially available databases of identified compounds. The probability of the efficient identification of unknown compound, using this approach, depends directly on the size and diversity of compound spectral databases. Generally, rare and unique biomarker compounds are not included in spectral databases, and other approaches are needed for their identification.

4.2.4 Metabolite quantitation

Among all metabolomic techniques, triple quadrupoles (QqQ) tandem mass spectrometry is considered a reference tool for the absolute quantitation of small molecules. It demonstrates high sensitivity and specificity when selected (or multiple) reaction monitoring (SRM or MRM) method is used, which can be applied for the quantitation of trace-level metabolites, with detection limit of ng/mL in sample matrices (e.g. plasma, serum, or cellular media). SRM or MRM can measure the real concentration of metabolites through absolute quantitation, by correlating analyte signal intensities with a calibration curve constructed using spiked stable isotope labelled analogues. However, QqQ-based quantitative methods are limited to targeted metabolites and often neglect information regarding other metabolites that are invisible due to the specificity of the analysis. Some analytes have non-specific transitions that are common for matrix interferences (e.g. the neutral loss of H_2O or CO_2). On the other hand, high-resolution mass spectrometry (HRMS), e.g. FT-MS (e.g. FT-ICR and Orbitrap) and TOF or QTOF, can provide global MS detection and offer better solutions compared with MRM analysis. HRMS, especially FT-MS in full-scan mode, can determine virtually all compounds present in a sample, because of its high resolving power, high mass accuracy, and broad dynamic range (Xiao *et al.*, 2012).

4.2.4.1 Physiological levels of metabolites

Identification of physiological levels of each metabolite in a defined biological system is one of most challenging issues connected with the absolute quantitation of metabolites. These levels need to be redefined for each individual system, depending on the experimental procedures. Physiological levels are measured using a non-stimulated sample control in the metabonomics approach, while in the metabolomic approach, the number of positive and negative control samples often needs to be higher, in order to obtain detailed information about the levels of the metabolites.

4.2.4.2 Interpretation and reporting of results

The key element in the long-term development of metabolomics as a science is standardization, including not only standardization of results, but also the unification of data file formats, data processing methods, sample preparation methods, and standardization of methods of supplementing the information to the databases. Properly collected, standardized data will provide the basis for future complex meta-analyses in the context of entire biological systems. With the aim of methodology standardization, the Metabolomics Standards Initiative of the Metabolomics Society formed the Chemical Analysis Working Group (CAWG) (http://cosmos-fp7.eu/msi), and within it, five working groups focused on biological context metadata, chemical analysis, data processing, the exchange format, and ontology (Sumner *et al.*, 2007) were formed. In recent standardization protocols, CAWG defined four levels of metabolite identification and reporting: level 1 – identified metabolites; level 2 – putatively annotated compounds; level 3 – putatively characterized compound classes; level 4 – unknown compounds (Salek *et al.*, 2013). To assign level 1 identification to a metabolite, it is necessary to compare two or more orthogonal properties of reference standards to experimental data, while level 2 and 3 annotations do not require reference standards. Chemical Analysis WG recommended the levels of identification, common names, and structural codes (e.g. InChI or SMILES) to be defined in publications and during database submission. Despite the intense work of the Metabolomics Society, published standards are still not universally respected. Six out of 20 randomly chosen metabolomics studies in 2013 defined in detail the annotation or identification methodology (Salek *et al.*, 2013). Preliminary recommendations of the Chemical Analysis WG for LC-MS/MS metabolite identification suggest a minimum of two independent domains relative to an authentic compound (retention time, unit, high-resolution mass spectrum, or MS/MS fragmentation pattern). As the complexity of metabolomic data grows, a strict observance of the recommendations for the standardization of results becomes even more important (Sumner *et al.*, 2007).

4.2.5 Bioinformatics and software tools

The basic tools are metabolic databases that allow the correct assignment of unknown compounds. Most of the available databases are free, with the possibility of the addition of the obtained experimental results. Depending on the application, they can be divided into four main groups:

- Comprehensive metabolomic databases – containing extensive transcriptomic, proteomic, and metabolomic data for creation of proper metabolic networks.
- Metabolic pathway databases – describing biological pathways in detail, including both metabolic and signalling pathways.
- Compound-specific databases – providing information about known or predicted compounds that could be possible metabolites.
- Spectral databases – providing mass spectra of metabolites for comparison with experimental results.

Databases useful for the metabolomics research are listed in Table 4.1.

Additionally, there are many bioinformatics tools allowing LC-MS data processing, identification of unknown metabolites, library searching, and statistical processing of the obtained results. Basic freeware used for metabolomic data analyses is listed in Table 4.2.

Table 4.1 Metabolomic freeware/open-source databases

Databases	Organism/matrix	Content	References
Kyoto Encyclopaedia of Genes and Genomes (KEGG): www.genome.jp/kegg	>700 different organism (from animals, plants and bacteria),	Contains over 15,000 compounds, 7742 drugs and nearly 11,000 glycan structures	Kanehisa *et al.* (2014)
BioCyc: http://biocyc.org	Contains collection of >3000 organism-specific Pathway/Genome Databases (PGDBs), each containing the full genome and predicted metabolic network.	Within the database have been separated organisms-dependent database: HumanCyc, EcoCyc and BsubCyc.	Caspi *et al.* (2014)
EcoCyc http://ecocyc.org	*Escherichia coli* K-12 MG1655	-	Keseler *et al.* (2013)
BsubCyc: http://bsubcyc.org	*Bacillus subtilis* (based on *B. subtilis* 168 genome sequence)	-	Caspi *et al.* (2014)
Reactome: http://www.reactome.org	19 species including *C. elegans*, *S. cerevisiae*, S. pombe, *P. falciparum* and *M. tuberculosis*	Contains information about over 20 000 metabolic pathways and around 70 000 reactions.	Croft *et al.* (2014)
Human Metabolome Database (HMDB): www.hmdb.ca	Human (*Homo sapiens*)	Contains over 7900 human metabolites in occurring in broad concentration range (1 µM > 1 nM)	Wishart *et al.* (2009)
Yeast Metabolome Database (YMDB): www.ymdb.ca	Yeast (*Saccharomyces cerevisae*)	Contains 2027 metabolites with 857 associated enzymes and 138 associated transporters	Jewison *et al.* (2011)
E. coli Metabolome Database (ECMDB): http://ecmdb.ca	*Escherichia coli* (strain K12, MG1655)	Contains 3755 metabolites with 1402 associated enzymes and 387 associated transporters. Includes also 1542 metabolic pathways linked to 3011 metabolites.	Guo *et al.* (2013)

Table 4.1 Continued

Databases	Organism/matrix	Content	References
BiGG: http://bigg.ucsd.edu	Integrates more than 70 published genome-scale metabolic networks into a single database	Contains 2,766 metabolites, and 3,311 metabolic and transport reactions	King *et al.* (2015)
MetaboLights: http://www.ebi.ac.uk/metabolights	Integrated cross-species and cross-technique metabolite database	Contains information about over 17 thousand compound derived from more than 2 000 organisms	Haug *et al.* (2013)
PubChem: http://pubchem.ncbi.nlm.nih.gov/	Information about structure, nomenclature and calculated physico-chemical data for compounds without their species-classification	Contains information about over 61 million chemical compounds	Kim *et al.* (2015)
ChEBI: http://www.ebi.ac.uk/chebi	Database mostly focused on natural products (metabolites) or synthetic products used to intervene in the processes of living organisms (drugs or toxins).	Contains information about over 47 000 chemical compounds	Hastings *et al.* (2013)
ChemSpider: http://www.chemspider.com	Contains information from around 500 diverse sources.	Contains information about over 35 million chemical compounds	Pence and Williams (2010)
In Vivo/In Silico Metabolites Database (IIMDB): http://metabolomics.pharm.uconn.edu/iimdb	Mammalian metabolites, drugs, secondary plant metabolites, and glycerophospholipids	Includes around 23,000 known compounds collected from existing biochemical databases plus more than 400,000 computationally generated human phase-I and phase-II metabolites of these known compounds.	Menikarachchi *et al.* (2013)
MassBank: http://www.massbank.jp	-	Contains over 41,000 mass spectra for over 35,000 different compounds	Horai *et al.* (2010)
METLIN: https://metlin.scripps.edu	-	Include over 240.000 MS spectra of different metabolites and over 80 000 MS/MS spectra (> 14,000 in high resolution)	Smith *et al.* (2005)
mzCloud: www.mzcloud.org	-	Includes over 640,000 mass spectra (MS and MS^n) collected in 5686 spectral trees for more than 44,000 compounds	-

Table 4.2 Basic freeware/open-source metabolomic bioinformatic environments

Software	Description	References
MetaboloAnalyst: www.metaboanalyst.ca	Web-based processing tool that accepts a variety of metabolomics data types including LC-MS. Allow to multipurpose sample processing such as statistical analysis (based on PCA and t-test), metabolite identification or metabolite pathway profiling	Xia *et al.* (2015)
OpenMS: http://open-ms.sourceforge.net	Open-source C++ library for LC/MS data management and analyses creating infrastructure for the rapid development of mass spectrometry-related software	Sturm *et al.* (2008)
XCMS Online: https://xcmsonline.scripps.edu	Cloud-based platform evolved from XCMS software for LC/MS-based data analysis. Software incorporates nonlinear retention time alignment, feature detection, and feature matching. Additionally allow automatically search MS/MS data against data from known metabolites in METLIN database	Tautenhahn *et al.* (2013)

In addition to these comprehensive software tools, many relatively simple and free software tools exist, such as:

- Fragment iDentificator (FiD) – dedicated to structural identification of product ions produced with tandem mass spectrometric measurements. It provides combinatorial search over all possible fragmentation paths and patterns, using unique algorithm for the ranking of potential structures (Heinonen *et al.*, 2008).
- Seven Golden Rules Software – Microsoft Excel-based software, for molecular formula calculation from accurate (HR) monoisotopic mas (Kind and Fiehn, 2007).

Furthermore, each of the mass spectrometer manufacturers offers their own chemometric tools. These software tools (or program packages) enable statistical processing, commercial databases search, as well as communication with open databases, such as ChemSpider or PubChem.

4.3 Metabolomics – degradation pathways and enzymes

4.3.1 General characterization of the enzymatic processes

Enzymes are biological catalysts that drive biochemical reaction rates forward. The conversion of substrates into products is conducted under conditions that lower the activation energy of the reaction. In comparison with most chemical catalysts, enzymes are usually substrate-selective, catalysing only specific reactions and remaining unchanged at the end of the reactions.

Most enzymes are proteins or glycoproteins, made up of at least one amino acid chain. The functionality of an enzyme depends on its conformation and alignment, its binding targets, and on the reaction targets. The enzyme regions directly involved in the catalytic process are called the active sites, and each enzyme may bind one or more protein or non-protein groups, called cofactors, that are crucial for its catalytic activity. These groups may be associated with the active sites through either covalent or non-covalent bonds. In a holoenzyme (enzyme combined with all necessary cofactors), the protein or glycoprotein moiety is called the apoenzyme, while the non-protein moiety is called the prosthetic group or cofactor.

A cofactor can be:

1. A coenzyme – organic molecules, usually vitamins or vitamin-derived compounds, which are loosely attached to the enzyme molecule combining with the enzyme–substrate complex.
2. A prosthetic group – an organic substance that is permanently attached to the protein or apoenzyme portion.
3. A metal ion-activator – positively charged metal ions, which temporarily bind to the active site of the enzyme, such as K^+, Fe^{2+}, Fe^{3+}, Cu^{2+}, Co^{2+}, Zn^{2+}, Mn^{2+}, Mg^{2+}, Ca^{2+} or Mo^{3+}.

In order for two molecules to react, it is necessary for them to collide with each other in the right orientation and with the proper energy. The energy level called activation energy must be sufficient to overcome the energetic barrier of the reaction. The enzyme-catalysed reaction is a fast, selective, and multi-step process. The key benefit of this reaction is that enzyme and substrate form a reaction intermediate with a lower activation energy compared with the reaction between reactants without a catalyst (Fig. 4.1).

Enzyme selectivity and specificity can be explained by two possible mechanisms: the lock and key and induced fit theories. In the first case, an enzyme and its substrate simply fit together, making an enzyme–substrate complex. In the second model, the enzyme molecule changes its shape as

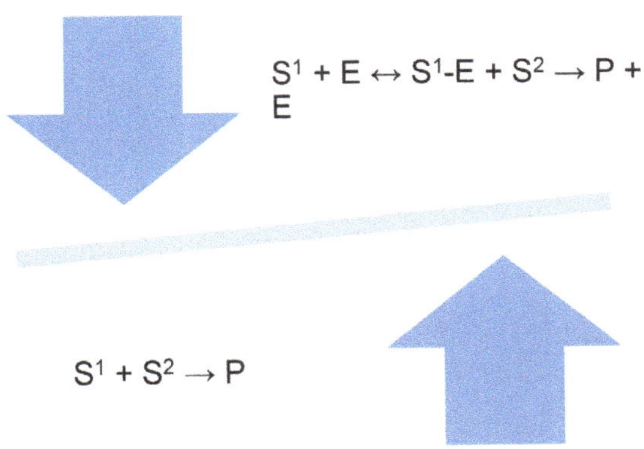

Figure 4.1 A simplified scheme of the energy consumption in the chemical and enzyme-catalysed reactions. S, substrate; E, enzyme; P, product.

a result of the approaching substrate molecule. This theory relies on the molecular flexibility, especially where single covalent bonds may freely rotate.

Major factors affecting enzyme activity are temperature and pH. As the temperature rises, a greater kinetic energy is applied to the reacting system, resulting in the higher chance of a successful collision leading to the increase in reaction rates. The enzyme structure begins to denature above the optimal temperature (Fig. 4.2). Furthermore, enzymes show the highest activities within relatively small optimal pH range, as the changes in pH can make and/or break intra- and intermolecular bonds, which leads to changes in the enzyme shape and its catalytic properties (Fig. 4.2). Additionally, the rate of an enzyme-catalysed reaction depends on enzyme and substrate concentrations, and it increases with the increase in these concentrations (Fig. 4.2). However, if only the substrate concentration is increased, the reaction rate rises to the point of saturation of the enzyme active sites. The enzyme–substrate complex has to dissociate, in order to release active sites for another portion of substrate to react (Fig. 4.2). If the substrate concentration is high and the temperature and pH is optimal and constant, the rate of catalysis is proportional to the concentration of the enzyme.

Enzyme activity may be reduced or stopped by numerous factors. Among them, most important ones are reaction conditions (see above), concentration of the substrate, cofactors or products, and the presence of various kinds of enzyme inhibitors. Inhibitors can block or distort the active site of the enzyme (active site-directed inhibitors) or can attach to the other parts of the enzyme and change its conformation and structure (non-active site-directed inhibitors).

4.3.2 General characterization of the enzymes involved in biodegradation

The removal of toxic compounds from the environment using microorganisms and plants are the most favoured forms of removal. Among all advantages of biodegradation, the selectivity of enzymatic reactions is the most important one. Purified enzymes are rarely applied for bioremediation, as many of the enzymes involved in the biodegradation processes are typical intercellular proteins that require special conditions, cofactors, and cellular machinery. However, the extracellular enzymes are typically very resistant to the environmental conditions and stay active in a wide range of temperatures, pH, humidity, redox factors, or different inhibitors. Therefore, these purified enzymes have been applied several times for the removal of xenobiotics or biodegradation of natural compounds.

The enzymes used for biodegradation belong to the six described categories (Lehninger *et al.*, 2004):

- oxidoreductases – catalyse the electron and proton transfer from a donor to an acceptor;
- transferases – transfer functional groups from a donor to an acceptor;

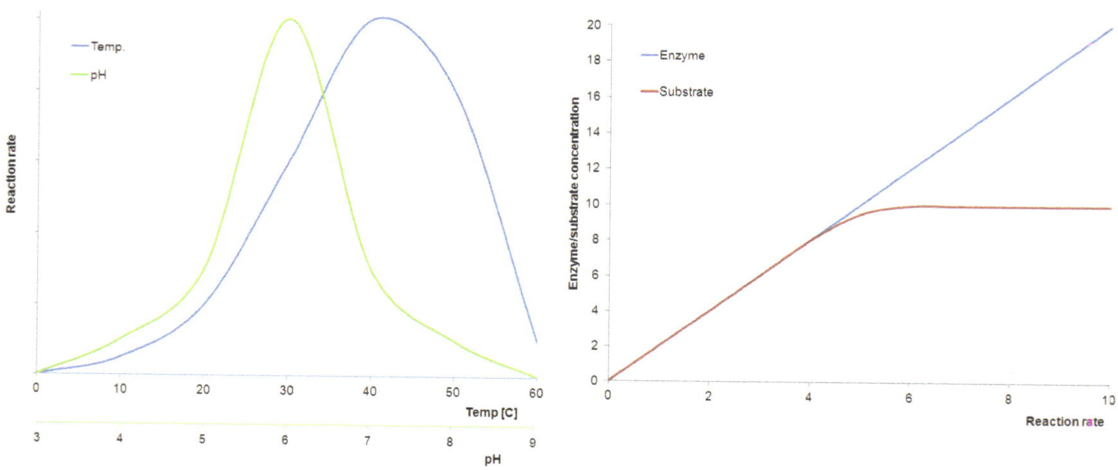

Figure 4.2 Factors affecting enzyme catalytic properties.

- hydrolases – catalyse the cleavage of C-C, C-O, C-N and other bonds using a water molecule;
- lyases – facilitate the cleavage of C-C, C-O, C-N and other bonds by elimination, leaving double bonds (in the reverse mode, they can catalyse the addition of groups across double bonds);
- isomerases – conduct geometric, structural rearrangements or isomerizations of the substrate;
- ligases (synthetases) – catalyse the joining of two molecules.

The most important enzyme classes involved in the biodegradation processes belong to oxidoreductases and hydrolases. Biodegradation pathways catalysed by different enzymes generally lead to the detoxification of the compounds. However, in some cases the reaction product may be more harmful than the original substrate. Biodegradation processes are divided into three major subgroups according to the toxicity of the final product:

1. Detoxification – the one- or multi-step reaction product is less toxic than the substrate, usually as a result of a simple modification (e.g. methylation or oxidation of the functional group) or conjugation (e.g. cysteine or glucuronic acid).
2. Activation – the one- or multi-step reaction product is more toxic than the substrate, as a result of a simple substrate modification (e.g. oxidation or dehydrogenation).
3. Mineralization – complete decomposition of the substrate to simple compounds (e.g. H_2O, CO_2, CH_4, and inorganic compounds). This process is always multi-step and incorporates several enzymatic reactions conducted by a single organism or a consortium of organisms.

All of these mechanisms can be either cometabolic or non-cometabolic. Cometabolism in the biodegradation reactions can be considered on cellular or population metabolism levels. Cellular cometabolism is a process in which the product formation takes place only in the presence of an additional compound acting as a sole source of energy (e.g. glucose) and nitrogen (e.g. amino acids). Cometabolism on the population level represents several organisms catalysing the reaction, and the by-products are used as substrates in the chain reactions conducted by different organisms within the population.

4.3.3 Oxidoreductase degradation pathways

Oxidoreductases catalyse the removal of toxic organic compounds in various organisms, including bacteria, fungi, and higher plants. The mechanisms of oxidoreductase catalytic activity relay on the cleavage of chemical bonds, together with a parallel transfer of electrons from a donor (reduced organic substrate) to an acceptor (chemical compound). Oxidoreductases include several kinds on enzymes, such as oxygenases, laccases and peroxidases.

4.3.3.1 Oxygenases

Oxygenases conduct the oxidation of reduced substrates by transferring oxygen from molecular oxygen (O_2), using FAD/NADH/NADPH as co-substrates. These enzymes are crucial in the metabolism of aromatic and aliphatic compounds, and have a broad substrate range, including the halogenated compounds, which form the largest group of environmental pollutants (e.g. pesticides, hydraulic and coolant fluids, plasticizers, chemical synthesis ingredients, or lignin bleaching agents in the paper industry). Based on the number of oxygen atoms necessary for oxygenation, two subcategories can be distinguished – monooxygenases and dioxygenases. The enzymatic activity of oxygenases results in the formation of various oxidized by-products (e.g. alcohols, ketones, or carboxylic acids), and it commonly ends with a ring cleavage, leading to the complete mineralization of the substrate via TCA cycle, in the case of aromatic compounds (Fetzner 2003; Arora *et al.*, 2009).

Monooxygenases incorporate one atom of the oxygen molecule into the substrate (Fig. 4.3). This class of enzymes is divided into two subclasses based on the presence of cofactor: flavin-dependent monooxygenases, containing flavin as a prosthetic group and requiring NADP or NADPH as a coenzyme, and haem-containing P450 monooxygenases, which are present in both eukaryotic and prokaryotic organisms (Karigar and Rao, 2011). Monooxygenases catalyse a very wide range of oxidative reactions with different substrates, including simple alkanes or complex intracellular molecules, such as steroids. For the biodegradation processes, monooxygenases are highly region- and stereoselective towards numerous substrates. The application of monooxygenases in the desulfurization, dehalogenation, denitrification,

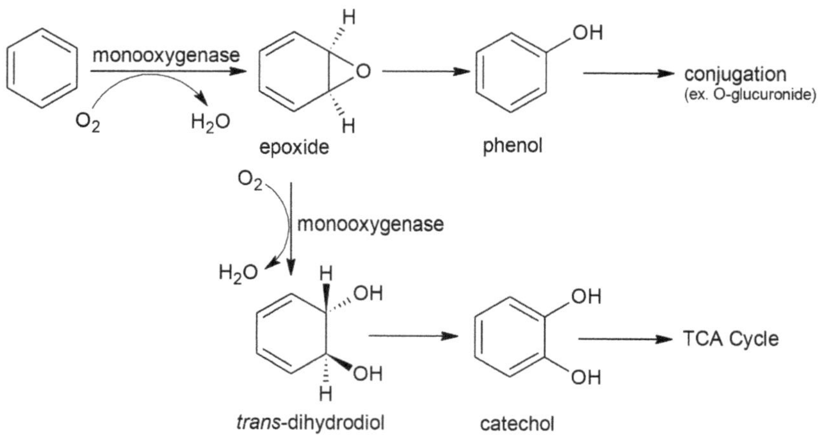

Figure 4.3 Monooxygenase mediated degradation of aromatic compounds.

ammonification, hydroxylation, biotransformation, and mineralization of various aromatic and aliphatic compounds is a key feature of their use in biodegradation reactions.

Numerous examples of the monooxygenase-mediated biodegradation pathways exist. They are specific for different substrates, including methane, alkanes, cycloalkanes, alkenes, haloalkenes, ethers, aromatic and heterocyclic hydrocarbons, halogenated-, sulfured-, or nitro-aromatics (Grosse *et al.*, 1999; Kulkarni and Chaudhari, 2007; Szewczyk and Długoński, 2009; Copley, 2009; Eshelli *et al.*, 2015). A good example of the monooxygenase catalytic activity is pentachlorophenol (and other chlorophenols), or the 2,4-D degradation pathway. This reaction is a key step for the formation of the further by-products (Szewczyk and Długoński, 2009) and may lead to complete mineralization of the compounds EAWAG Biocatalysis/Biodegradation Database (http://eawag-bbd.ethz.ch/index.html). As a result of oxidative dehalogenation, the chlorine atom is removed from the position 4 of the aromatic ring and substituted with a hydroxyl group, which finally leads to the formation of 2,3,5,6-tetrachlorohydroquinone (TeCH) (Fig. 4.4) or 2,4-dichlorophenol (Fig. 4.5).

Dioxygenases are multicomponent enzyme systems that introduce two oxygen atoms into substrate, and primarily oxidize aromatic compounds. Aromatic hydrocarbon dioxygenases specifically catalyse the oxygenation of numerous structures of substrates, and belong to a large family of Rieske non-haem iron oxygenases (Rieske (2Fe–2S) cluster and mononuclear iron in each α subunit) (Dua *et al.*, 2002). Catechol dioxygenases act as natural compounds for the degradation of environmental aromatics, and transform aromatic precursors into aliphatic products (Fig. 4.6). They have been found in the soil bacteria, Actinomycetales, fungi, and higher organisms (Heller *et al.*, 2010; Saito *et al.* 2000; Zhou *et al.*, 2006; Yuasa and Ball, 2011). The intradiol cleaving enzymes utilize Fe(III), while in some cases extradiol cleaving enzymes utilize Fe(II) and Mn(II) (Karigar and Rao, 2011). Similar to monooxygenase pathways, dioxygenase involvement may lead to a complete decomposition of the molecules via multi-component reactions, where some steps are catalysed by dioxygenases (Figs. 4.4 and 4.5).

4.3.3.2 Laccases

Laccases are copper-containing oxidoreductases that catalyse the mono-oxidation of various substrates (e.g. phenols and aromatic or aliphatic amines) to the corresponding radicals, using molecular oxygen as the final electron acceptor. These enzymes are primarily involved in the depolymerization of lignin, and represent very important nutrient enzymes for microorganisms (Karigar and Rao, 2011). The enzymes are particularly widespread in ligninolytic basidiomycetes, but they can be found in certain prokaryotes, insects, and plants as well, indicating that the laccase redox process is ubiquitous in nature (Baldrian, 2006). Laccases play important roles in several biometabolic steps, including those involved in fungal pigmentation, plant lignification, lignin biodegradation, humus turnover, and cuticle sclerotization, wherein

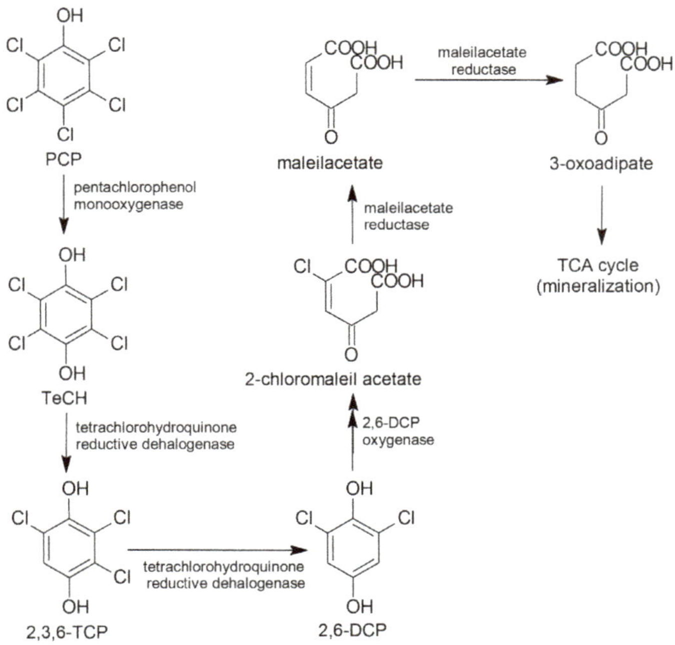

Figure 4.4 Pentachlorophenol degradation pathway.

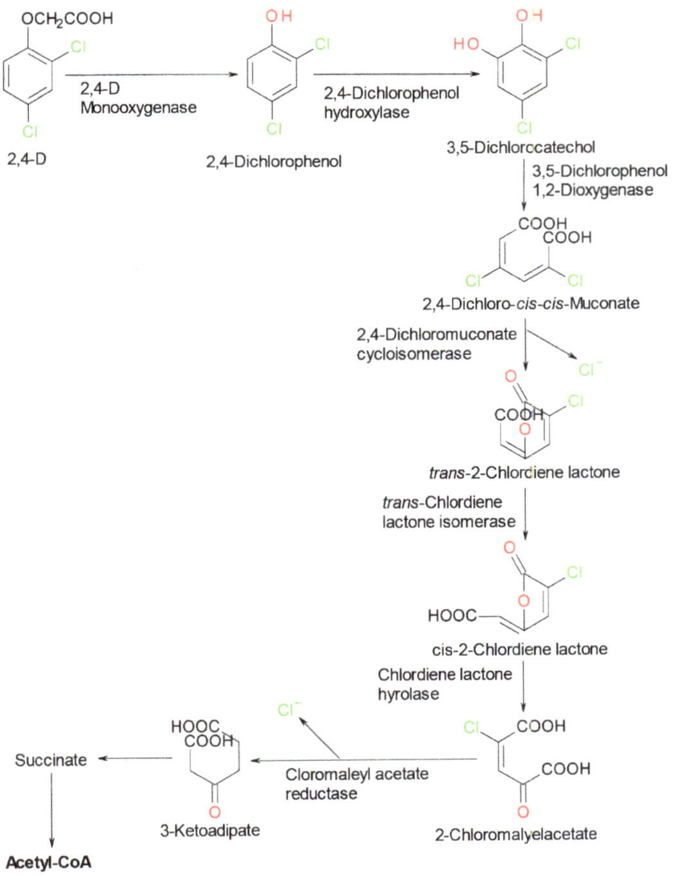

Figure 4.5 2,4-dichlorophenol biodegradation pathway.

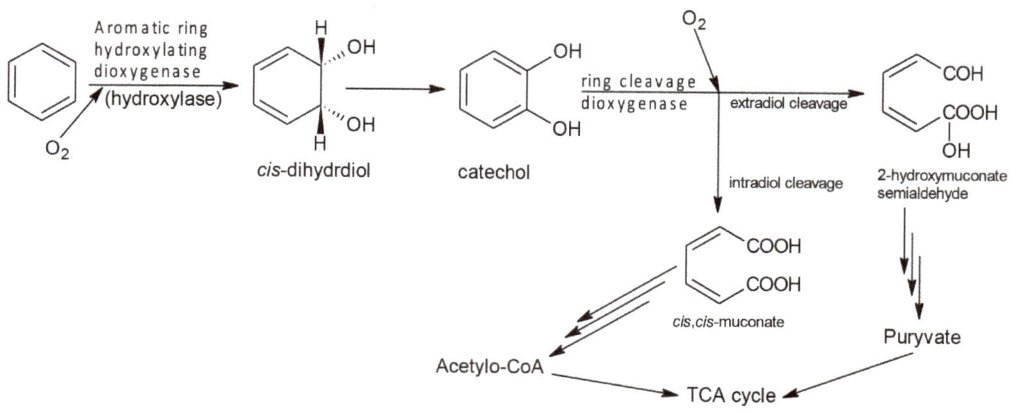

Figure 4.6 Dioxygenase mediated degradation of aromatic compounds.

naturally occurring low-molecular-weight phenolic compounds and natural fibre polymers are utilized as substrates (Ruiz-Duenas and Martinez, 2009; Jeon *et al.*, 2012). The oxidation of the phenolic and methoxyphenolic acids may result in the decarboxylation or demethylation, and in some cases laccase-mediated reactions result in the cleavage of the complex structures situated on certain C-C or C–N bonds. Intra and extracellular laccases were identified in different microorganisms.

Of the laccases available from various species, fungal laccases are of particular commercial interest, because these enzymes have relatively high redox potentials. Therefore, the enzymes are more suitable for use in oxidative processes compared with the other forms of laccases. Moreover, fungal laccases are secreted extracellularly and enzyme purification is very simple, which represents a particular advantage for biotechnological applications (Baldrian, 2006). Fungal laccases have proven to be useful in several areas of biotechnology, including organic syntheses, pulp/textile bleaching, bioremediation, chemical grafting, and polymer surface modification (Kunamneni *et al.*, 2008; Mikolasch and Schauer, 2009; Witayakran and Ragauskas, 2009; Jeon *et al.*, 2012). The general reaction mechanisms during phenolic substrate oxidation by laccase are presented in Fig. 4.7.

Many laccases are used in the biodegradation studies and for bioremediation. Among them, recently, a pathway for malachite green (MG) decolorization was presented by Yang *et al.* (2015). The optimized MG decolorization by a novel laccase, LacA, isolated from the white-rot fungus *Cerrena* sp., achieved the maximum decolorization efficiency of 91.64% in the absence of a laccase mediator. Based on the intermediates identified by LC-MS, LacA catalysed MG transformation

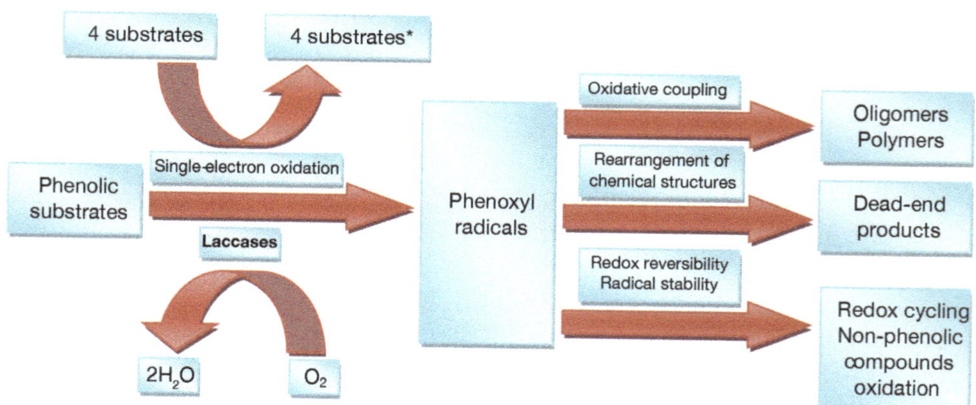

Figure 4.7 Representative oxidative reactions of phenolic substrates catalysed by laccase enzymes. Sub and Sub* indicate phenolics and phenoxyl radicals, respectively (Joan *et al.*, 2012).

via two separate, co-existing pathways (Fig. 4.8). Together with the decolorization, LacA was able to reduce MG toxicity.

Biodegradation of bis-azo dye, Reactive Black 5 (RB5), conducted by ascomycete fungus *Trichoderma atroviride* F03 (Adnan *et al.*, 2015) represent a different example. Fig. 4.9 shows the proposed pathway catalysed by extracellular laccase. Additionally, the dependency of laccase on the molecular oxygen suggests that it is a very simple process, and the characterization of metabolites demonstrated that this biodegradation process does not generate toxic aromatic amines, and therefore represents an eco-friendly natural treatment.

4.3.3.3 Peroxidases

Peroxidases are a group of oxidoreductases that catalyse the reduction of peroxides, such as hydrogen peroxide, and the oxidation of a variety of organic and inorganic compounds. Peroxidases can be haem-containing and non-haem proteins. In mammals, they are involved in various biological processes, such as immune system or hormone regulation. In plants, they are involved in the metabolism of auxins, formation of lignin or suberin (Wakamatsu and Takahama, 1993), cross-linking of cell wall components, pathogen defence systems (Biles and Martyn, 1993), or cell elongation (Hiner *et al.*, 2002; Koua *et al.*, 2009).

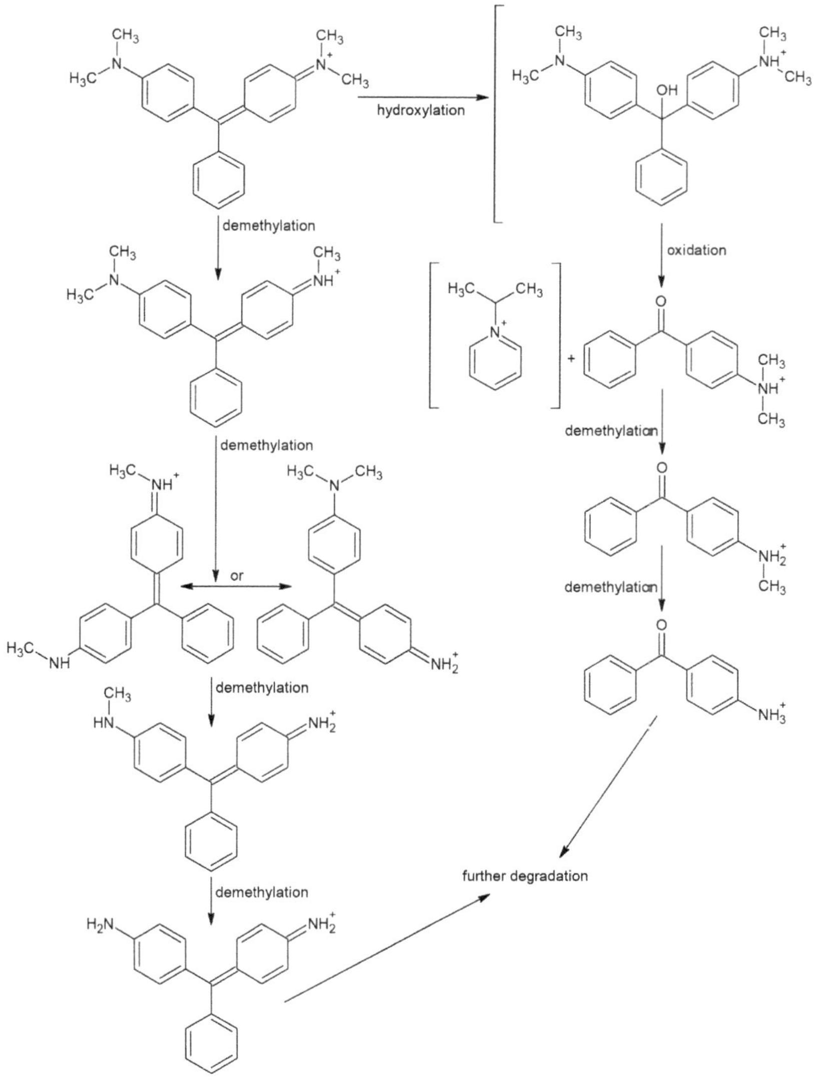

Figure 4.8 Proposed mechanism of LacA-mediated MG degradation. Probable (non-identified) metabolites in parenthesis (Yang *et al.*, 2015).

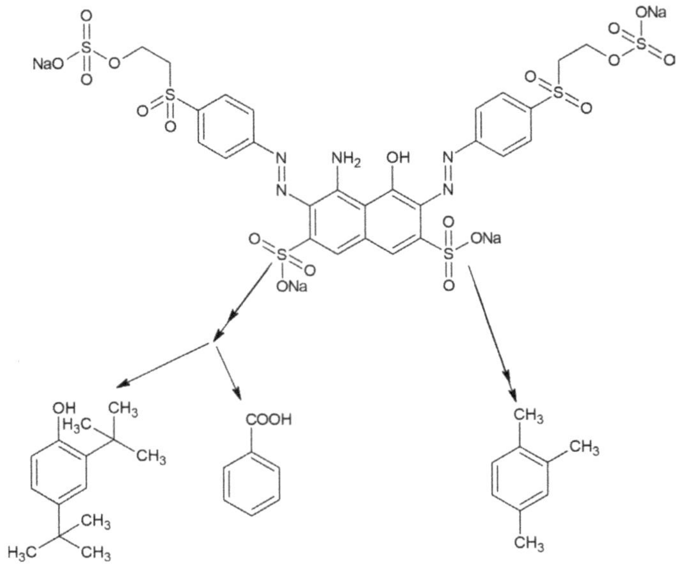

Figure 4.9 The proposed pathway of RB5 biodegradation by ascomycete fungus T. atroviride F03 (Adnan *et al.*, 2015).

Peroxidases were widely used for many years in clinical biochemistry procedures and enzyme immunoassays, and they have a potential for soil detoxification, while horseradish peroxidase (HRP), together with soybean and turnip peroxidases, have been used for the bioremediation of wastewaters contaminated with phenols, cresols, and chlorinated phenols. Lignin peroxidase (LiP) and manganese peroxidase (MnP) can be successfully used for biopulping and biobleaching in the paper industry, and can produce oxidative breakdown of synthetic azo dyes (Hamid and Rehman, 2009). Some newly suggested peroxidase applications include treatment of wastewaters containing phenolic compounds, synthesis of various aromatic chemicals, and the removal of peroxide from materials such as food items and industrial waste (Agostini *et al.*, 2002).

Haem peroxidases have been divided into two groups, peroxidases found only in animals, and those found in plants, fungi, and prokaryotes. The latter group has been subdivided into three classes based on the sequence comparison:

- Class I – intracellular enzymes (yeast cytochrome c peroxidase, ascorbate peroxidase (APX) from plants, and bacterial catalase peroxidases).
- Class II – extracellular fungal peroxidases, the most widely known ones are lignin peroxidase (LiP) and manganese peroxidase (MnP) from *Phanerochaete chrysosporium*, and *Coprinus cinereus* peroxidase or *Arthromyces ramosus* peroxidase (ARP).
- Class III – secretory plant peroxidases, such as HRP and peroxidases from barley or soybean.

Non-haem peroxidases form five independent families – thiol peroxidase (glutathione peroxidases and peroxiredoxins), alkylhydroperoxidase, non-haem haloperoxidase, manganese catalase, and NADH peroxidase (Koua *et al.*, 2009). A majority of the potential application studies was focused on the treatment of phenolic contaminants in the presence of H_2O_2 (Kim *et al.*, 2005; Mohan *et al.*, 2005; Cheng *et al.*, 2006).

The beneficial effects of HRP in the treatment of contaminants, including anilines, hydroxyquinoline, and arylamine carcinogens, such as benzidines and naphthylamines, have been demonstrated. Additionally, it has the ability to co-precipitate recalcitrant contaminants, by inducing the formation of mixed polymers of toxic and natural components that behave similar to the polymeric products of easily removable contaminants (Hamid 2009). In addition to HRP, and due to a high potential for the degradation of toxic substances, the most studied peroxidases are lignin peroxidase (LiP), manganese-dependent peroxidise (MnP), and versatile peroxidase (VP).

Microbial lignin peroxidases

Lignin peroxidase (LiP) was first identified in 1983 (Aitken *et al.*, 1994). Lignin peroxidases are haem proteins secreted mainly by the white-rot fungus *P. chrysosporium* during secondary metabolism. In the presence of co-substrate H_2O_2 and a mediator, such as veratryl alcohol, LiP degrades lignin and other phenolic compounds. Lignin peroxidase (LiP) plays a central role in the biodegradation of the plant cell wall constituent, lignin. It was shown that it mineralizes a variety of recalcitrant aromatic compounds and oxidizes a number of polycyclic aromatic and phenolic compounds (Fig. 4.10). The role of LiP in lignin depolymerization has also been confirmed, and its mechanism is very similar to that of HRP (Hamid 2009).

Microbial manganese peroxidases

Manganese peroxidase (MnP) is an extracellular haem enzyme secreted by the lignin-degrading fungi that oxidizes Mn(II) to the oxidized Mn(III) in a multi-step reaction. MnP catalyses this reaction in the presence of Mn(III)-stabilizing ligands. The resulting Mn(III) complexes can afterwards perform the oxidation of organic substrates. MnP produced by *P. chrysosporium* has also been shown to catalyse the oxidation of several monoaromatic phenols (Fig. 4.11) and aromatic dyes, but these reactions depend on the presence of both divalent manganese and certain types of buffers (Hamid, 2009).

Microbial versatile peroxidase

Microbial versatile peroxidase (VP) enzymes are able to directly oxidize Mn(II), methoxybenzenes, phenolic aromatic substrates like MnP, LiP and HRP. VP has extraordinary broad substrate specificity and ability to oxidize the substrates in the absence of manganese. It has also been demonstrated that VP is able to oxidize both phenolic and non-phenolic lignin model dimers (Ruiz-Duenas et al., 2007).

4.3.4 Hydrolase degradation pathways

Hydrolytic enzymes break chemical bonds in the xenobiotic molecules, which leads to a change in their toxic properties, resulting in the detoxification or activation of the substrate. The most favourable reaction is detoxification, which was successfully used for the biodegradation of oil spills, organophosphate, organochlorine, and carbamate insecticides. Hydrolases are able to catalyse

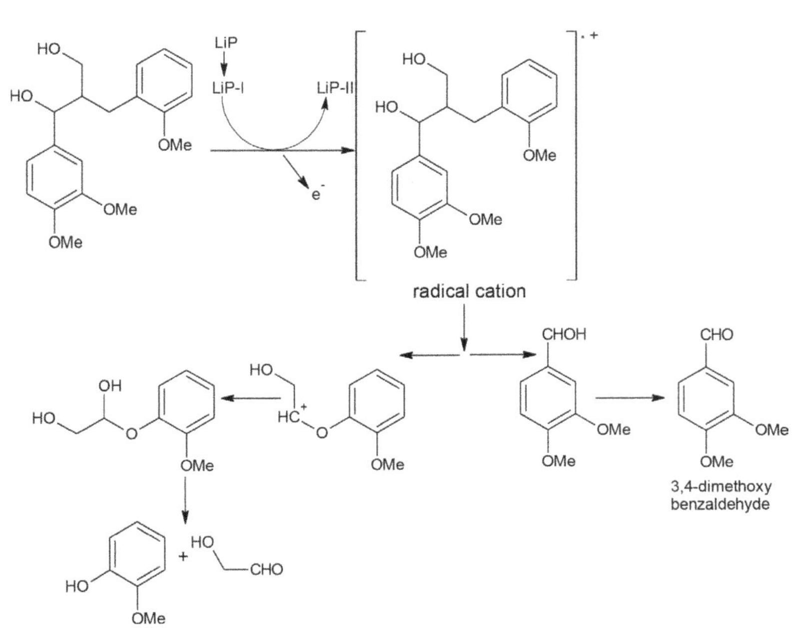

Figure 4.10 Lignin peroxidase (LiP)-catalysed oxidation of nonphenolic β-O-4 lignin model compound (Karigar and Rao, 2011).

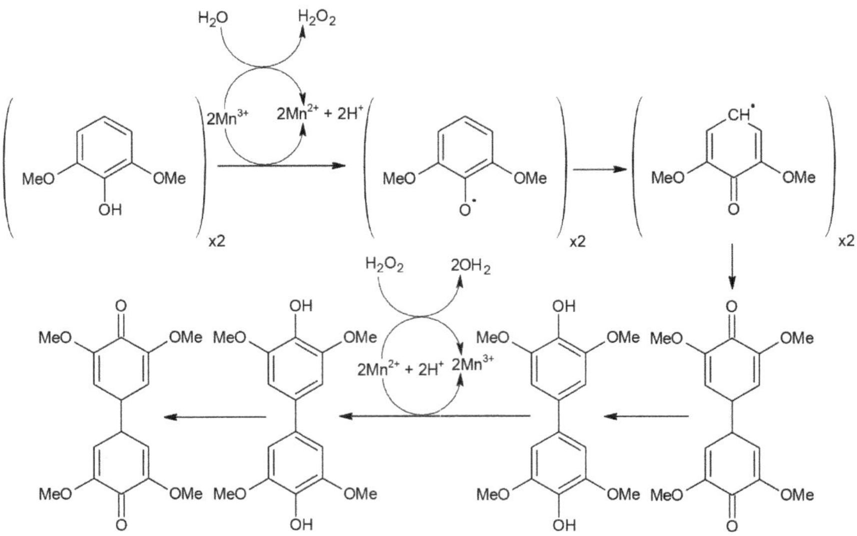

Figure 4.11 Proposed mechanism for the oxidation of 2,6-dimethoxyphenol by the MnP system (Karigar and Rao, 2011).

different types of reactions including condensations and alcoholyses. The main advantages of this class of enzymes are:

- availability (ubiquitous enzymes)
- lack of cofactor stereoselectivity
- resistance to water-miscible solvents.

Hydrolases are classified as EC 3 in the EC number classification of enzymes, and they can be further classified into several subclasses, based on the bonds they act upon:

- EC 3.1: ester bonds (esterases: nucleases, phosphodiesterases, lipase, phosphatase);
- EC 3.2: sugars (DNA glycosylases, glycoside hydrolase);
- EC 3.3: ether bonds;
- EC 3.4: peptide bonds (proteases/peptidases);
- EC 3.5: carbon–nitrogen bonds, other than peptide bonds;
- EC 3.6: acid anhydrides (acid anhydride hydrolases, including helicases and GTPase);
- EC 3.7: carbon–carbon bonds;
- EC 3.8: halide bonds;
- EC 3.9: phosphorus–nitrogen bonds;
- EC 3.10: sulfur–nitrogen bonds;
- EC 3.11: carbon–phosphorus bonds;
- EC 3.12: sulfur–sulfur bonds;
- EC 3.13: carbon–sulfur bonds.

Many kinds of the enzymes belonging to hydrolase group are extracellular hydrolytic enzymes. Among them, amylases, proteases, lipases, DNases, pullulanases, and xylanases have a potential for different uses in numerous areas including food industry, biomedical sciences, feed additive, and chemical industries. Enzymes of major importance are hemicellulase, cellulase, and glycosidase, because of their applications in the degradation of biomass (Sánchez-Porro *et al.*, 2003; Schmidt, 2006).

Biodegradation pathways catalysed by hydrolases differ based on the bonds and substrates with which they interact. A few examples of their widespread activity are described in the following text, focused on the pathways involving lipases, esterases, cellulases, and proteases.

4.3.4.1 Lipases

Lipases catalyse the hydrolysis of triacylglycerols to glycerol and free fatty acids. Lipolytic reactions typically take place at the lipid–water interface, where lipolytic substrates usually form equilibrium between monomeric, micellar, and emulsified states. One of the examples of lipase activity is triolein hydrolysis conducted by *Candida rugosa* lipase (Fig. 4.12), where consecutive breaks of ester bonds result in diolein, monoolein, and glycerol formation with parallel release of oleic acid (Karigar and Rao, 2011). Another example is the degradation of palm oil mill effluent (POME)

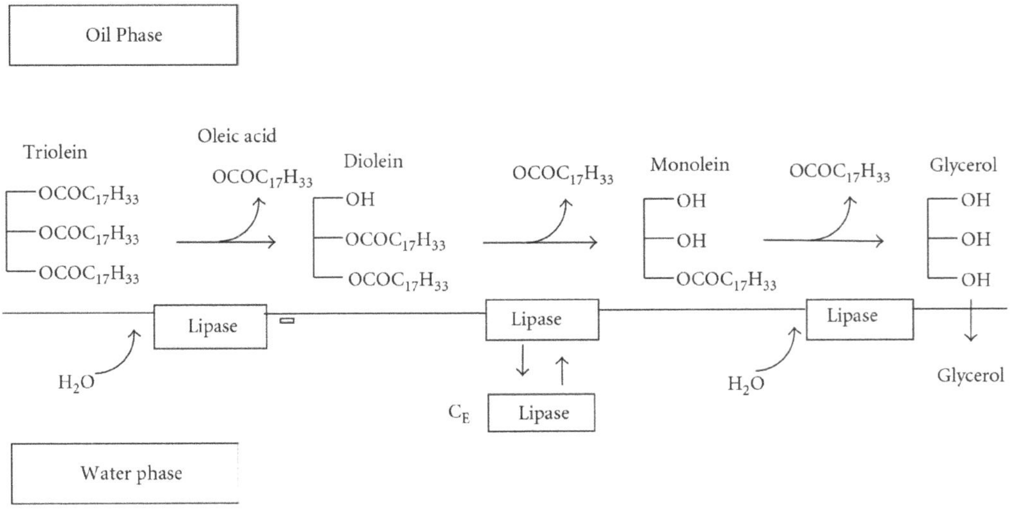

Figure 4.12 Proposed mechanism for triolein hydrolysis by *Candida rugosa* lipase in biphasic oil–water system. C_E represents the enzyme concentration in the bulk of the water phase (reproduced from Karigar and Rao, 2011).

by *Pseudomonas aeruginosa, Bacillus subtilis,* and *Candida albicans* lipases. Although some organic compounds were not completely degraded (oleic acid, palmitic acid, and octadecanoic acid), as they persisted throughout the period of analysis, the potential industrial applications of the enzyme, as well as the application in biodegradation of POME or different types of oil-polluted habitats, were postulated (Okwute *et al.*, 2015). The lipase from *Pseudomonas fluorescens* and *B. subtilis* was proved to be a highly effective degrader of poly(butylene succinate) (PBS) at pH 7 and 37°C (Asheeba *et al.*, 2011). An interesting application is the degradation of castor oil and lipase production by *P. aeruginosa* where production of castor oil bioproducts and lipase itself adds biotechnological and economical values to the overall process, as castor oil has recently become a promising candidate for biodiesel production (Amara and Salem, 2009).

4.3.4.2 Esterases

Esterases play a very important role in the degradation of natural materials and industrial pollutants, e.g. cereal wastes, plastics, and different toxic chemicals. They can be useful in the synthesis of optically pure compounds, perfumes, and antioxidants (Panda and Gowrishankar, 2005). Esterases can catalyse three types of reactions: esterification, interesterification, and transesterification reactions, with very good chemo-, regio- and/or enantioselectivity. These enzymes have been isolated from plants, animals, and microorganisms, and it was shown that all classes of microorganisms, including bacteria, fungi, and actinomycetes produce esterases – either constitutive or inducible (Sayali *et al.*, 2013).

A good example of esterase activity is polyurethane (PU) biodegradation by *Pseudomonas chlororaphis* ATCC 55729 (Howard, 2012). Analysis of the breakdown products of PU revealed that the main metabolites were derived from the polyester segment of the polymer. The metabolites produced were diethylene glycol, trimethylolpropane, and dimethyladipic acid (Fig. 4.13).

4.3.4.3 Cellulases

Cellulases are enzymes that hydrolyse β-1,4 bonds in cellulose chains. They are produced by fungi, bacteria, protozoans, plants, and animals. Cellulases contain non-catalytic carbohydrate-binding modules (CBMs) and/or other functionally known or unknown modules, which may be located at the N- or C-terminus of a catalytic module. In nature, complete cellulose hydrolysis is mediated by a combination of three main types of cellulases.

- endoglucanases (EC 3.2.1.4)
- exoglucanases, including cellobiohydrolases (CBHs) (EC 3.2.1.91)
- β-glucosidase (BG) (EC 3.2.1.21).

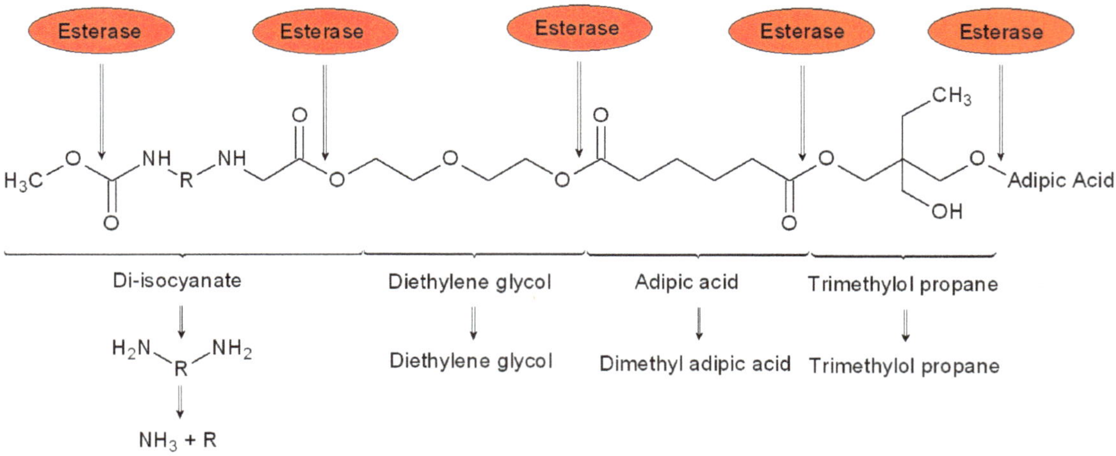

Figure 4.13 Theoretical degradative pathway of polyester polyurethane by esterase activity of *Pseudomonas* (Howard, 2012).

In order to hydrolyse and metabolize insoluble cellulose, microorganisms must secrete cellulases (with the possible exception of BG) that are either free or cell surface-bound. Cellulases are increasingly being used for various industrial purposes – in the textile industry, pulp and paper industry, and food industry, and as the additives in detergents, and for the improvement of digestibility of animal feeds (Zhang and Zhang, 2013). Together with major enzymes, some ancillary enzymes are also present. During the enzymatic hydrolysis, cellulose is degraded by the cellulases to reducing sugars that can be fermented by yeasts or bacteria to ethanol (Sun and Cheng, 2002).

4.3.4.4 Proteases

Proteases are a class of enzymes that occupy a pivotal position with respect to their applications in both physiological and commercial fields. Proteolytic enzymes catalyse the cleavage of peptide bonds in other proteins, and they perform both degradative (total hydrolysis of proteins in an aqueous environment) and synthetic functions (in non-aqueous environment) (Fig. 4.14). Proteases are divided into exo- and endopeptidases, based on their action at or away from the termini, respectively (Rao *et al.*, 1998). Endopeptidases are grouped based on the position of the active site (e.g. serine endopeptidase, cysteine peptidase, metallopeptidases). Proteases that act on free amino or carboxyl terminal amino acids are called aminopeptidase or

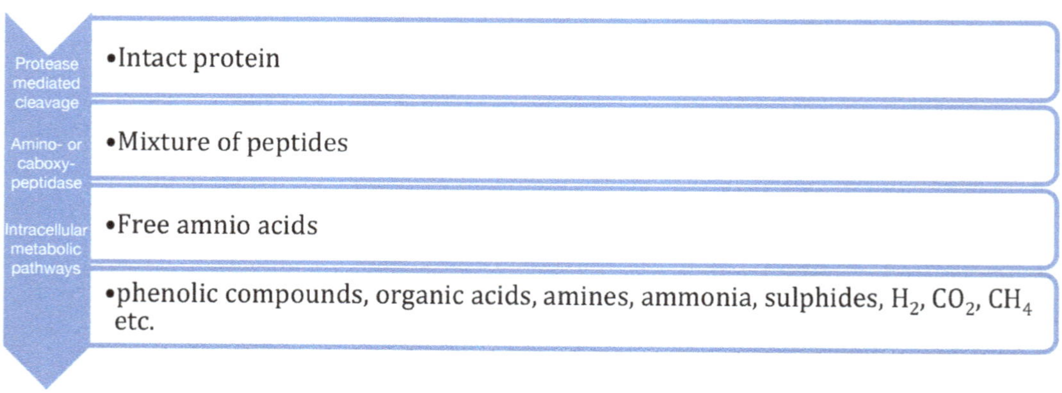

Figure 4.14 The schematic pathway of protein degradation conducted by proteases at initial steps.

carboxypeptidase, respectively. Proteases have wide range of applications in food, leather, detergent, and pharmaceutical industries.

4.3.5 ROS enzymes role in biodegradation

Reactive oxygen species (ROS) are generated during mitochondrial oxidative metabolism as well as in the cellular response to xenobiotics, cytokines and bacterial invasion. ROS presence induces oxidative stress, which refers to the excess of oxidants over the cell capability to activate an effective antioxidant response. Oxidative stress can result in macromolecular damage but accumulating evidence indicates that ROS may also serve as critical signalling molecules in both cell proliferation and survival (Ray *et al.*, 2012). Most of the reactive oxygen species are generated as by-products during mitochondrial electron transport or metal-catalysed oxidation reactions. The sequential reduction of oxygen through the addition of electrons leads to the formation of a number of ROS, including superoxide, hydrogen peroxide, hydroxyl radical, hydroxyl ion, and nitric oxide.

Cells have a variety of defence mechanisms against the harmful effects of ROS. The two most important strategies are enzymatic and non-enzymatic ones, involving small molecule antioxidants that play a role in detoxification (Held, 2014). The major enzymes involved in the ROS scavenging are (Ray *et al.*, 2012):

- superoxide dismutase
- catalase
- ascorbate and glutathione peroxidase
- peroxiredoxins
- thioredoxins
- kinases and phosphatases (involved in cell proliferation and survival).

The biodegradation of xenobiotics under aerobic conditions may result in a ROS overproduction, especially in the pathways involving oxidoreductases as major degradation agents. One of the toxic mechanisms of xenobiotics is ROS generation that leads to the self-defence mechanism activation.

The up-regulation of ROS enzymes, represented by the manganese superoxide dismutase and catalase, was observed in *Paecilomyces marquandii* cultures during alachlor biodegradation (Szewczyk *et al.*, 2015). Similar up-regulation of ROS enzymes was observed in *Metarhizium robertsii* cultures, exposed to 4-n-nonylphenol (Szewczyk *et al.*, 2014). During 2-ethylhexyl nitrate biodegradation in *Mycobacterium austroafricanum* IFP 2173 12, stress response proteins were exclusively found in the cultures containing xenobiotics, and chloride peroxidase was identified among them (Nicolau *et al.*, 2009). A recent microevolutionary study investigated the process of *Pseudomonas aeruginosa* PAO1 mechanisms of resistance to a recently developed novel antibacterial zinc Schiff-base (ZSB) compound, at the proteome level. One of the mechanisms of the resistance was effective removal (biodegradation) of ZSB. A dramatic change in the proteome was observed for ZSB treated *P. aeruginosa* PAO1 P0. Outer membrane proteins and enzymes connected to oxidative stress were found to be significantly differentially expressed, and thioredoxin and aconitase were found to be up-regulated in the ZSB-containing cultures (Cierniak *et al.*, 2013).

4.4 Cell metabolism during biodegradation

Exposure to xenobiotics leads to multilevel stress caused by their toxicity. Their influence on the cell metabolism affects completely or partly, directly or indirectly, cellular converging catabolism, diverging anabolism, and cyclic pathways (Fig. 4.15). The targeted monitoring of metabolic pathways during the process of biodegradation can be a very fruitful source of data regarding the cellular strategies involved in detoxification, energy uptake and consumption, or self-defence and cell death mechanisms. The multilevel metabolomic data may be fundamental for a deeper understanding of the biodegradation pathways, which may lead to a successful environmental process optimization. Different small or large molecular markers can be used for the monitoring of the xenobiotic influence on cell metabolism – free amino acids, organic acids, cofactors, signalling molecules, simple and poly sugars, phospholipids, proteins, or mixed biological polymers (e.g. glycoproteins).

The most common method applied in the targeted and untargeted profiling of compounds is MS and MS/MS scanning which allow the collection of the quantitative, relative quantitative, trend change,

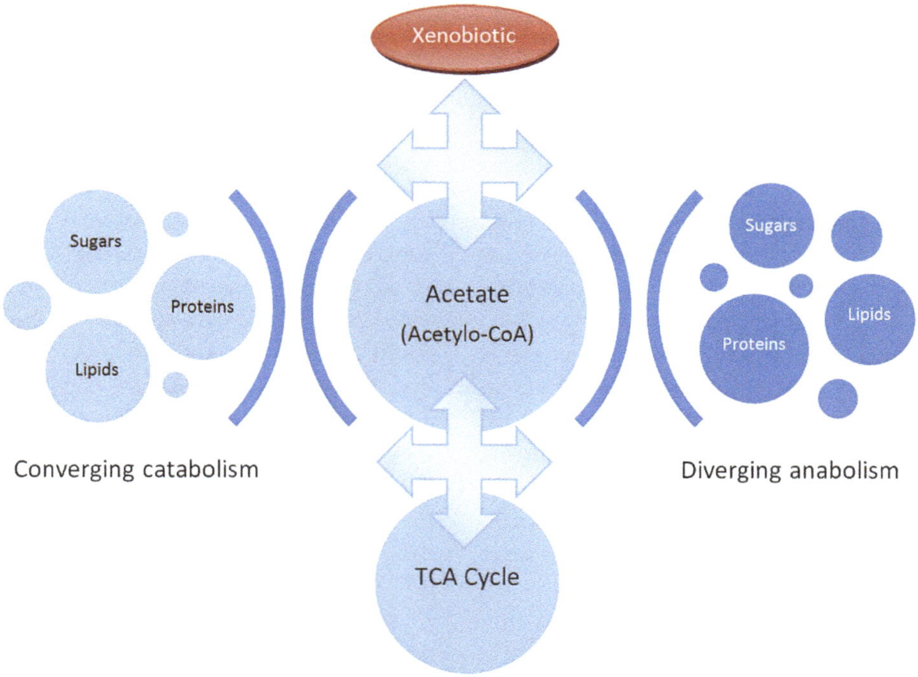

Figure 4.15 Schematic representation of xenobiotic influence on cellular metabolism.

and qualitative data, for a large number of compounds in one run. Typical post-processing of the data involves PCA analysis of the series of control and test samples.

Targeted metabolomic study of the white-rot fungus *P. chrysosporium* exposed to benzoic acid (BA), included numerous compounds: glycolysis intermediates, TCA cycle intermediates, amino acids, and lipids. The effective removal of BA altered fungal regulation of TCA, mannitol cycles, and the regulation and utilization of trehalose as a storage sugar. The majority of the TCA cycle metabolite concentrations decreased in the presence of BA (Matsuzaki *et al.*, 2008). A comprehensive and untargeted metabolomics study of the degradation of phenanthrene by bacterium *Sinorhizobium* sp. C4 included numerous compounds as well. High accumulation of some intermediates of glycolysis (e.g. 2-phosphoglycerate), TCA cycle (e.g. oxalate) and others (2-hydroxyglutarate, malonate, and 3-hydroxyisovalerate) were found in the phenanthrene-supplemented cultures. This increase was paralleled by the reduction in the levels of other metabolites in the same metabolic pathway (e.g. 3-phosphoglycerate, malate, and α-ketoglutarate) (Keum *et al.*, 2008). The degradation of alachlor by *Paecilomyces marquandii* leads to the xenobiotic removal, with several identified by-products, and the up-regulation of oxidative stress markers. These up-regulated markers in alachlor-containing cultures include ROS enzymes (manganese superoxide dismutase and catalase), ascorbate, and GSH. It was observed that the addition of alachlor drives carbon metabolism towards higher and more efficient consumption during the initial stages of growth (increased glycolysis and TCA cycle rate), but it also changes the glucose utilization to form polysaccharides, supplementary materials, lipopolysaccharides, glycosphingolipids (UDP-glucose/galactose) (Szewczyk *et al.*, 2015; Słaba *et al.*, 2015). The lipidomic approach was applied in the study on the influence of nickel (Ni) and selenium (Se) on phospholipid composition in shoots and roots of wheat seedlings. Ni toxicity was associated with a higher level of phosphatidic acid species in the roots while in the shoots, the phosphatidylcholine/phosphatidylethanolamine ratio was about five-fold higher than in roots and decreased in Ni-containing samples. The authors concluded that the changes in the phospholipids profile may be used as an indicator of Ni stress (Bernat *et al.*, 2014).

References

Adnan, L.A., Sathishkumar, P., Yusoff, A.R.M., and Hadibarata, T. (2015). Metabolites characterisation of laccase mediated Reactive Black 5 biodegradation by fast growing ascomycete fungus *Trichoderma atroviride* F03. Inter. Biodet. Biodeg. *104*, 274–282.

Agostini, E., Hernandez-Ruiz J., Arnao, M.B., Milrand, S.R., Tigier, H.A., and Acosta, M. (2002). A peroxidase isoenzyme secreted by turnip (*Brassica napus*) hairy-root culture inactivation by hydrogen peroxide and application in diagnostic kits. Biotechnol. Appl. Biochem. 35, 1–7.

Aitken, M.D., Venkatadri, R., and Irvine, R.L. (1989). Oxidation of phenlic pollutants by a lignin degrading enzyme from the white-rot fungus *Phanerochaete chrysosporium*. Water Res. 23, 443–450.

Amara, A.A., and Salem, S.R. (2009). Degradation of castor oil and lipase production by *Pseudomonas aeruginosa*. American-Eurasian J. Agric. Environ. Sci. 5, 556–563.

Arora, P.K., Kumar, M., Chauhan, A., Raghava, G.P., and. Jain, R.K. (2009). OxDBase: a database of oxygenases involved in biodegradation. BMC Res.Notes 2, 67.

Asheeba, S.T., Raman, M.D., Begum, T.S., and Sathishkumar, D. (2011). Effect of Lipase in Biodegrading Polybutylene Succinate. Int. J. Environ. Sci. 2, 1021–1029.

Baldrian, P. (2006). Fungal laccases – occurrences and properties. FEMS Microbial. Rev. *30*, 215–242.

Bernat, P., Gajewska, E., Bernat, T., and Wielanek, M. (2014). Characterisation of the wheat phospholipid fractionin the presence of nickel and/or selenium. Plant. Growth. Regul. 72, 163–170.

Biles, C.L., and Martin, R.D. (1993). Peroxidase, polyphenoloxidase, and shikimate dehydrogenase isozymes in relation to the tissue type, maturity and pathogen induction of watermelon seedlings. Plant. Physiol. Biochem. *31*, 499–506.

Biocatalysis/Biodegradation database – available online: http://eawag-bbd.ethz.ch/index.html. Accessed 27 April 2016.

Caspi, R., Altman, T., Billington, R., Dreher, K., Foerster, H., Fulcher, C.A., Holland, T.A., Keseler, I.M., Kothari, A., Kubo, A., *et al.* (2014). The MetaCyc database of metabolic pathways and enzymes and the BioCyc collection of Pathway/Genome Databases, Nucleic Acids Res. *42*, 459–471.

Chen, G., Cui, L., Teo, G.S., Ong, C.N., Tan, C.S., and Choi, H. (2015). MetTailor: dynamic block summary and intensity normalization for robust analysis of mass spectrometry data in metabolomics. Bioinformatics *31*, 3645–3652.

Cheng, J., Yu, S.M., and Zuo, P. (2006). Horseradish peroxidase immobilized on aluminum pillared interlayer clay for the catalytic oxidation of phenolic water. Water Res. *40*, 283–290.

Chu, Y., Jiang, H., Ju, J., Li, Y., Gong, L., Wang, X., Yang, W., and Deng, Y. (2015). A metabolomic study using HPLC-TOF/MS coupled with ingenuity pathway analysis: Intervention effects of *Rhizoma alismatis* on spontaneous hypertensive rats. J. Pharm. Biomed. Anal. *117*, 446–452.

Cierniak, P., Jübner, M., Müller, S., and Bender, K. (2013). Insights into mechanisms and proteomic characterisation of *Pseudomonas aeruginosa* adaptation to a novel antimicrobial substance. PLoS ONE *8*, e66862.

Copley, S.D. (2009). Evolution of efficient pathways for degradation of anthropogenic chemicals. Nat. Chem. Biol. *5*, 559–566.

Croft, D., Mundo, A.F., Haw, R., Milacic, M., Weiser, J., Wu, G., Caudy, M., Garapati, P., Gillespie, M., Kamdar, M.R., *et al.* (2014). The Reactome pathway knowledgebase. Nucleic Acids Res. *42*, 472–477.

Dua, M., Singh, A., Sethunathan, N., and Johri, A. (2002). Biotechnology and bioremediation: successes and limitations. Appl. Microbiol. Biotechnol. *59*, 143–152.

Eshelli, M., Harvey, L., Edrada-Ebel, R.A., and McNeil, B. (2015). Metabolomics of the biodegradation process of aflatoxin B1 by *Actinomycetes* at an initial pH of 6.0. Toxins 7, 439–456.

Fetzner, S. (2003). Oxygenases without requirement for cofactors or metal ions. Appl. Microbiol. Biotechnol. *60*, 243–257.

Grosse, S., Laramee, L., Wendlandt, K.-D., McDonald, I.R., Miguez, C.B., and Kleber, H.-P. (1999). Purification and characterization of the soluble methane monooxygenase of the type II methanotrophic bacterium *Methylocystis* sp. strain WI 14. Appl. Environ. Microbiol. *65*, 3929–3935.

Guerrasio, R., Haberhauer-Troyer, C., Steiger, M., Sauer, M., Mattanovich, D., Koellensperger, G., and Hann, S. (2013). Measurement uncertainty of isotopologue fractions in fluxomics determined via mass spectrometry. Anal. Bioanal. Chem. *405*, 5133–5146.

Guo, A.C., Jewison, T., Wilson, M., Liu, Y., Knox, C., Djoumbou, Y., Lo, P., Mandal, R., Krishnamurthy, R., and Wishart, D.S. (2013). ECMDB: the *E. coli* Metabolome Database. Nucleic Acids Res. *41*, 625–630.

Hamid, M., and Rehman, K. (2009). Potential applications of peroxidases. Food Chem. *115*, 1177–1186.

Hanke, T., Nöh, K., Noack, S., Polen, T., Bringer, S., Sahm, H., Wiechert, W., and Bott, M. (2013). Combined fluxomics and transcriptomics analysis of glucose catabolism via a partially cyclic pentose phosphate pathway in *Gluconobacter oxydans* 621H. Appl. Environ. Microbiol. *79*, 2336–2348.

Hastings, J., de Matos, P., Dekker, A., Ennis, M., Harsha, B., Kale, N., Muthukrishnan, V., Owen, G., Turner, S., Williams, M., and Steinbeck, C. (2013). The ChEBI reference database and ontology for biologically relevant chemistry: enhancements for 2013. Nucleic Acids Res., *41*, 456-D63.

Haug, K., Salek, R.M., Conesa, P., Hastings, J., de Matos, P., Rijnbeek, M., Mahendraker, T., Williams, M., Neumann, S., Rocca-Serra, P., *et al.* (2013). MetaboLights – an open-access general-purpose repository for metabolomics studies and associated meta-data. Nucleic Acids Res. *41*, 781–786.

Heinonen, M., Rantanen, A., Mielikäinen, T., Kokkonen, J., Kiuru, J., Ketola, R.A., and Rousu, J. (2008). FiD: a software for ab initio structural identification of product ions from tandem mass spectrometric data. Rapid Commun. Mass. Spect. *22*, 3043–3052.

Held, P. (2014). An introduction to reactive oxygen species. measurement of ROS in cells. Available online: http://www.biotek.com/resources/articles/reactive-oxygen-species.html. Accessed 27 April 2016.

Heller, M.C., Drew, C.P., Jackson, K.A., Griffey, S., and Watson, J.L. (2010). A potential role for indoleamine 2,3-dioxygenase (IDO) in *Rhodococcus equi* infection. Vet. Immunol. Immunopathol. *138*, 174–182.

Hiner, A.N., Ruiz, J.H., López, J.N.R., Cánovas, F.G., Brisset, N.C., Smith, A.T., Arnao, M.B., and Acosta, M. (2002). Reactions of the class II peroxidases, lignin peroxidase and *Arthromyces ramosus* peroxidase, with hydrogen peroxide. Catalase-like activity, compound III formation, and enzyme inactivation. J. Biol. Chem. *277*, 26879–26885.

Horai, H., Arita, M., Kanaya, S., Nihei, Y., Ikeda, T., Suwa, K., Ojima, Y., Tanaka, K., Tanaka, S., Aoshima, K., *et al.* (2010). MassBank: a public repository for sharing mass spectral data for life sciences. J. Mass. Spec. *45*, 703–714.

Howard, G.T. (2012). Polyurethane biodegradation. In Microbial Degradation of Xenobiotics, S.N. Singh, ed. (Springer, Berlin), pp. 371–394.

Howlett, R.M., Davey, M.P., Paul, Q.W., and Kelly, D.J. (2014). Metabolomic analysis of the food-borne pathogen *Campylobacter jejuni*: application of direct injection mass spectrometry for mutant characterisation. Metabolomics *10*, 887–896.

Jeon, J.R., Baldrian, P., Murugesan, K., and Chang, Y.S. (2012). Laccase-catalysed oxidations of naturally occurring phenols: from in vivo biosynthetic pathways to green synthetic applications. Microb Biotechnol. *5*, 318–332.

Jewison, T., Knox, C., Neveu, V., Djoumbou, Y., Guo, A.C., Lee, J., Liu, P., Mandal, R., Krishnamurthy, R., Sinelnikov, I., *et al.* (2012). YMDB: the Yeast Metabolome Database. Nucleic Acids Res. doi: 10.1093/nar/gkr916

Kanehisa, M., Goto, S., Sato, Y., Kawashima, M., Furumichi, M., and Tanabe, M. (2014). Data, information, knowledge and principle: back to metabolism in KEGG. Nucleic Acids Res. *42*, 199–205.

Karigar, C.S., and Rao, S.S. (2011). Role of microbial enzymes in the bioremediation of pollutants: a review. Enzyme Res. *2011*, 11.

Keseler, I.M., Mackie, A., Peralta-Gil, M., Santos-Zavaleta, A., Gama-Castro, S., Bonavides-Martínez, C., Fulcher, C., Huerta, A.M., Kothari, A., Krummenacker, M., *et al.* (2013). EcoCyc: fusing model organism databases with systems biology. Nucleic Acids Res. *41*, 605–612.

Keum, Y.S., Seo, J.S., Li, Q.X., and Kim, J.H. (2008). Comparative metabolomic analysis of *Sinorhizobium* sp. C4 during the degradation of phenanthrene. Appl. Microbiol. Biotechnol. *80*(5), 863–872.

Kim, G.Y., Lee, K.B., Cho, S.H., and Shim, J., and Moon, S.H. (2005). Electroenzymatic degradation of azo dye using an immobilized peroxidase enzyme. J. Hazard. Mat. *126*, 183–188.

Kim, S., Thiessen, P.A., Bolton, E.E., Chen, J., Fu, G., Gindulyte, A., Han, L., He, J., He, S., Shoemaker, B.A., *et al.* (2015). PubChem Substance and Compound databases. Nucleic Acids Res. *44*(D1), D1202–D1213.

Kind, T., and Fiehn, O. (2007). Seven Golden Rules for heuristic filtering of molecular formulas obtained by accurate mass spectrometry. BMC Bioinformatics *8*, 105.

King, Z.A., Lu, J.S., Dräger, A., Miller, P.C., Federowicz, S., Lerman, J.A., Ebrahim, A., Palsson, B.O., and Lewis, N.E. (2015). BiGG Models: A platform for integrating, standardizing, and sharing genome-scale models. Nucleic Acids Res. *44*(D1), D515–D522.

Koua, D., Cerutti, L., Falquet, L., Sigrist, C.J., Theiler, G., Hulo, N., and Dunand, C. (2009). PeroxiBase: a database with new tools for peroxidase family classification. Nucleic Acids Res. 37, D261-D266.

Kulkarni, M., and Chaudhari, A. (2007). Microbial remediation of nitro-aromatic compounds: an overview. J. Environ. Man. *85*, 496–512.

Kunamneni, A., Camarero, S., Garcia-Burgos, C., Plou, F.J., Ballesteros, A., and Alcalde, M. (2008). Engineering and applications of fungal laccases for organic synthesis. Microb. Cell Fact 7, 32–48.

Lehninger, A.L., Nelson, D.L., and Cox, M.M. (2004). Lehninger's Principles of Biochemistry, 4th edn (New York: W.H. Freeman).

Martin, J.C., Maillot, M., Mazerolles, G., Verdu, A., Lyan, B., Migné, C., Defoort, C., Canlet, C., Junot, C., Guillou, C., *et al.* (2015). Can we trust untargeted metabolomics? Results of the metabo-ring initiative, a large-scale, multi-instrument inter-laboratory study. Metabolomics *11*, 807–821.

Matsuzaki, F., Shimizu, M., and Wariishi, H. (2008). Proteomic and metabolomic analyses of the white-rot fungus *Phanerochaete chrysosporium* exposed to exogenous benzoic acid. J. Proteome Res. *7*, 2342–2350.

Menikarachchi, L.C., Hill, D.W., Hamdalla, M.A., Mandoiu, I.I., and Grant, D.F. (2013). In silico enzymatic synthesis of a 400,000 compound biochemical database for nontargeted metabolomics. J. Chem. Inf. Model. *53*, 2483–2492.

Mikolasch, A., and Schauer, F. (2009). Fungal laccases as tools for the synthesis of new hybrid molecules and biomaterials. Appl. Microbiol. Biotechnol. *82*, 605–624.

Mohan, S.V., Prasad, K.K., Rao, N.C., and Sarma, P.N. (2005). Acid azo dye degradation by free and immobilized horseradish peroxidase (HRP) catalyzed process. Chemosphere *58*, 1097–1105.

Nicholson, J.K., and Lindon, J.C. (2008). Systems biology: Metabonomics. Nature *455*, 1054–1056.

Nicolau, E., Kuhn, L., Marchal, R., and Jouanneau, Y. (2009). Proteomic investigation of enzymes involved in 2-ethylhexyl nitrate biodegradation in *Mycobacterium austroafricanum* IFP 2173. Res. Microbiol. *160*, 838–847.

Okwute, O.L., Stephen, E., and Anyanwu, P.I. (2015). Biodegradation of palm oil mill effluent (POME) and lipase activity by *Pseudomonas aeruginosa*, *Bacillus subtilis* and *Candida albicans*. Br. Microbiol. Res. J. 9, 1–10.

Panda, T., and Gowrishankar, B.S. (2005). Production and applications of esterases. Appl. Microbiol. Biotechnol. *67*, 160–169.

Pence, H.E., and Williams, A. (2010). ChemSpider: an online chemical information resource. J. Chem. Educ. In J Chem Educ *87*, 1123–1124.

Rao, M.B., Tanksale, A.M., Ghatge, M.S., and Deshpande, V.V. (1998). Molecular and biotechnological aspects of microbial proteases. Microbiol. Mol. Biol. Rev. *62*, 597–635.

Ray, P.D., Huang, B.W., and Tsuji, Y. (2012). Reactive oxygen species (ROS) homeostasis and redox regulation in cellular signaling. Cell. Signal. *24*, 981–990.

Roessner, U., and Bowne, J. (2009). What is metabolomics all about? BioTechniques *46*, 363–365.

Ruiz-Duenas, F.J., Morales, M., Pérez-Boada, M., Choinowski, T., Martínez, M.J., Piontek, K., and Martínez, Á.T. (2007). Manganese oxidation site in *Pleurotus eryngii* versatile peroxidase: a site-directed mutagenesis, kinetic, and crystallographic study. Biochemistry *46*, 66–77.

Ruiz-Duenas, F.J., and Martinez, A.T. (2009). Microbial degradation of lignin: how a bulky recalcitrant polymer is efficiently recycled in nature and how we can take advantage of this. Microb. Biotechnol. *2*, 164–177.

Ryffel, F., Helfrich, E.J., Kiefer, P., Peyriga, L., Portais, J.C., Piel, J., and Vorholt, J.A. (2015). Metabolic footprint of epiphytic bacteria on *Arabidopsis thaliana* leaves. ISME J. 1–12.

Saito, A., Iwabuchi, T., and Harayama, S. (2000). A novel phenanthrene dioxygenase fromnocardioides sp. strain kp7 expression in *Escherichia coli*. J Bacteriol. *182*, 2134–2141.

Salek, R.M., Steinbeck, C., Viant, M.R., Goodacre, R., and Dunn, W.B. (2013). The role of reporting standards for metabolite annotation and identification in metabolomic studies. Gigascience 2, 13.

Sánchez-Porro, C., Martin, S., Mellado, E., and Ventosa, A. (2003). Diversity of moderately halophilic bacteria producing extracellular hydrolytic enzymes. J. Appl. Microbiol. *94*, 295–300.

Sayali, K., Sadichha, P., and Surekha, S. (2013). Microbial esterases: an overview. Int. J. Curr. Microbiol. Appl. Sci. *2*, 135–146.

Schmidt, O. (2006). Wood and tree Fungi: Biology, damage, protection, and use (Springer, New York).

Scholz, M., and Selbig, J. (2007). Visualization and analysis of molecular data. In Metabolomics – Methods and Protocols, W. Weckwerth, eds. (Humana Press Inc., Totowa, NJ), pp. 87–104.

Słaba, M., Różalska, S., Bernat, P., Szewczyk, R., Piątek, M.A., and Długoński, J. (2015). Efficient alachlor degradation by the filamentous fungus *Paecilomyces marquandii* with simultaneous oxidative stress reduction. Bioresource technology *197*, 404–409.

Smith, C.A., O'Maille, G., Want, E.J., Qin, C., Trauger, S.A., Brandon, T.R., Custodio, D.E., Abagyan, R., and Siuzdak, G. (2005). METLIN: a metabolite mass spectral database. Ther Drug Monit. 27, 747–751.

Sturm, M., Bertsch, A., Gröpl, C., Hildebrandt, A., Hussong, R., Lange, E., Pfeifer, N., Schulz-Trieglaff, O., Zerck, A., Reinert, K., and Kohlbacher, O. (2008). OpenMS – an open-source software framework for mass spectrometry. BMC Bioinformatics *9*, 163.

Sue, T., Obolonkin, V., Griffiths, H., and Villas-Bôas, S.G. (2011). An exometabolomics approach to monitoring microbial contamination in microalgal fermentation processes by using metabolic footprint analysis. Appl. Environ. Microbiol. 77, 7605–7610.

Sumner, L.W., Amberg, A., Barrett, D., Beale, M.H., Beger, R., Daykin, C.A., Fan, T.W., Fiehn, O., Goodacre, R., Griffin, J.L., *et al.* (2007). Proposed minimum reporting standards for chemical analysis Chemical Analysis Working Group (CAWG) Metabolomics Standards Initiative (MSI). Metabolomics 3, 211–221.

Sun, Y., and Cheng, J. (2002). Hydrolysis of lignocellulosic materials for ethanol production: a review. Bioresour. Technol. *83*, 1–11.

Sysi-Aho, M., Katajamaa, M., Yetukuri, L., and Oresic, M. (2007). Normalization method for metabolomics data using optimal selection of multiple internal standards. BMC Bioinformatics *8*, 1–17.

Szewczyk, R., and Długoński, J. (2009). Pentachlorophenol and spent engine oil degradation by *Mucor ramosissimus*. Inter. Biodet. Biodeg. *63*, 123–129.

Szewczyk, R., Soboń, A., Sylwia, R., Dzitko, K., Waidelich, D., and Długoński, J. (2014). Intracellular proteome expression during 4-n-nonylphenol biodegradation by the filamentous fungus *Metarhizium robertsii*. Int. Biodeterior. Biodegrad. *93*, 44–53.

Tautenhahn, R., Patti, G.J., Rinehart, D., and Siuzdak, G. (2012). XCMS Online: A Web-Based Platform to Process Untargeted Metabolomic Data Analytical Chemistry. Anal. Chem. *84*, 5035–5039.

Szewczyk, R., Soboń, A., Słaba, M., and Długoński, J. (2015). Mechanism study of alachlor biodegradation by *Paecilomyces marquandii* with proteomic and metabolomic methods. J. Hazard. Mater. *291*, 52–64.

Vázquez-Fresno, R., Llorach, R., Marinic, J., Tulipani, S., Garcia-Aloy, M., Espinosa-Martos, I., Jiménez, E., Rodríguez, J.M., and Andres-Lacueva, C. (2014), Urinary metabolomic fingerprinting after consumption of a probiotic strain in women with mastitis. Pharmacol. Res. *87*, 160–165.

Wakamatsu, K., and Takahama, U. (1993). Changes in peroxidase activity and in peroxidase isozymes in carrot callus. Physiol. Plant, *88*, 167–171.

Wishart, D.S., Knox, C., Guo, A.C., Eisner, R., Young, N., Gautam, B., Hau, D.D., Psychogios, N., Dong, E., Bouatra, S., *et al.* (2009). HMDB: a knowledgebase for the human metabolome. Nucleic Acids Res. 37, 603–610.

Witayakran, S., and Ragauskas, A.J. (2009). Synthetic applications of laccase in green chemistry. Adv. Synth. Catal. *351*, 1187–1109.

Xia, J., Sinelnikov, I.V., Han, B., and Wishart, D.S. (2015). MetaboAnalyst 3.0 – making metabolomics more meaningful. Nucleic Acids Res. *43*, 251–257.

Xiao, J.F., Zhou, B., and Ressom, H.W. (2012). Metabolite identification and quantitation in LC-MS/MS-based metabolomics. Trends Anal. Chem. 32, 1–14.

Yang, J., Yang, X., Lin, Y., Ng, T.B., Lin, J., and Ye, X. (2015). Laccase-catalyzed decolorization of malachite green: performance optimization and degradation mechanism. PLoS ONE *10*, e0127714.

Yuasa, H.J., and Ball, H.J. (2011). Molecular evolution and characterization of fungal indoleamine 2,3-dioxygenases. J. Mol. Evol. 72, 160–168.

Zhang, X.Z., and Zhang, Y.H.P. (2013). Cellulases: characteristics, sources, production, and applications. In Bioprocessing Technologies in Biorefinery for Sustainable Production of Fuels, Chemicals and Polymers, First Edition. S. Yang, H.A. El-Enshasy, and N. Thongchul, eds. (John Wiley & Sons, Inc., Hoboken, NJ), pp. 131–146.

Zhou, H.W., Guo, C.L., Wong, Y.S., and Tam, N.F. (2006). Genetic diversity of dioxygenase genes in polycyclic aromatic hydrocarbon-degrading bacteria isolated from mangrove sediments. FEMS Microbiol. Lett. *262*, 148–157.

Proteomics as a Tool for Metabolic Pathways Inspection in Microbial Cells 5

Rafał Szewczyk and Konrad Kowalski

Contents

Abstract

Proteomics has become an important part of the characterization of biological systems or single organisms. In this chapter, background relating to proteomics studies including definitions, instrumentation, and basic as well as more complex workflows for protein extraction and identification with different instrumental techniques and bioinformatics tools are described. The application of proteomics in biodegradation is described in terms of selected research studies covering typical approaches: gel-free and gel-based as well as qualitative and quantitative comparative proteomics workflows and metaproteomics.

5.1 Introduction

Proteomics is a rapidly growing field of molecular biology that is concerned with systematic, high-throughput protein expression analyses of cells or organisms. The results of proteomics studies typically consist of inventories of the protein content of differentially expressed proteins across multiple conditions. Cells respond to internal and external changes by regulating protein levels and activities; and changes in the proteome of a cell thus provide a snapshot of the cell in action. Proteomics data enable a greater understanding of the structure, function, and interactions of the entire protein content in a specific organism.

5.1.1 Definitions

Proteomics is the branch of life sciences describing the methods of proteins analysis – the proteome of a certain biological system. Currently, there are three main branches of proteomics: structural, functional, and quantitative. Structural proteomics is mostly focused on protein structure analysis and structure–function relationships. This kind of analysis can be performed using a wide range of methods typically used for biomacromolecule analysis, such as X-ray crystallography, NMR spectroscopy, electron microscopy, or circular dichroism (Roy *et al.*, 2014). Bioinformatics currently offers various ways of modelling and predicting protein structures using portals such as Protein Model Portal (http://www.proteinmodelportal.org) that provides access to different protein models created by comparative modelling methods provided by various affiliate sites. The RCSB protein data bank (PDB) already contains more than 100,000 protein structures (http://www.rcsb.org/pdb). Mass spectrometry-based techniques now also offer many new possibilities for in-depth structural analysis of macromolecules. Affinity purification-mass spectrometry (AP-MS) is a method involving sample enrichment and protein selection based on protein affinities for certain ligands, such as DNA or RNA fragments, lipids, peptides, or other proteins, especially antibodies. Coupling AP-MS with standard mass spectrometry-based analysis can help to identify potential protein interaction partners in whole analysed proteomes based only on protein–ligand interaction dynamics (Dunham *et al.*, 2012; Walzthoeni *et al.*, 2013). Another useful tool in structural proteomics is proton-deuterium exchange (HDX) combined with liquid chromatography and mass spectrometry. This technique makes use of the unique properties of amide hydrogens for rapid exchange on deuterium, during which H-D exchange occurs under native conditions in solution such that the exchange rate is directly dependent on the exposure of the amide bond to the environment. While HDX-MS yields only basic information about protein structure (such as conformation and solvent accessibility), it also provides precise information about the structure of the analysed protein–ligand complexes and the dynamics of their formation (Chalmers *et al.*, 2011; Kan *et al.*, 2013). Electrospray ionization-type MS methods use soft ionization and have therefore been used as a simple method of detecting protein complex formation for many years (Jurneczko *et al.*, 2011). The development of new ion mobility mass spectrometers (IM-MS) now also allows for ions to be separated based on their cross-section in the gaseous phase. Differences in biomacromolecule cross-sections relate directly to protein structure and shape and can therefore provide information in both native protein analyses and protein–ligand interactions studies (Konijnenberg *et al.*, 2013; Jurneczko *et al.*, 2011). Protein complexes can also be analysed by chemical cross-linking MS (CX-MS). This sample preparation method involves a crosslinking reaction based on absolute distances between polypeptide chains and yields detailed structural information (Walzthoeni *et al.*, 2013).

Functional proteomics explore not only the interactions between proteins and the dynamics of the formation and dissolution of protein complexes, but also variations in the activity and biological function of a protein depending on its interactions. This kind of analysis can be performed mostly using methods of molecular biology, such as electrophoretic mobility shift assays (EMSA), DNase I footprinting or surface plasmon resonance (SPR) (Cai *et al.*, 2012).

Currently, the most popular proteomics approach is to utilize quantitative proteomics to identify and quantify proteins. The simplest quantitative analyses are based primarily on comparisons of the intensities of stained proteins separated by 2-DE electrophoresis. This method, however, is limited by the fact that different proteins (depending on the amino acid sequence and structure) have different dye incorporation capacities, and thus despite such analyses being conducted within a single experiment to allow for direct comparison between test and control samples, the results must be interpreted with caution. A similar but certainly more accurate measure of relative protein concentration can be achieved by sample derivatization with isotope-coded affinity tags (ICAT). Relative concentrations are calculated as the intensity ratio between the same peptide signal labelled with different isotopes tags. Absolute protein concentration determination by mass spectrometry-liquid chromatography techniques is conducted in the manner described in detail for small molecules in Chapter 4.

5.1.2 Current proteomics

It is extremely challenging to identify and quantify proteins, as well as to determine the posttranslational modification states and interaction partners for all proteins in the cell. For this reason, proteomics has become a technology-driven field. The proteome changes constantly and is characterized by a huge dynamic concentration range, therefore representing an extremely complex system. The major impact in the proteomics field has been driven by dramatic improvements in mass spectrometry in recent years. Systems biology seeks to achieve a global understanding of the mechanisms by which a system, as a whole, adapts to its environment; and the 'ome' and 'omics' terms (genomics, transcriptomics, proteomics, and metabolomics) indicate totality. Recent advances in all omics approaches involve integration with mathematics and the engineering sciences, resulting in an in-depth understanding of complex native biological systems in terms of not only structure, but also dynamics. The development of a systems level approach to biotechnological investigations may enable the design of biological systems with desired properties (Wright *et al.*, 2012).

5.2 Instrumentation, methods and software

5.2.1 Sample preparation workflows

Sample preparation is one of the most crucial processes in proteomics research: the results of a proteomics experiment depend on the conditions under which the starting materials were obtained (e.g. growth time, culture conditions), the origin of the starting materials (cell culture vs. soil consortium), and various other factors. In contrast to gene expression levels, protein content may vary markedly between time points during an analysis. Considering these variations, the answer to the question of how many samples and repetitions are required in an experiment is – as many as possible. For this reason, an appropriate experimental workflow and careful sample preparation are some of the most important factors in obtaining significant and reliable results. The information presented below was sourced from a review article by Bodzoń-Kulakowska *et al.* (2007).

Proteins included in a proteomics study may originate from one of many sources. Major protein sources are:

- cultured cells (intracellular proteome and secretome);
- animal and plant tissues;
- body fluids (blood, serum, plasma, cerebrospinal fluid, saliva);
- environmental samples such as water, soil, sludge, etc. (metaproteomics approach).

The general workflow for protein extraction includes several major steps depending on the protein source used:

- Cell disruption: mechanical (ultrasonic, pressure, freeze–thaw); osmotic; or detergent lysis. Homogenization of a sample involves processing a sample such that its composition and structure is uniform throughout the whole volume (i.e., the sample's physical properties should change, but not the chemistry of its components).
- Protein solubilization: once isolated, proteins in their native state are often insoluble and a protein solubilization thus has a marked effect on the quality of the final results. Considering the great diversity and heterogeneity among proteins and sample source-derived interfering contaminants in biological extracts, simultaneous solubilization of all proteins remains a challenge. To avoid protein modification, aggregation, or precipitation resulting in the occurrence of artefacts and subsequent protein loss, the sample solubilization process involves the inclusion of chaotropes (e.g. urea and/or thiourea), detergents {e.g. 3-[(3-Cholamidopropyl)-dimethyl-ammonio]-1-propane sulfonate (CHAPS) or Triton X-100}, reducing agents [dithiothreitol/dithioerythritol (DTT/DTE) or tributylphosphine (TBP)], and protease inhibitors in sample buffers. Each sample and each set of conditions requires unique, experimentally determined treatment conditions.
- Contaminant removal: buffers, salts, and detergents are included in sample solutions to aid protein solubility; however, often these components interfere with further protein separation steps, with protein digestion, or with mass spectrometry analysis and must therefore be removed.
- Salts: removal may be achieved by dialysis, ultrafiltration, gel filtration, precipitation with trichloroacetic acid (TCA) or organic solvents, or solid-phase extraction. Alternatively, commercially available clean-up kits can be used for salt removal.

- Detergents: the most common detergent removal methods include dialysis, gel filtration chromatography, hydrophobic adsorption chromatography, ion-exchange chromatography, and protein precipitation.
- High-abundance proteins: various strategies for the removal of high-abundance proteins have been reported, most of which are based on affinity chromatography using dye-ligands or their derivatives and antibodies (immunoaffinity depletion).
- Lipids: lipids are widely present in biological fluids. Numerous proteins are complexed with lipids, and this interaction reduces their solubility and may therefore affect the pI and MW of the proteins. Most often, in cases where 2-DE separation will be conducted, the use of a centrifugal filter device and the inclusion of CHAPS in the sample buffer allow for efficient lipid and salt removal.
- Polysaccharides and nucleic acids: the presence of polysaccharides and nucleic acids in sample buffers can result in viscous solutions, clogging the pores of the polyacrylamide gels, thus causing either precipitation or extended focusing times, which in turn result in background smearing. To eliminate polysaccharides from protein preparations, they can be precipitated using TCA, acetone, ammonium sulfate, or phenol/ammonium acetate. A subsequent centrifugation step may be beneficial. High-speed ultracentrifugation is used when larger contaminating polysaccharides are present. Digestion with protease-free DNase and RNase is often carried out for the removal of DNA and RNA from protein preparations.
- Other contaminants (including small ionic molecules, nucleotides, metabolites, phospholipids, and insoluble material e.g. organelles): Non-proteinaceous impurities may form complexes with proteins, thereby hampering their solubilization. Precipitation with TCA/acetone or other salt-excluding techniques is effective in removing these contaminants, and high-speed centrifugation may also be an effective alternative.
- Protein enrichment: it may be advantageous to reduce the complexity of a sample by pre-fractionation or to enrich it for proteins of our interest prior to analysis. The most common protein enrichment and purification methods rely on selective precipitation using acetone, TCA, ethanol, isopropanol, diethylether, chloroform/methanol, ammonium sulfate, or polyethylene glycol (PEG); alternatively, a number of commercially available affinity precipitation kits can be used. Other methods of protein enrichment and purification include electrophoresis, centrifugation, membrane processing, and chromatographic or solid phase extraction (SPE) techniques. The fundamental idea of pre-fractionation is to separate a sample into distinguishable fractions containing restricted numbers of molecules, which can be achieved fractionated using a variety of approaches including precipitation, centrifugation, liquid chromatography and electrophoresis-based methods, filtration, and velocity- or equilibrium sedimentation.

5.2.2 Gel-based proteomics

One-dimensional (1-DE) and two-dimensional (2-DE) electrophoresis are the most common techniques currently applied in proteomics. While 1-DE is used only for fast screening due to its limitations in terms of resolution, 2-DE remains the most widely used proteomics tool for resolving protein mixtures from all kinds of protein sources (Fig. 5.1). Depending on the gel size and pH gradient used, 2-DE has been shown to resolve more than 5000 proteins in a single gel with a sensitivity of <1 ng protein per gel spot. Immobilized pH gradient techniques for isoelectric point focusing (IEF) separation were a major development in 2-DE: IEF stripes improved gel reproducibility, rendering, and quantitative comparison between gels. The introduction of narrow range pH gradient zoom gels furthermore enhanced the capacity and detection sensitivity of 2-DE. The efficacy of 2-DE can be further enhanced by sample pre-fractionation methods such as sequential extraction with increasingly stronger solubilization solutions and preparative IEF separation (Loh and Cao, 2008).

Electrophoretic separation of proteins requires appropriate staining for visualization and quantitation purposes. Coomassie Brilliant Blue and silver staining are the most widely used protein detection methods in 2-DE; however, their low sensitivity (especially in the case of Coomassie Brilliant Blue) or limited linear range may lead to inaccurate

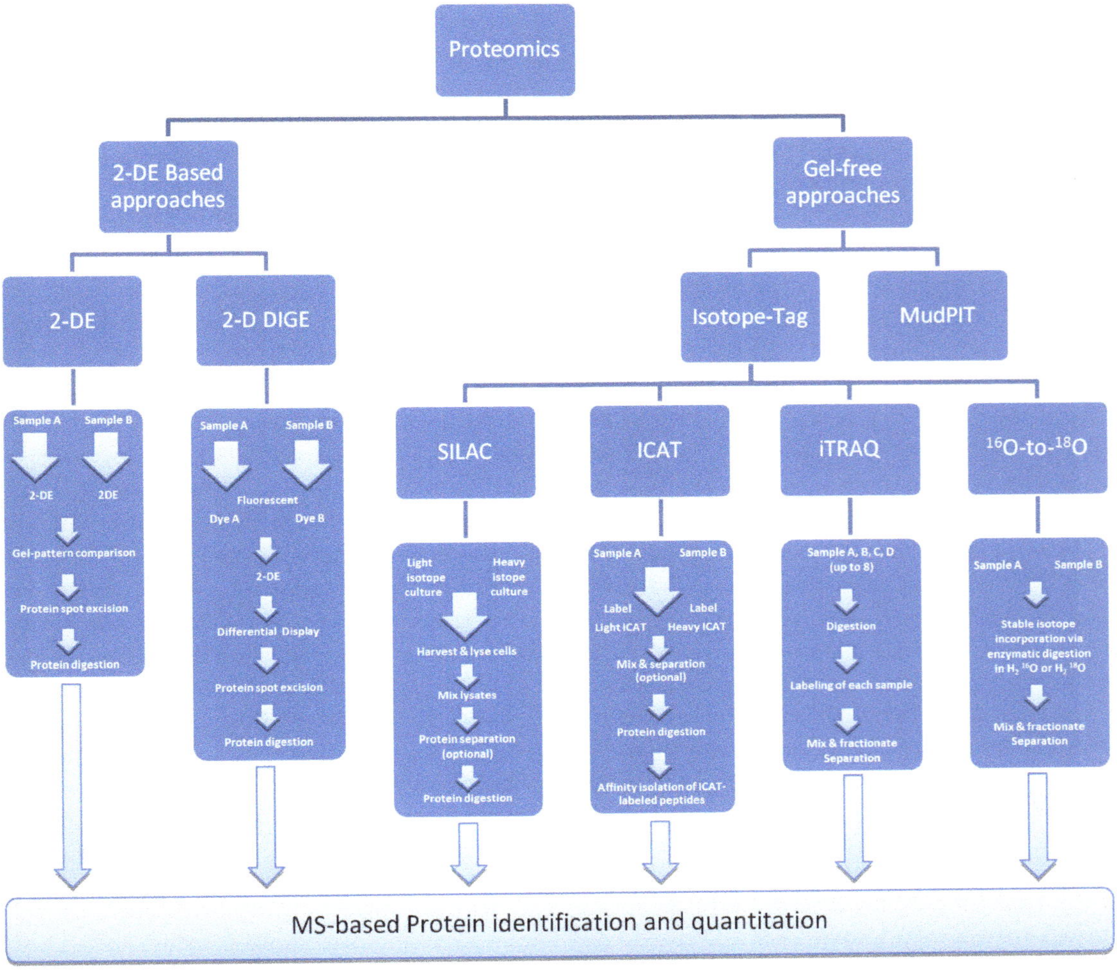

Figure 5.1 Schematic illustration of representative proteomics approaches.

quantitative analysis. The application of radioactive labelling or the use of novel fluorescent dyes, on the other hand, has been demonstrated to improve protein quantitation (Lanne and Panfilov, 2005; Berkelmann, 2006). Multiplexing technology using two-dimensional difference gel electrophoresis (2-DE DIGE) is another approach to overcoming the limitations of quantitative analysis. In 2-DE DIGE, the simultaneous separation, detection, and quantification of multiple samples tagged with fluorescent dyes can be performed in a single gel. Compared with conventional 2-DE with fewer gel replicates, quantitative statistical data with high-confidence levels can be obtained for expression profiling using 2-DE DIGE (Loh and Cao, 2008).

Beside the limitations of 2-DE, this technique is widely used in proteome analysis. The major advantages of 2-DE include an extremely high-resolving power for complex protein mixtures; simple visualization of separation and quantification results; as well as easy identification of post-translational modifications (PTMs) in separated 2-DE spots. Other interesting parameters that can be determined by 2-DE are M_r and pI, which can be estimated from 2-DE maps. Such data may be helpful in positively identifying a protein in contrast to non-gel-based methods in which M_r and pI information are lost.

Although very popular, 2-DE has technical limitations that have led to the development of gel-free strategies, which have been shown to be useful in proteomics studies that cannot be addressed using gel-based strategies alone.

5.2.3 Gel-free proteomics

Liquid chromatography (LC) is an essential tool for the separation of proteins after proteolytic digestion for proteomic analysis. Individual peptides must be separated using LC to avoid ion suppression and the limitations of mass spectroscopy (MS), which requires a certain amount of time to collect high-quality fragmentation spectra. Good separation of low abundance peptides is particularly important in the case of 'shotgun' proteomics.

The most common LC technique used for LC-MS analysis in proteomics is reversed phase chromatography (RP), which is based on hydrophobic interactions and functions to separate non-polar compounds in water-miscible solvents. The stationary phase can vary considerably in terms of hydrophobicity and selectivity depending on the length of the aliphatic chains (C_2 to C_{18}). After proteolytic digestion, peptides are usually separated using linear gradients of water and acetonitrile, while acids [e.g. formic acid (FA) and acetic acid (AA)] or volatile ammonium salts (e.g. ammonium formate, ammonium acetate, and ammonium carbonate) are used to maintain the pH of the mobile phase.

An alternative method for separating mixtures of peptides is hydrophilic interaction chromatography (HILIC). This chromatography technique is similar to normal phase (NP) chromatography and is used exclusively for the separation of polar compounds. In contrast to NP chromatography, HILIC is compatible with MS detection by utilizing the same mobile phases as RP chromatography (water, acetonitrile, methanol) (Guo and Gaiki, 2011). It has been shown that HILIC yields good results for the analysis of post-translational modifications, particularly phosphorylation and glycosylation (Boersema *et al.*, 2008).

In addition to one-dimensional chromatography, multidimensional peptide separation is becoming more widely used. The classic approach to peptide multivariate analysis in proteomics is preceded by LC-MS separation with 2-DE electrophoresis; however, considerable time can be saved by using multidimensional chromatography online coupling different stationary phase columns, e.g. HILIC, RP, size exclusion (SEC), or strong cation exchange (SCX) (Horvatovich *et al.*, 2010). This approach has been shown to improve the robustness of proteomic analyses not only by improving the selectivity but also the sensitivity of peptide detection (Wu *et al.*, 2012). An example of a multidimensional chromatography method by which proteomic study efficiency has been improved is multidimensional protein identification technology (MudPIT) (Wolters *et al.*, 2001; Washburn *et al.*, 2001), which is based on cation exchange pre-fractionation and RP LC separation of tryptic peptides and has been shown to improve the identification of unique low-abundance proteins (Loh and Cao, 2008; Gilany *et al.*, 2015).

To increase separation efficiency for peptide analysis, chromatographic techniques can be applied in the micro- (μLC) or nano- (nLC) scale. Both these separation techniques provide not only better chromatographic separation at a corresponding lower consumption of mobile phases, but also make use of micro- or nano-electrospray ion sources to increase the ionization intensity. These low flow-rate and high-pressure techniques are usually carried out with flow rates and maximum pressure limits, respectively, of 1–200 μl/min and 10,000 psi for μLC and 0.01–50 μl/min and 15,000 for nLC. Compared with conventional LC and UHPLC methods, these are often associated with problems with the stability and reproducibility of proteomic analysis since high working pressures and small diameter capillaries multiply technical problems such as dead volume and increase susceptibility to clogging or contamination of the system (Nazario *et al.*, 2015).

The gel-free proteomic techniques described here are used to achieve the best possible separation of the peptides obtained by proteolytic digestion of proteins before mass spectrometric detection and fragmentation (Fig. 5.1). Better separation always provides a better quantitative and qualitative proteomic result; however, the choice of techniques is dependent on the complexity of the test material as well as the expected outcome defined in the experimental design.

5.2.4 Mass spectrometry

Mass spectrometry (MS), a modern analytical technique that allows for the accurate measurement of the ratio between the molecular weight and the electric charge of an ion (m/z), is commonly used in investigations into or confirmation of the structures of organic compounds as well as in the elucidation of qualitative and quantitative relationships in the mixture.

Regardless of this construction and purpose, a mass spectrometer must consist of:

- an ion source or a device that produces ions (ESI, APCI, MALDI, Photospray);
- an analyser wherein the pre-ionized particles undergo separation on the basis of their m/z;
- A detector that counts the number of ions from the analyser.

Depending on the separation technique used (gel-based or gel-free proteomics), peptides can be introduced into the mass spectrometer usually via liquid chromatography coupled with electrospray ion source ESI-MS/MS or can be analysed directly from spots by matrix-assisted laser desorption/ionization MALDI-MS/MS. Each qualitative MS/MS experiment consists of two types of scans: general scan MS, that is carried out over a wide m/z range and fragmentation MS/MS scan in which collision-induced dissociation (CID) of selected mass m/z peptides allocated by general scan MS is performed. Based on a fully characterized mechanism of peptide CID fragmentation (Fig. 5.2), bioinformatics tools (e.g. Mascot) can fully or partially restore the peptide sequences with parallel identification and localization of post-translational modifications (PTMs), allowing for protein identification (Fig. 5.3).

Quantitative peptide analyses are usually performed using triple-quadrupole mass spectrometers (QqQ) and based on a scan mode termed multiple reaction monitoring (MRM), which is a scan based on the selection of pseudo molecular ions and generated a result from CID, selected fragmentation ions. Peptide quantification using MRM pairs can

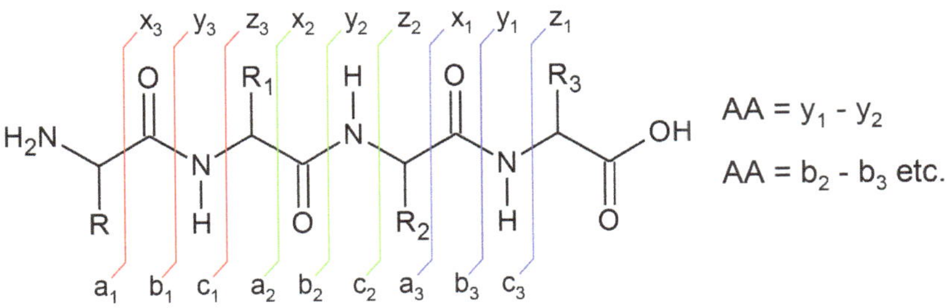

Figure 5.2 MS/MS-based fragmentation and sequencing of peptides.

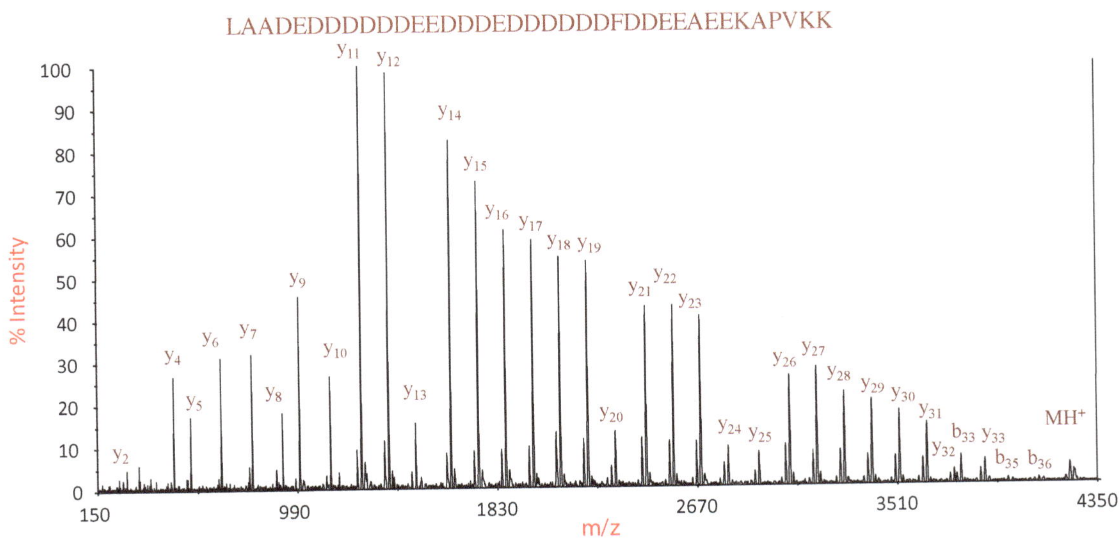

Figure 5.3 Example of sequencing of the peptide on the basis of MS/MS spectrum.

be optimized experimentally or predicted using dedicated software tools such Skyline (https://brendanx-uw1.gs.washington.edu/labkey/project/home/software/Skyline/begin.view).

The huge diversity in mass spectrometer types and scanning modes provides many options for different scanning experiments and the right choice depends on the specific requirements of each proteomic analysis.

5.2.5 Identification and quantitation

Protein identification (protein ID) was the first mass spectrometry application in proteomics. There are two different strategies for MS protein ID: top down, i.e. sequence analysis of the whole (intact) protein; and bottom-up, i.e. recreation of a protein sequence from the sequences of the peptides resulting from enzymatic degradation of the protein. From fragmentation spectra, peptide sequences are defined and confirmed using a library search (e.g. Mascot). The bottom-up strategy is currently the most popular proteomic approach for the identification of both pure proteins (e.g. spots on two-dimensional gels) as well as proteins in complex mixtures (using so called 'shotgun' proteomics). In the case of pure proteins or mixtures of several proteins, identification can be performed with MALDI TOF/TOF or by means of tandem LC-MS/MS systems. More complex mixtures of proteins require higher resolution systems (e.g. Orbitrap or QqTOF) for proper identification.

Relative quantitation can be performed using gel-free isotope labelling methods such as stable isotope labelling by amino acids in cell culture (SILAC), isotope-coded affinity tags (ICAT), isobaric tag for relative and absolute quantitation (iTRAQ), or using ^{16}O:^{18}O ratios.

The SILAC method of quantitative protein analysis is based on the proteomic MS/MS analysis of two or more isotopically labelled samples. Isotope labelling of proteins is performed using culture media containing amino acids rich in heavy isotopes (e.g. arginine or lysine containing carbon ^{13}C or ^{15}N). The incorporation of 'heavy' acids is carried out at the certain cell-culture stage. This method is not often applied in quantitative proteomics because of the limitations of *in vitro* experimental models and the amount of isotope labelled standard required (Chen *et al.*, 2015; Chahrour *et al.*, 2015).

The ICAT and iTRAQ methods are similar quantitative approaches based on the simultaneous chemical labelling of samples with different isotope tags. Depending on the protocol used, the isotope labelled reagent will react with the amino acid side chains (e.g. cysteine) of the undigested protein (ICAT) or with N-terminus primary amino group in a peptide mixture after proteolytic digestion (iTRAQ). The ICAT method allows for the simultaneous labelling of a maximum of two samples (suitably heavy and light isotope), while iTRAQ allows for the use of up to four different isotope tags. These analyses yield relative concentrations of each peptide (and therefore the corresponding concentrations of the proteins) calculated from the intensity ratio of species labelled with different signal tags (Chahrour *et al.*, 2015). Most proteomic software tools include ICAT or iTRAQ functionality at both the data interpretation levels (jTraqX – http://sourceforge.net/projects/protms/files/jTraqX) and at the level of experimental design (Skyline).

The ^{16}O:^{18}O exchange methodology is based on the naturally occurring (in $H_2^{18}O$ solution) endoprotease-catalysed oxygen exchange process at the C-terminal carboxylic acid (Heller *et al.*, 2003). As in the case of ICAT, ^{16}O:^{18}O allows only two samples to be labelled simultaneously; however, this technique tends to be used as a simpler alternative to ICAT and iTRAQ.

5.2.6 Post-translational modification analysis

Several different post-translational modifications (PTMs) of proteins are currently known and described. These include:

- Native irreversible modifications: modifications necessary for the native biological function of a protein that occur during or immediately after protein translation: glycosylation.
- Reversible modification for regulation of protein activity or function: phosphorylation and acetylation.
- Irreversible modifications leading to protein degradation: poly-ubiquitination.

The effects of PTMs are strongly protein-dependent: while glycosylation often modulates protein activity, it facilitates maturation and protects against proteolytic degradation in certain proteins. Proteomic MS/MS analysis of PTMs is performed by the methods described previously (Figs. 5.2 and 5.3). Key in these analyses is the suitable enrichment of the test sample for peptides/proteins with the modification of interest. Such enrichment may be performed by any kind of affinity chromatography including antibody affinity, lectin affinity, or proteolytic enzyme activity (Doll and Burlingame, 2015; Ren *et al.*, 2014).

5.2.7 Software tools and databases

Software tools are an essential part of modern proteomics. Proteomic studies performed using mass spectrometers generate huge amounts of both MS and MS/MS spectra that cannot be interpreted manually, and without search engines or prediction/simulation software tools, it would not be possible to process and interpret proteomic data.

Mascot is the most versatile and widely used search engine for the interpretation of qualitative mass spectrometry proteomic data. The algorithm used in Mascot was created based on the earlier MOWSE algorithm (MOlecular Weight Search) and indicates the probability of the occurrence of the unknown protein based on the sequence of the peptide resulting from the proteolytic digestion method used (e.g. tryptic). Databases and search engines are freely available on the Matrix Science website (www.matrixscience.com); however, to perform multiple searches and to access additional databases, a license must be purchased. The Mascot search engine generates information on the probability of a protein occurrence (protein ID) based on a database search and a parameter termed False Discovery Rate (FDR) by searching against a randomized sequence of the same length for every sequence in the target database (decoy database)(Wang *et al.*, 2009). Besides Mascot, there are many other algorithms and databases for qualitative proteomic analysis, including open source tools such as X!Tandem (www.thegpm.org) and ProteinProspector (www.prospector.ucsf.edu). Most mass spectrometer manufactures also offer their own search engines and commercially available databases.

As in the case of qualitative analysis, many different software tools have been developed for quantitative proteomics over the years. Currently, the most commonly used tool for this purpose is Skyline (https://brendanx-uw1.gs.washington.edu), which is a freely available, open-source software for building SRM/MRM, targeted MS/MS, DIA/SWATH, and targeted DDA with MS1 quantitative methods and for analysing mass spectrometer data. Software tools used for analyses based on known protein sequences are compatible with most software packages used by mass spectrometer manufacturers. The Skyline community offers full support to their users via useful manuals and tutorials as well as webinars. Project Skyline is currently being expanded by MacCoss Lab with additional expansions and projects (freeware and open-source), also including elements of qualitative proteomics: Panorama, Topograph, and BiblioSpec.

In addition to these highly advanced software tools for managing proteomic data, there are many other basic but useful tools. The greatest diversity of tools is offered by the Swiss Institute of Bioinformatics on their ExPASy Bioformatics Resources Portal (www.expasy.org/proteomics), where users have access to tools ranging from protein sequence databases for the prediction of post-translational modifications to structural proteomics tools.

5.2.8 Other methods

Currently, quantitative proteomics is based primarily on mass spectrometry techniques. In contrast, structural proteomics also relies on other techniques for detailed descriptions of protein structure and interactions. The most accurate information about secondary and tertiary structure of proteins is obtained using spectrophotometric methods such as circular dichroism (CD) and fluorescence spectroscopy.

Circular dichroism spectra provide information about changes in protein secondary structures caused by a number of physical and chemical factors. Such measurements allow for the simple and precise identification of structural changes that lead to protein denaturation and aggregation. In the far UV range, CD furthermore provides information about the content of individual elements of secondary structure in the spatial structure of the protein. In an experimental CD spectrum for a protein with dominant α-helices, there are three specific extremes: two minima for $\lambda = 222\,nm$ (associated with an $n \rightarrow \pi^*$ shift) and $\lambda = 208\,nm$ (associated with a $\pi \rightarrow \pi^*$ transition) and one maximum for $\lambda = 192\,nm$ (associated with the $\pi \rightarrow \pi^*$ transition). The β-structure of a protein is associated with two $\pi \rightarrow \pi^*$ shifts on a CD spectrum: a minimum for $\lambda = 218\,nm$ and a maximum for $\lambda = 195\,nm$. In the near-UV range, CD can provide additional information about the tertiary structure of a protein. In this regard, the amino acids that affect circular dichroism are cysteine (S–S bond), tryptophan, phenylalanine, and tyrosine. In contrast to far-UV CD spectra, near-UV (260–350 nm range) CD spectral analysis lacks well-defined rules for signals (Park and Raines, 2004).

The second spectrophotometric method used in structural proteomic studies is based on intrinsic protein fluorescence. The intrinsic fluorescent properties of proteins are the result of the presence of three aromatic amino acids: tryptophan (280 → 348 nm), tyrosine (274 → 303 nm), and phenylalanine (257 → 282 nm). The fluorescence intensity depends on the number of fluorophores but also on the tertiary structure of the protein, and thus is dependent on the environmental conditions and interactions with other molecules. Quenching of fluorescence intensity is directly related to the interaction of a fluorophore with other molecules or protein side-chains in the fluorophore microenvironment. Studies on both fluorescence quenching and anisotropy provide detailed information about protein interactions and may therefore provide a means of describing the dynamics of such interactions (Greenfield, 2004).

5.3 Proteomics of biodegradation

5.3.1 Comparative untargeted and targeted proteomics

The large majority of proteomic investigations of environmental microorganisms are focused on model microorganisms cultured in the laboratory. These species have been studied for their interesting traits such as the ability to tolerate, degrade, or precipitate toxic compounds or for their versatility in the use of electron donors, electron acceptors, or carbon and energy sources. These qualities deem such organisms attractive for environmental biotechnology applications, and proteomics may lead to a better understanding of the functions of these organisms in specific habitats. A technical advantage to using laboratory-cultured model organisms is that their genomes can be sequenced, which almost always improves the quality of protein identifications. The use of such model organisms allows for flexibility during method development for more complete proteome profiling (Lacerda and Reardon, 2009).

Comparative proteomics represents a large portion of the research data associated with the human genome project and cancer and other disease biomarker research; however, there are also many other application of comparative proteomics in research fields such as plant physiology or microbiology. Proteomic studies have been carried out on the organisms involved in the biodegradation of different substrates, both natural and anthropogenic; however, only very few such studies have been reported. Data from selected comparative proteomics studies in the field of biodegradation are presented in this chapter.

A good example of proteomic characterization of biodegradation by a single organism is a study on the removal of alachlor by *Paecilomyces marquandii* (Szewczyk *et al.*, 2015). In this study, targeted and untargeted LC-MS/MS methods were applied in the characterization of changes in glucose concentration and intermediates of glycolysis and the TCA cycle as well as to monitor alachlor removal (complete depletion after 168 h of culture) and xenobiotic metabolite levels. Basic data on changes in culture parameters such as pH and dry weight were also collected. Data were collected at 24- to 48-hour intervals during the 7 days of the experiment both in the presence (test) and absence (control) of alachlor as well as from abiotic controls (where applicable). From all these data, but in particular from the alachlor metabolite profiling data (Fig. 5.4), the fifth day of incubation was chosen for a comparative proteomic investigation.

Protein separation by 2-DE yielded 247 and 265 protein spots in the control and alachlor-containing samples, respectively. The gel analysis resulted in the selection of 58 proteins from the control cultures and 42 proteins from the alachlor-containing cultures for MS/MS sequencing followed by Mascot

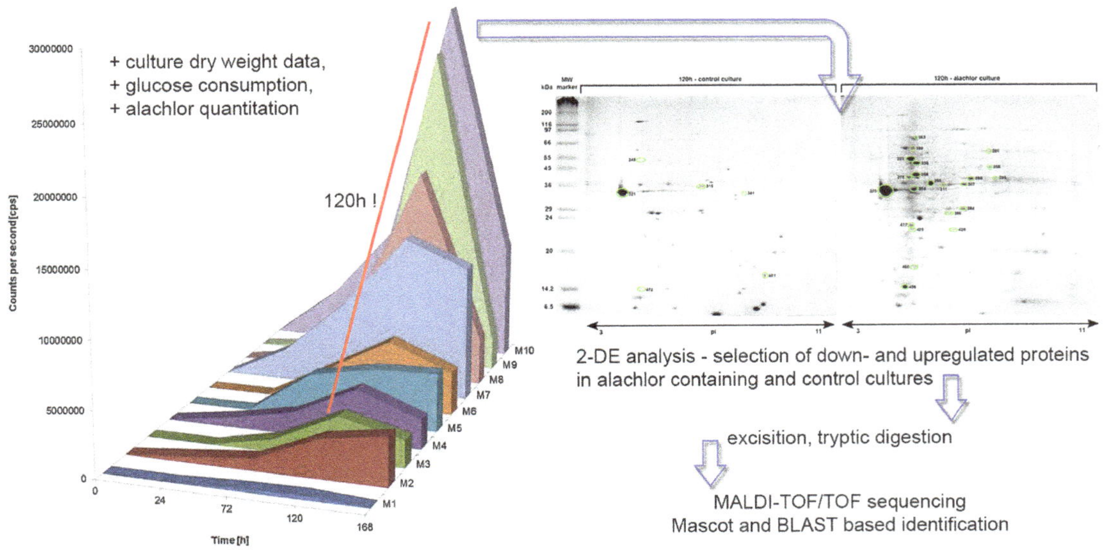

Figure 5.4 Trend analysis of the identified metabolites during alachlor (50 mg/l) biodegradation by *Paecilomyces marquandii* on Sabouraud medium. M1–M10 are identified metabolites (Szewczyk *et al.*, 2015).

algorithm identification using the NCBI non-redundant database (taxonomy set to fungi) and functional alignments (BLAST searches with the delta-blast algorithm). The data revealed that, in the presence of a strong xenobiotic, enzymes associated with carbon source metabolism (kinases, transaldolases, reductases, dehydrogenases, mutases, and isomerases), energy (ATP synthase), and ROS (manganese superoxide dismutase, catalase) were up-regulated. Furthermore, the unique overexpression of cyanide hydratase, which hydrolyses nitriles to ammonia or amide and the corresponding carboxylic acid, was identified in alachlor-containing cultures. A comparison of the identified metabolite structures indicated that this enzyme as the key component in the alachlor removal process.

Physiological and proteomic approaches complement each other, and a combination of the two can lead to a better understanding of bacterial responses under different growth conditions. A comparative physiological and gel-free proteomic study of *Pseudomonas putida* growth on citrate and benzoate revealed the presence of various cellular mechanisms used for changing from one to another carbon source (Fig. 5.5) (Mandalakis *et al.*, 2013). Apart from the identification of the enzymes associated with the peripheral catabolism of benzoate via the catechol ortho-cleavage pathway, two other enzymes were found to be up-regulated, which led to the discovery that the protocatechuate ortho-cleavage pathway may also play a role in benzoate degradation. Moreover, the up-regulation of an arylsulfatase involved in the catabolism of sulfated aromatic hydrocarbons co-existing in nature with benzoate also reported. Another group of up-regulated proteins was shown to be associated with the uptake of respective exogenous C-substrates (i.e. specific benzoate and citrate transporters), while several other groups of proteins were found to be responsible for various physiological adjustments made by *P. putida* F1 in response to changes in C-source. Among these adjustments were (a) the induction of the stress response as a result of benzoate toxicity; (b) the activation of redox-balancing due to the higher C:O ratio of benzoate versus citrate; (c) the induction of metabolic pH regulation as a result of alkalization due to the presence of trisodium citrate; and (d) an increase in cellular motility and chemosensory mechanisms as a result of benzoate toxicity.

The results of this study clearly demonstrate the power of proteomics in clarifying bacterial responses to various carbon sources including organic pollutants. Such data may be applied in environmental engineering to clarify the molecular background of certain processes.

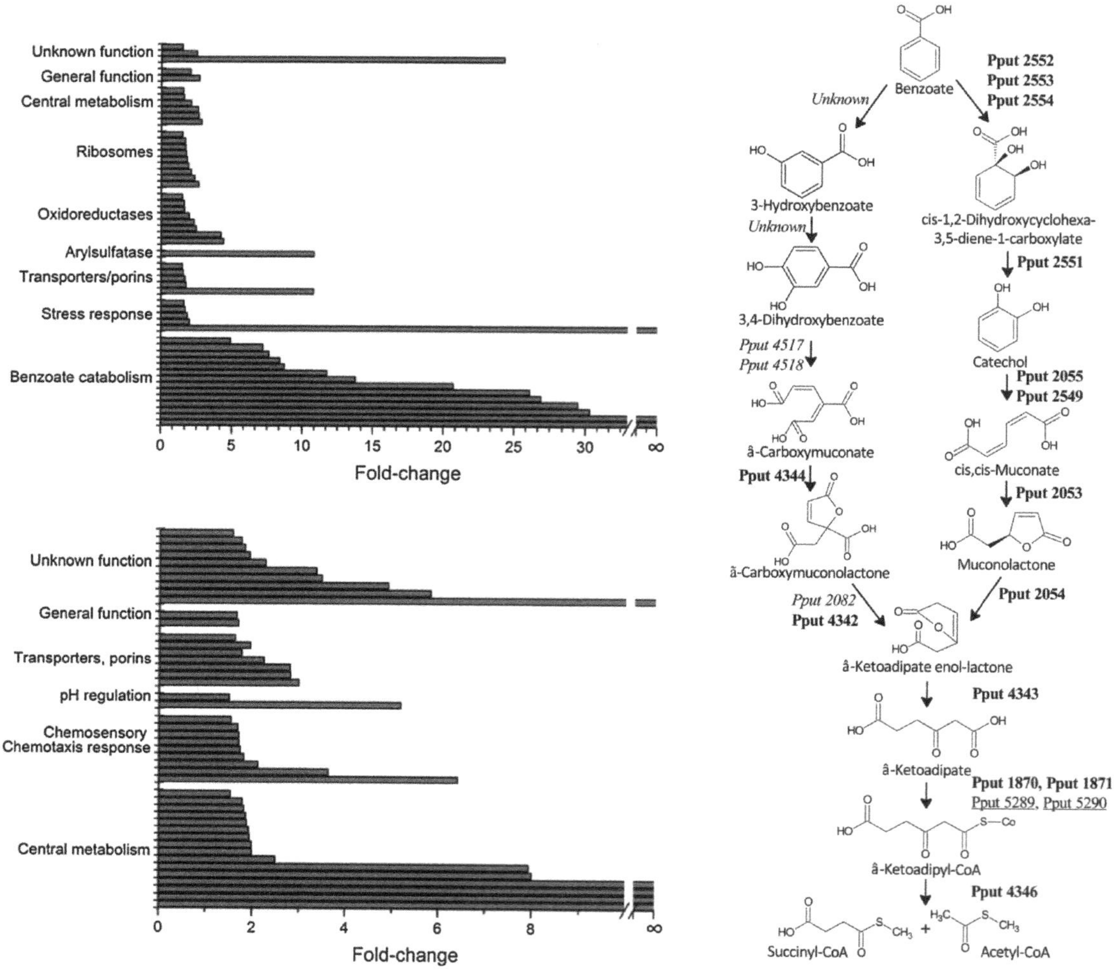

Figure 5.5 (Left) Fold-change of the functional groups that were preferentially expressed in *P. putida* F1 growth on a) benzoate and b) citrate. (Right) KEGG pathway map for benzoate degradation in *P. putida* F1. The proteins identified as up-regulated by one (underlined) or at least two peptides (in bold) are highlighted. Undetected proteins are also shown in italics (reproduced from Mandalakis *et al.*, 2013, SpringerOpen licence).

A quantitative proteomics study was conducted on 17β-oestradiol degradation by *Stenotrophomonas maltophilia* ZL1 (Li *et al.* 2012). Relative quantitation of the proteins was conducted using a label-free LC-MS/MS method of spectral counting with normalized spectral counts. The number of identified proteins in each of the four samples ranged from 193 to 263. To characterize the biodegradation process, the proteomics of glycolysis, the TCA cycle, oxidative phosphorylation, amino acid metabolism, and lipid metabolism were assessed. On the basis of these data, the authors identified associations between cellular metabolism pathways and characterized the involvement of certain enzymes (phenolpuryvate tautomerase and aromatic amino-acid transaminase) in 17β-oestradiol degradation (Fig. 5.6). Summarizing their findings, the authors proposed that E2 can be converted to E1 by *S. maltophilia* ZL1 and that E1 and its degradation intermediates can be utilized in aromatic amino acid biosynthesis.

5.3.2 Metaproteomics

Metaproteomics refers to the proteomic study of communities as metaorganisms, where a metaorganism is defined as a collection of organisms evolving as a whole, sharing genes and metabolic capacities. For this reason, metaproteomic studies

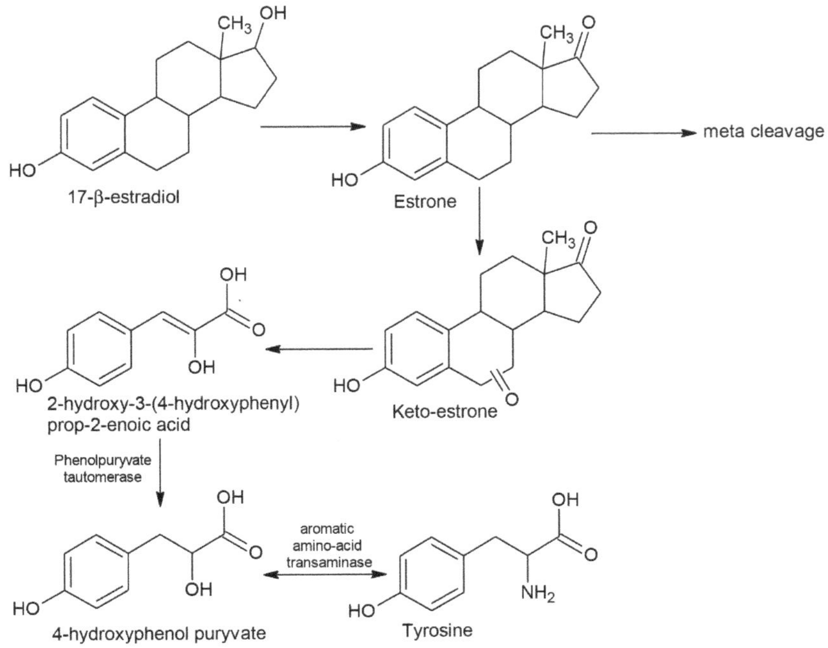

Figure 5.6 Proposed biotransformation pathway of 17β-oestradiol by *S. maltophilia* ZL1.

are far more complex than proteomic studies of pure cultures. Laboratory-based investigations of microbial communities represent an intermediate level of complexity between isolates and complex environmental samples, allowing information to be obtained about a community rather than a single microorganism without the challenges associated with protein extraction from environmental samples. Proteomic investigations of microbial communities in their native environments provide the most realistic information about their function; however, these also pose the greatest experimental and bioinformatic challenges (Lacerda and Reardon, 2009). Proteins reflect actual functionality with respect to metabolic reactions and regulatory cascades and therefore provide more direct information about microbial activity than functional genes and even the corresponding messenger RNAs (Wilmes and Bond, 2006). The use of proteins furthermore bears the potential to reveal the identity of active microorganisms via database analysis using the level of homology to other species (Benndorf *et al.*, 2007). Thus, the presence of specific proteins in environmental samples is a potentially reliable indicator for microbial function (Benndorf *et al.*, 2007).

Thus far, metaproteome data have been reported for various environmental systems including soil (Benndorf *et al.*, 2007; Chourey *et al.*, 2010), activated sludge (Wilmes and Bond, 2006), water (Benndorf *et al.*, 2007), leaf litter (Schneider *et al.*, 2012), and a composting plant (Liu *et al.*, 2015). Metaproteomics has also been used to define key catabolic players at contaminated sites to predict pollutant degradation networks in the environment (Guazzaroni *et al.*, 2013).

An interesting example of the application of metaproteomics is a study on major microbial characterization and biodegradation functions in a large-scale aerobic composting plant (Liu *et al.* 2015). The authors characterized the main biodegradation pathways in a municipal solid waste composting system across the main phases. Of 640 proteins, 449 exhibited the highest sequence identity to bacterial proteins and 191 with fungal proteins. The microbial succession investigation revealed that Bacillales, Actinobacteria, and *Saccharomyces* strains increased significantly in abundance during composting process. The Gammaproteobacteria were identified as the largest group, accounting for 40.3% of the total bacterial proteins.

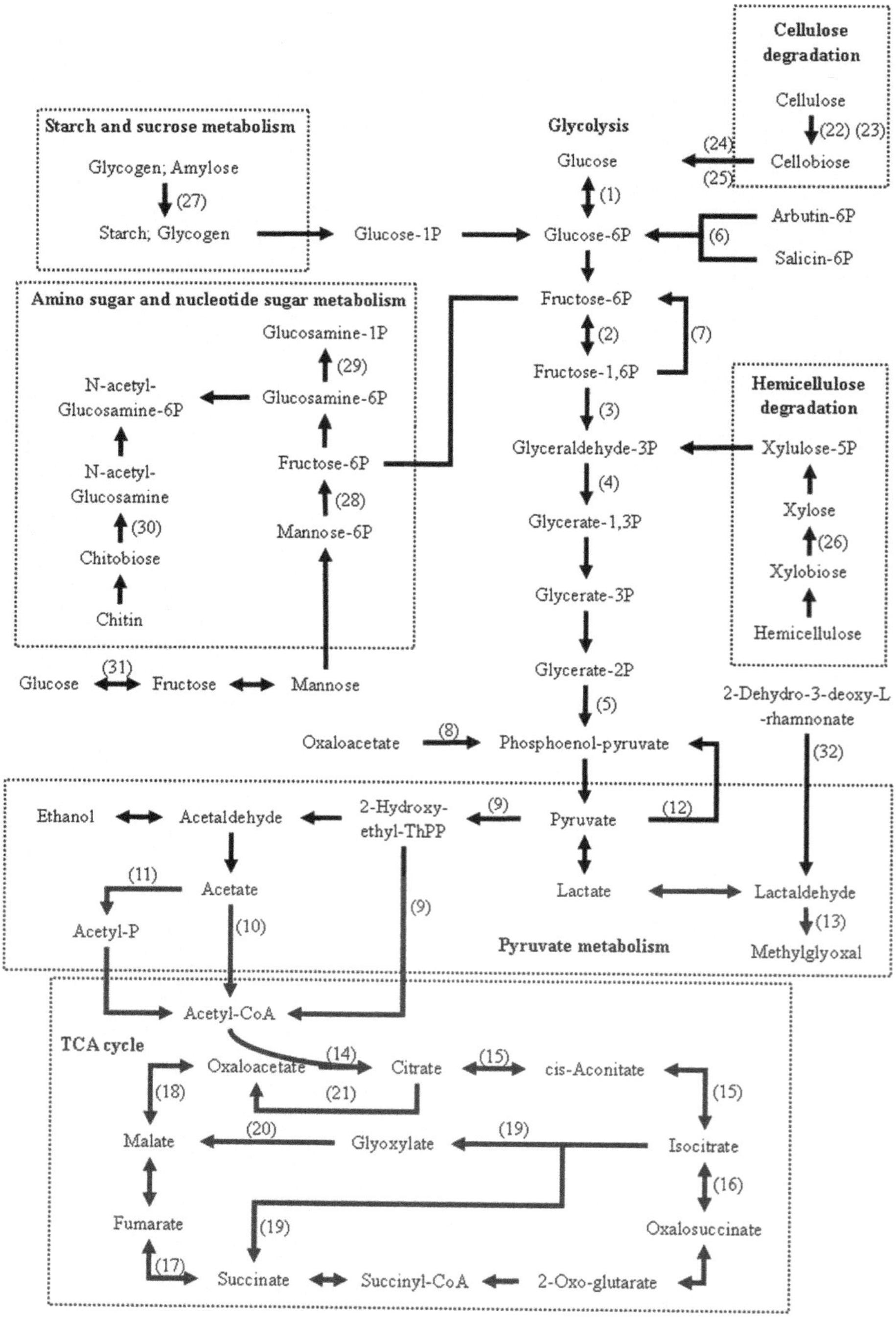

Figure 5.7 Depiction of the carbohydrate metabolic characteristics of microbial communities inferred from the composting plant metaproteome. Proteins are: (1) hexokinase; (2) 6-phosphofructokinase; (3) triosephosphate isomerase; (4) glyceraldehyde-3-phosphate dehydrogenase; (5) enolase; (6) aryl-phospho-beta-D-glucosidase; (7) fructose-1,6-bisphosphatase; (8) Phosphoenolpyruvate carboxykinase; (9) Pyruvate dehydrogenase; (10) acetyl-coenzyme A synthetase; (11) acetate kinase; (12) pyruvate, phosphate dikinase; (13) glyoxylate/hydroxypyruvate reductase; (14) citrate synthase; (15) Aconitate hydratase; (16) Isocitrate dehydrogenase; (17) succinate dehydrogenase; (18) malate dehydrogenase; (19) isocitrate lyase; (20) malate synthase; (21) Citrate lyase; (22) Exoglucanase; (23) endoglucanase; (24) cellobiose dehydrogenase; (25) betaglucosidase; (26) endo-1,4-beta-xylanase B; (27)1,4-alpha-glucan branching enzyme GlgB; (28) mannose-6-phosphate isomerase; (29) Phosphoglucosamine mutase; (30) beta-N-acetylhexosaminidase; (31) 2-keto-3-deoxy-L-rhamnonate aldolase; (32) xylose isomerize (reproduced from Liu *et al*., 2015).

Carbohydrate metabolism was identified as the main metabolic pathway in the composting process and cellulose and hemicelluloses were found to be the main carbon sources, since three types of cellulases essential for cellulose degradation were identified (Fig. 5.7). Fungi were identified as the main producers of cellulases in the earlier phase; however, only bacterial and no fungal cellulolytic enzymes were detected at the end of the active phase. This finding supported the conclusion that the key microbiological population for cellulose degradation in different composting stages differed, while it did not support thermophilic fungal activity through the thermophilic phase. In the curing phase, a mixed community of bacteria and fungi was shown to enhance cellulose decomposition, which suggested that the curing phase might be a rate-limiting period in the cellulose decomposition process.

Although some denitrifying bacteria were detected in the samples, only *Pseudomonas stutzeri* was shown to produce denitrification enzymes. Denitrification enzyme production by other denitrifying bacteria was not evident, suggesting that *Pseudomonas* species act as key players in the denitrification process during composting.

As highlighted by the authors of the compost plant study, metaproteomics-based investigations are limited in terms of protein extraction efficiency and data analysis. Nevertheless, the data obtained from the metaproteome analysis provided deep insight into molecular details of the composting process.

Another example of the use of metaproteomics in the field of biodegradation is a study on microbial metabolism in aromatic-contaminated environments (Guazzaroni *et al.*, 2012). The authors applied an omic approach that included both metagenomics and metaproteomics of anthropogenically influenced and polyaromatic hydrocarbon (PAHs)-contaminated soil with (Nbs) or without (N) bio-stimulation with calcium ammonia nitrate, NH_4NO_3, and KH_2PO_4 as well as the commercial surfactant Iveysol and two naphthalene-enriched communities derived from both soils (CN2 and CN1, respectively).

This investigation led to the identification of distinct biodegradation capabilities for the utilization of potential growth-supporting aromatics, which results in bio-stimulated communities being extremely fit for naphthalene utilization and non-stimulated communities exhibiting a greater metabolic window than previously predicted. Beside the characterization of microbial communities on the basis of genetic and proteomic data, interactions among members were hypothesized and degradation networks for complex microbial communities were proposed.

A total of 1116 proteins were identified and quantified from cells harvested from enrichment cultures CN1 and CN2. A total of 582 (52%) of the proteins were identified in both samples, with the relative intensities of 123 and 227 proteins being significantly higher in CN1 and CN2, respectively. A total of 132 proteins were furthermore found to be specific to CN1, most of which were affiliated with *Azospirillum* (B71%) and *Comamonas* (B17%) species, whereas 402 proteins (B94% associated with Gammaproteobacteria of the *Pseudomonas* genus) were found exclusively in CN2. On the basis of their probable functions, the proteins were sorted into five major groups, which were associated with roles in aerobic degradation, electron acceptors, energy metabolism, transport, and folding/stress proteins.

Based on the data obtained from the metagenomic and metaproteomic studies, four consortia metabolic networks were proposed (Fig. 5.8). The study revealed that the PAHs-contaminated soil ecosystem was delicate and that its metabolic capability may be easily lost as a result of bio-stimulation, which could have a significant ecological impact. Finally, the curated database containing protein sequences with biochemical functions shown to be involved in biodegradation of aromatics via di- and tri-hydroxylated intermediates may be used in future investigation related to microbial biodiversity, ecology, and function in response to aromatics.

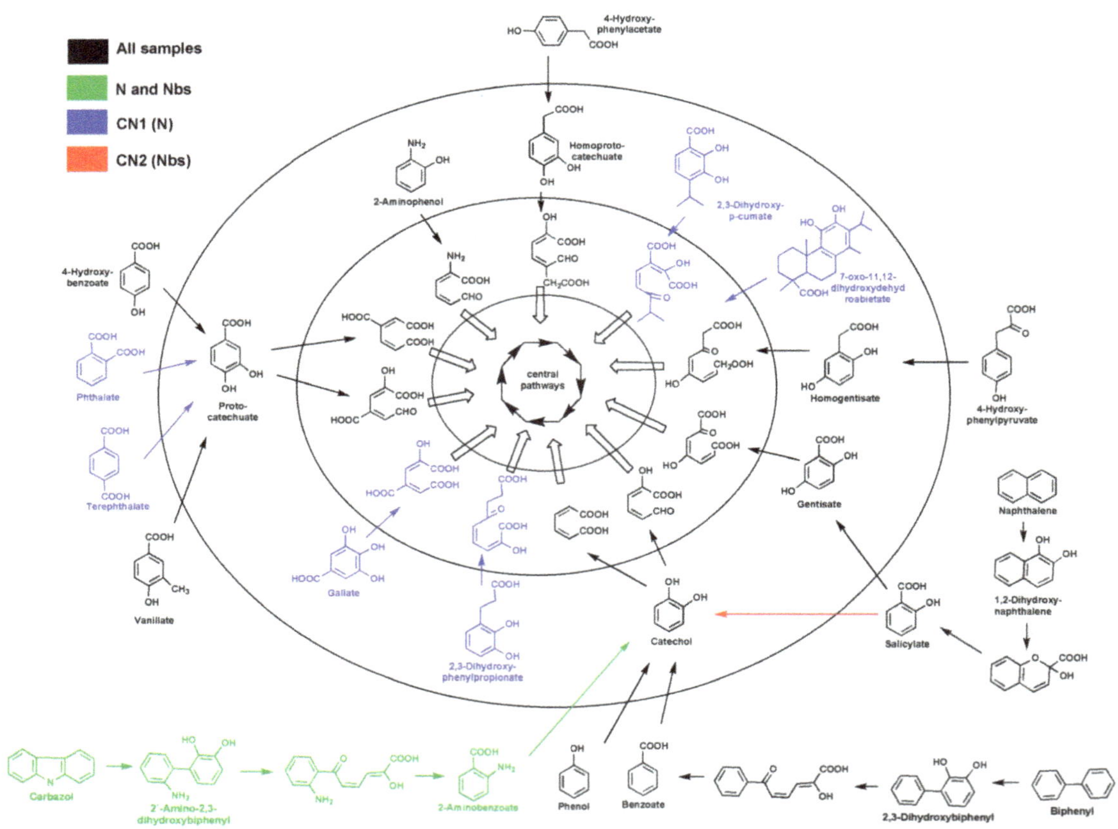

Figure 5.8 Potential aerobic degradation networks of aromatics via di- and tri-hydroxylated intermediates in the four investigated communities. The colour code used for the respective pathways is as follows: black, all samples; green, N and Nbs (specific for soil communities); blue, CN1 (N) (specific for non bio-stimulated communities]; red, CN2 (Nbs) (specific for bio-stimulated communities). Reprinted by permission from Macmillan Publishers Ltd: The ISME Journal (Guazzaroni *et al.*, 2012), copyright (2012).

References

Benndorf, D., Balcke, G.U., Harms, H., and Von Bergen, M. (2007). Functional metaproteome analysis of protein extracts from contaminated soil and groundwater. ISME J. *1*, 224–234.

Berkelman, T. (2006). Sensitivity and protein-to-protein consistency of Flamingo fluorescent gel stain compared to other fluorescent stains. Mol. Cell. Proteom. *5*, S47-S147.

Bodzoń-Kulakowska, A., Bierczyńska-Krzysik, A., Dylag, T., Drabik, A., Suder, P., Noga, M., Jarzębińska, J., and Silberring, J. (2007). Methods for samples preparation in proteomic research. J. Chromatogr. B Analyt. Technol. Biomed. Life Sci. *849*, 1–31.

Boersema, P.J., Mohammed, S., and Heck, A.J. (2008). Hydrophilic interaction liquid chromatography (HILIC) in proteomics. Anal. Bioanal. Chem. *391*, 151–159.

Cai, Y.H., and Huang, H. (2012). Advances in the study of protein–DNA interaction. Amino Acids *43*, 1141–1146.

Chahrour, O., Cobice, D., and Malone, J. (2015). Stable isotope labelling methods in mass spectrometry-based quantitative proteomics. J. Pharm. Biomed. Anal. *113*, 2–20.

Chalmers, M.J., Busby, S.A., Pascal, B.D., West, G.M., and Griffin, P.R. (2011). Differential hydrogen/deuterium exchange mass spectrometry analysis of protein–ligand interactions. Expert Rev. Proteomics. *8*, 43–59.

Chen, X., Wei, S., Ji, Y., Guo, X., and Yang, F. (2015). Quantitative proteomics using SILAC: Principles, applications, and developments. Proteomics, *15*, 3175–3192.

Chourey, K., Jansson, J., VerBerkmoes, N., Shah, M., Chavarria, K.L., Tom, L.M., Brodie, E.L., and Hettich, R.L. (2010). Direct cellular lysis/protein extraction protocol for soil metaproteomics. J. Proteome Res. *9*, 6615–6622.

Doll, S., and Burlingame, A.L. (2014). Mass spectrometry-based detection and assignment of protein posttranslational modifications. ACS Chem. Biol. *10*, 63–71.

Dunham, W.H., Mullin, M., and Gingras, A.C. (2012). Affinity-purification coupled to mass spectrometry: basic principles and strategies. Proteomics, *12*, 1576–1590.

Gilany, K., Minai-Tehrani, A., Savadi-Shiraz, E., Rezadoost, H., and Lakpour, N. (2015). Exploring the Human Seminal Plasma Proteome: An Unexplored Gold Mine of Biomarker for Male Infertility and Male Reproduction Disorder. J Reprod Infertil. *16*, 61.

Greenfield, N.J. (2004). Circular dichroism analysis for protein–protein interactions. In Protein–protein Interactions: Methods and Applications. H. Fu, ed. (Humana Press Inc., Totowa, NJ), *261*, pp. 55–77.

Guazzaroni, M.E., Herbst, F.A., Lores, I., Tamames, J., Peláez, A.I., López-Cortés, N., Alcaide, M., Del Pozo, M.V., Vieites, H.M., von Bergen, M., *et al.* (2013). Metaproteogenomic insights beyond bacterial response to naphthalene exposure and bio-stimulation. ISME J. *7*, 122–136.

Guo, Y., and Gaiki, S. (2011). Retention and selectivity of stationary phases for hydrophilic interaction chromatography. J. Chromatogr. A. *1218*, 5920–5938.

Heller, M., Mattou, H., Menzel, C., and Yao, X. (2003). Trypsin catalyzed 16 O-to-18 O exchange for comparative proteomics: tandem mass spectrometry comparison using MALDI-TOF, ESI-QTOF, and ESI-ion trap mass spectrometers. J. Am. Soc. Mass Spectrom. *14*, 704–718.

Horvatovich, P., Hoekman, B., Govorukhina, N., and Bischoff, R. (2010). Multidimensional chromatography coupled to mass spectrometry in analysing complex proteomics samples. J. Sep. Sci. *33*, 1421–1437.

Jurneczko, E., and Barran, P.E. (2011). How useful is ion mobility mass spectrometry for structural biology? The relationship between protein crystal structures and their collision cross sections in the gas phase. Analyst *136*, 20–28.

Kan, Z.Y., Walters, B.T., Mayne, L., and Englander, S.W. (2013). Protein hydrogen exchange at residue resolution by proteolytic fragmentation mass spectrometry analysis. Proc. Natl. Acad. Sci. U.S.A. *110*, 16438–16443.

Konijnenberg, A., Butterer, A., and Sobott, F. (2013). Native ion mobility-mass spectrometry and related methods in structural biology. Biochim. Biophys. Acta *1834*, 1239–1256.

Lacerda, C.M., and Reardon, K.F. (2009). Environmental proteomics: applications of proteome profiling in environmental microbiology and biotechnology. Brief. Funct. Genomic. Proteomic. *8*, 75–87.

Lanne, B., and Panfilov, O. (2005). Protein staining influences the quality of mass spectra obtained by peptide mass fingerprinting after separation on 2-D gels. A comparison of staining with coomassie brilliant blue and sypro ruby. J. Proteome Res. *4*, 175–179.

Li, Z., Nandakumar, R., Madayiputhiya, N., and Li, X. (2012). Proteomic analysis of 17β-estradiol degradation by *Stenotrophomonas maltophilia*. Environ. Sci. Technol. *46*, 5947–5955.

Liu, D., Li, M., Xi, B., Zhao, Y., Wei, Z., Song, C., and Zhu, C. (2015). Metaproteomics reveals major microbial players and their biodegradation functions in a large-scale aerobic composting plant. Microb. Biotechnol. *8*, 950–960.

Loh, K.C., and Cao, B. (2008). Paradigm in biodegradation using *Pseudomonas putida* – a review of proteomics studies. Enzyme Microb. Technol. *43*, 1–12.

Mandalakis, M., Panikov, N., Dai, S., Ray, S., and Karger, B.L. (2013). Comparative proteomic analysis reveals mechanistic insights into *Pseudomonas putida* F1 growth on benzoate and citrate. AMB Express, *3*, 64.

Nazario, C.E., Silva, M.R., Franco, M.S., and Lanças, F.M. (2015). Evolution in miniaturized column liquid chromatography instrumentation and applications: An overview. J. Chromatogr. A. *1421*, 18–37.

Park, S.H., and Raines, R.T. (2004). Fluorescence polarization assay to quantify protein–protein interactions. In Protein–protein Interactions: Methods and Applications. H. Fu, ed. (Humana Press Inc., Totowa, NJ), *261*, pp. 161–165.

Ren, R.J., Dammer, E.B., Wang, G., Seyfried, N.T., and Levey, A.I. (2014). Proteomics of protein post-translational modifications implicated in neurodegeneration. Transl. Neurodegener. 3, 23.

Roy, U., Woods, A.G., Sokolowska, I., and Darie, C.C. (2014). Utility of computational structural biology in mass spectrometry. In Advancements of Mass Spectrometry in Biomedical Research, A.G. Woods, and C.C. Darie, eds. (Springer International Publishing, Switzerland), *806*, 107–128.

Schneider, T., Keiblinger, K.M., Schmid, E., Sterflinger-Gleixner, K., Ellersdorfer, G., Roschitzki, B., Richter, A., Eberl, L., Zechmeister-Boltenstern, S., and Riedel, K. (2012). Who is who in litter decompositionandquest; metaproteomics reveals major microbial players and their biogeochemical functions. ISME J. *6*, 1749–1762.

Szewczyk, R., Soboń, A., Słaba, M., and Długoński, J. (2015). Mechanism study of alachlor biodegradation by *Paecilomyces marquandii* with proteomic and metabolomic methods. J. Hazard. Mater. *291*, 52–64.

Walzthoeni, T., Leitner, A., Stengel, F., and Aebersold, R. (2013). Mass spectrometry supported determination of protein complex structure. Curr. Opin. Struct. Biol. 23, 252–260.

Wang, G., Wu, W.W., Zhang, Z., Masilamani, S., and Shen, R.F. (2008). Decoy methods for assessing false positives and false discovery rates in shotgun proteomics. Anal. Chem. *81*, 146–159.

Washburn, M.P., Wolters, D., and Yates, J.R. (2001). Large-scale analysis of the yeast proteome by multidimensional protein identification technology. Nat Biotechnol. *19*, 242–247.

Wilmes, P., and Bond, P.L. (2006). Metaproteomics: studying functional gene expression in microbial ecosystems. Trends Microbiol. *14*, 92–97.

Wolters, D.A., Washburn, M.P., and Yates, J.R. (2001). An automated multidimensional protein identification technology for shotgun proteomics. Anal. Chem. 73, 5683–5690.

Wright, P.C., Noirel, J., Ow, S.Y., and Fazeli, A. (2012). A review of current proteomics technologies with a survey on their widespread use in reproductive biology investigations. Theriogenology 77, 738–765.

Wu, Q., Yuan, H., Zhang, L., and Zhang, Y. (2012). Recent advances on multidimensional liquid chromatography–mass spectrometry for proteomics: From qualitative to quantitative analysis – a review. Anal. Chim. Acta *731*, 1–10.

Lipidomics in Studies on Adaptation Mechanisms of Microorganisms to the Toxic Effects of Hazardous Compounds

6

Przemysław Bernat

Contents

Abstract

Anomalies in environmental conditions activate a series of processes in microbial cells that allow them to minimize the negative impact of the environmental stress. These environmental factors, such as temperature, nutrient shortage, or dangerous pollutants, may result in changes in the cytoplasmic membrane. Since membranes constitute the main target for the action of toxic compounds, most adaptive mechanisms are concerned with maintenance of the membrane fluidity and lipid-phase stability. These processes can be modulated by altering the lipid composition of the membrane. In order to understand the contribution of membrane lipid composition to the functionality of membranes, comprehensive structural and quantitative information on the organellar lipidomics is essential. Improved lipidomic technologies, which are described in this review, have greatly enhanced our knowledge about the lipid biology of microorganisms at the level of individual species.

6.1 Introduction

Lipids occur in all living organisms and are major components of biological membranes. Their function was originally relegated to composing the membrane bilayer and energy storage, but in recent years, it has been recognized that lipids play diverse roles as molecules involved in cell regulation and signalling. Lipids are defined as small hydrophobic or amphiphilic molecules that originate either entirely or in part from two distinct types of building blocks: ketoacyl and isoprene groups (Oresic *et al*, 2008). Owing to the many possible variations of lipid building blocks and how those blocks are linked, lipids are structurally highly diverse. Oresic *et al.* (2008) estimated that the number of lipids covering the major lipid classes is close to 200,000.

According to the LIPID MAPS consortium and others, lipids can be classified into eight major categories (namely fatty acyls, glycerolipids, glycerophospholipids, sphingolipids, sterol lipids, prenol lipids, saccharolipids, and polyketides) and dozens of lipid classes and subclasses (Köfeler *et al.*, 2012). Their presence and functionality are greatly influenced by the surrounding environment and many factors such as temperature, osmotic concentration, nutrient contents, hormonal concentrations, and health conditions (Wang *et al.*, 2015).

6.2 Lipidomics: a distinct field of metabolomics

Metabolomics is the study of small molecules in biological samples. Conventional metabolomic methods are used to extract water-soluble molecules, such as amino acids, vitamins, and carbohydrates (Layre *et al.*, 2013). On the other hand, lipids are insoluble in aqueous solutions and need special attention. In contrast to the term 'metabolome', the name 'lipidome' was introduced and first appeared in the literature in 2001 (Loizides-Mangold, 2013). It describes the complete lipid composition within a cell, a tissue, or an organism. Subsequently, Rilfors and Lindblom (2002) introduced the term 'functional lipidomics', which describes the role played by membrane lipids. With the advancement of technology, particularly in methods of mass spectrometry (MS), the field of lipidomics has emerged. In 2003, Han and Gross defined lipidomics as a global analysis of cellular lipidomes by a comprehensive MS approach (Loizides-Mangold, 2013). Nowadays, lipidomics is seen in a broader context. It involves the identification and quantification of all individual cellular lipid species and the characterization of their interactions with neighbouring lipids and proteins. This also includes the expression of proteins involved in lipid metabolism (Loizides-Mangold, 2013). It can be said that lipidomics is a branch of metabolomics.

6.3 An increasing interest in microbial lipidomics

Temperature changes, nutrient shortage, or the presence of dangerous pollutants in the surrounding environment will activate a series of processes in microbial cells that allow them to minimize the negative impact of these factors (Murínová and Dercová, 2014). Such factors may result in changes in the cytoplasmic membrane, which is a dynamic structure made up of a stable phospholipid bilayer interspersed with motile proteins. The lipid components of the membrane form a barrier to the transport of molecules, whereas the protein components act as transport structures. Even subtle changes in the lipid composition can have a huge impact on the membrane properties. Moreover, the flexibility and adaptation capacity of the membrane largely determines the survival of the cell (Murínová and Dercová, 2014). Since membranes constitute the main target for the action of toxic compounds, most adaptive mechanisms are concerned with maintenance of the membrane fluidity and lipid-phase stability (Murínová and Dercová, 2014). These processes can be modulated by alteration of the lipids that make up the membrane. In order to understand the contribution of membrane lipid composition to the functionality of membranes, comprehensive structural and quantitative information on the organellar lipidomics is essential (Klose *et al.*, 2013). Improved lipidomic technologies have greatly enhanced our knowledge about lipid biology at the level of the individual species (Köfeler *et al.*, 2012). It can be said that microbial lipidomics represents an innovative tool in the study of microbial ecology (Lopalco *et al.*, 2011).

6.4 Molecular species of phospholipids

The main constituents of microbial membranes are phospholipids. The phospholipid bilayer is the environment in which sterols, sphingolipids, and proteins carry out their vital functions, such as energy transduction, signal transduction, solute transport, DNA replication, protein targeting and trafficking, cell–cell recognition, and secretion (Frega *et al.*, 2012). There are two general classes of phospholipids: glycerophospholipids and sphingolipids.

Glycerophospholipids have a three-carbon glycerol backbone with different substitutions at the three-carbon systematic nomenclature (sn) positions (Xia and Jemal 2009) (Fig. 6.1). The sn-1 position can contain a fatty acid ester, an alkyl ether, or a vinyl ether, giving rise to phosphatidyl, plasmanyl, or plasmenyl (plasmalogen) glyc-

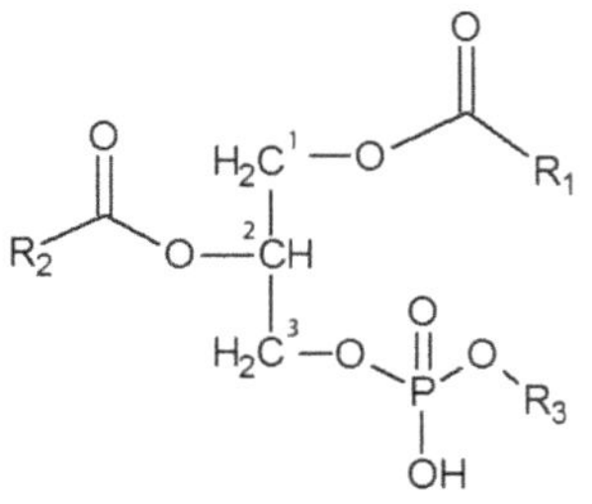

R_1, R_2 - fatty acid alkyl chain
R_3 - polar head group

Figure 6.1 Scheme of glycerophospholipids.

erophospholipids, respectively. The sn-2 position contains a fatty acid group, except in lysophospholipids where the sn-2 hydroxyl group remains free without being esterified (Xia and Jemal, 2009). The polar head groups are attached to the sn-3 positions and distinguish the different phospholipid classes: phosphatidic acid (PA), phosphatidylcholine (PC), phosphatidylethanolamine (PE), phosphatidylglycerol (PG), phosphatidylinositol (PI), phosphatidylserine (PS), and diphosphatidylglycerol (DPG; a dimeric form of PG, also known as cardiolipin (CL)) (Fig. 6.2). In addition, the phospholipid classes are distinguished according to their

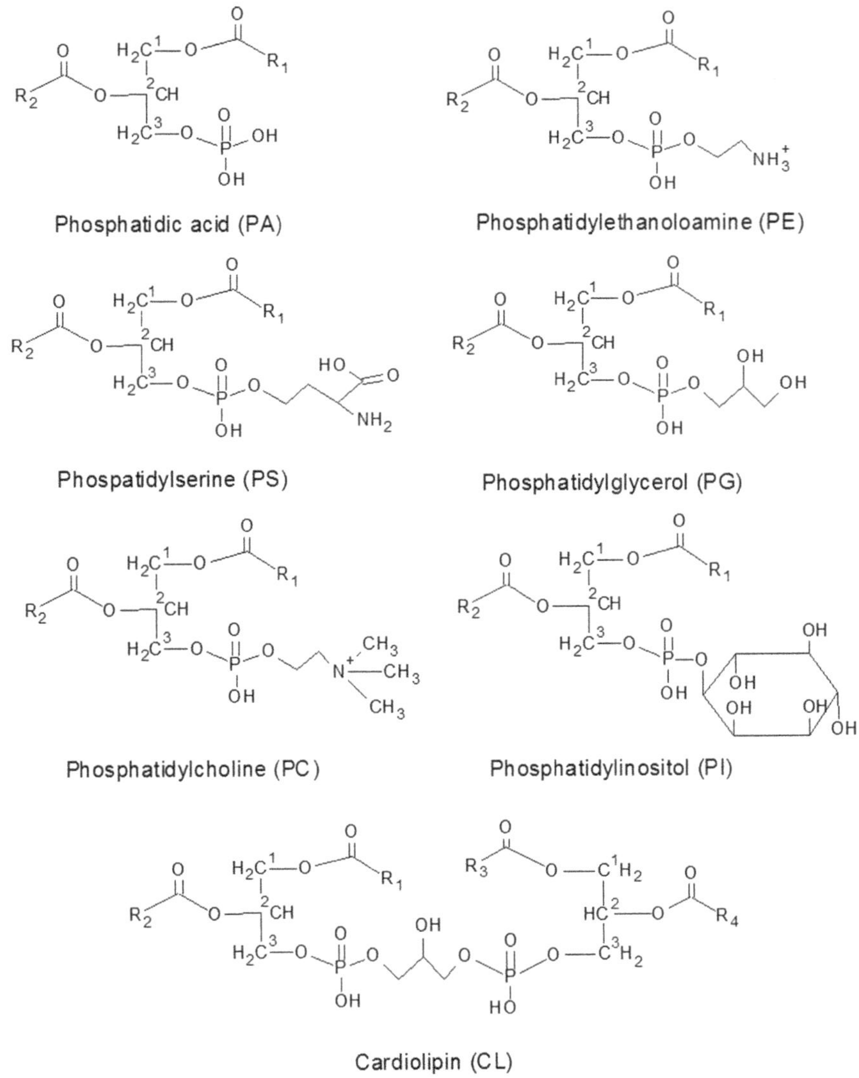

Figure 6.2 Chemical formula for glycerophospholipids.

anionic (PG, DPG, PS, PI) or zwitterionic (PE, PC) properties (Mazzella *et al.*, 2004). Sphingomyelins (SMs) are not derived from glycerol (Fig. 6.3). The backbone in sphingomyelin is sphingosine, an amino alcohol that contains a long, unsaturated hydrocarbon chain. In sphingomyelin, the amino group of the sphingosine backbone is linked to a fatty acid by an amide bond (Berg *et al.*, 2002). The sn-3 position contains the phosphate-choline polar head group, the same group found in the PC class of glycerophospholipids.

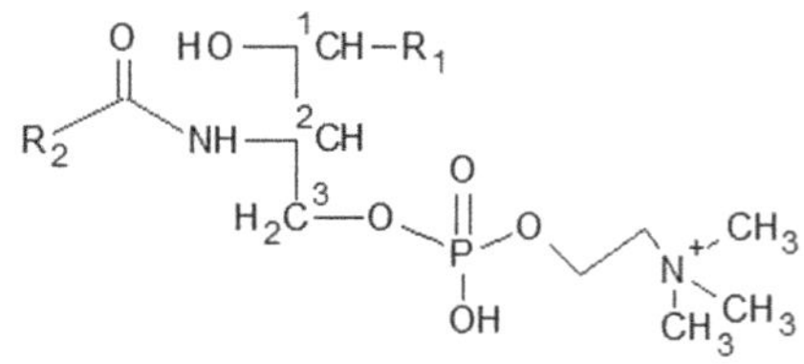

Figure 6.3 Chemical structure of sphingomyelin.

PA is an essential phospholipid that is involved in membrane biosynthesis and signal transduction in living cells (Testerink and Munnik, 2005; Yao and Rock, 2013) (Table 6.1). PC is the most abundant phospholipid in eukaryotic membranes (Sohlenkamp *et al.*, 2003). Besides having a structural role in membranes and lipoproteins, PC plays an important role in signal transduction, as it is a major source of lipid secondary messengers in eukaryotes. Based on genomic data, it was found that more than 10% of all bacteria possess PC (Sohlenkamp *et al.*, 2003). PE is one of the most common bacterial lipids and the second most abundant in eukaryotic cells, behind PC. It plays a major role in membrane architecture and function. It is able to form reversed non-lamellar structures such as the hexagonal phase. PE can stabilize membrane proteins in their optimal conformation and plays a pivotal role during cell division to ensure the proper progression of cytokinesis (Gibellini and Smith, 2010). PG is one of the major constituents of bacterial membranes. Although its concentration in eukaryotic membranes is low, it is nevertheless present, such as in mitochondria (Zhao *et al.*, 2008). Whereas PI is

Table 6.1 Functions of glycerophospholipids in microorganisms

Glycerophospholipid	Functions in microbial cells	References
Phosphatidylcholine (PC)	1. Interactions symbiotic/pathogenic bacteria-eukaryotic hosts 2. Membrane structure formation 3. Resistance to heavy metals 4. The major membrane-forming phospholipid in eukaryotes	Sohlenkamp *et al.* (2003), Boeris and Lucchesi (2012)
Phosphatidylethanolamine (PE)	1. Membrane architecture 2. Adaptation to organic solvents and temperature stress	Mrozik *et al.* (2004), Murzyn *et al.* (2005); Rühl *et al.* (2012)
Phosphatidic acid (PA)	1. Basis in the synthesis of all membrane phospholipids 2. Signal transduction	Testerin and Munnik (2005); Yao and Rock (2013)
Phosphatidylserine (PS)	1. Precursor for PE synthesis 2. Protein activation and localization	Kay and Grinstein (2011); Parsons and Rock (2013)
Phosphatidylglycerol (PG)	1. Membrane architecture 2. Adaptation to organic solvents and temperature stress	Mrozik *et al.* (2004); Rühl *et al.* (2012)
Phosphatidylinositol (PI)	1. Signal transduction 2. Metabolism of proteins, nucleic acids, and carbohydrates	Chi *et al.* (1999)
Cardiolipin (CL)	1. Formation of dynamic protein–lipid membrane domains 2. Cell division 3. Resistance to organic compounds 4. Mitochondrial function	Romantsov *et al.* (2007); Minkler and Hoppel (2010); Seydlová *et al.* (2013)

rarely found in bacteria (except in the class *Actinobacteria*), high amounts have been determined in eukaryotic cells (Hawthorne, 1982). It plays important roles in a variety of cellular processes, such as cell growth, metabolism of proteins, nucleic acids, and carbohydrates as well as lipids themselves, and signal transduction (Chi *et al.*, 1999). PS, which is a minor phospholipid, plays important roles within the intracellular environment of eukaryotic cells. Indeed, a number of important intracellular proteins require PS for their proper localization and/or activation (Kay and Grinstein, 2011). In bacteria, PS is known to be a precursor of PE and serves as an obligatory intermediate in its formation (Parsons and Rock, 2013). Cardiolipin is a unique phospholipid present in mitochondria and is a component of bacterial membranes. In eukaryotic cells, it is present almost exclusively in the inner mitochondrial membrane, where it is required for normal mitochondrial function (Minkler and Hoppel, 2010). SMs have significant signalling and regulatory roles within cells. They are one of the main phospholipid components of the lipid rafts that affect the membrane fluidity and membrane protein trafficking (Frega *et al.*, 2012).

6.5 Methods of phospholipid analysis

6.5.1 Extraction

The isolation of phospholipids is usually accomplished by solvent extraction of the total lipids of a sample and further purification of the phospholipid fraction. Introduced more than 58 years ago by Folch *et al.* (1957) and Bligh-Dyer (1959), most studies still rely on these general extraction procedures, often in modified versions. Folch *et al.* (1957) were one of the first groups of researchers to recognize that a solvent mixture must be adequately polar in order to dissociate the lipids from cell membranes but also not so polar that the solvent does not readily dissolve all triacylglycerols and other non-polar lipids (Iverson *et al.*, 2001). The so-called 'Folch' method employs a ratio of 1 part sample to 20 parts mixture of chloroform/methanol (MeOH) (2:1, v/v) followed by several washings of the crude extracts (Iverson *et al.*, 2001). The Bligh-Dyer method is also based on a mixture of chloroform/MeOH (1:2, v/v), but uses a subsequent addition of 1 volume of chloroform and 1 volume of water. These methods are still in use because the extraction yield is practically quantitative (Frega *et al.*, 2012). As a less-toxic alternative, chloroform has been replaced with dichloromethane. In 2008, Matyash *et al.*, introduced a novel sample extraction procedure, employing MeOH and methyl tert-butyl ether (MTBE) (1.5:5, v/v), which were added to the sample together with water. The advantage of MTBE extraction over a chloroform-containing solvent mixture came from the low density of the lipid-containing organic phase that forms the upper layer during phase separation. This greatly simplified the lipid collection. Furthermore, compared with chloroform, MTBE is non-toxic and non-carcinogenic (Cajka and Fiehn 2014).

6.5.2 Liquid chromatography separation, detection, and quantitation

The separation of phospholipids into different classes can be accomplished either with normal-phase (NP) or with reversed-phase (RP) high-performance liquid chromatography (HPLC). Normal-phase liquid chromatography (LC) uses the differences of polar moieties to separate analytes. Phospholipids are eluted in order of increasing polarity of the head group: phosphatidylglycerol, phosphatidylinositol, phosphatidylethanolamine, phosphatidylserine, phosphatidylcholine, and sphingomyelin. The stationary phase in NP chromatography typically consists of unmodified silica particles, with free silanol groups being the interacting functional groups. Silanol groups modified by amino or diol functional groups are also commonly used in the separation of phospholipids (Brouwers, 2011). Highly non-polar solvents with a low ionization capacity have to be used during separation (Cajaka and Fiehn, 2014). A good separation of all phospholipid classes is made possible by using gradient elution, starting with a solvent of low polarity (e.g. heptane, chloroform, and hexane) and ending with a solvent mixture of higher polarity. Unfortunately, the separation of species within a given class is usually very limited (Buré *et al.*, 2013). Nowadays, the reversed phase is the most widely used for the analysis of complex lipids. The separation is based on the lipophilicity of the combined chain length and the number of double bonds present in the fatty acid side chains (Peterson and Cumming, 2006). Using a RP gradient, phospholipids belonging to the same class, but with different fatty-acid chain lengths, could be introduced into the mass spectrometer source with different eluent times. Various types of C_{18}, C_8 and phenyl columns with mobile phases containing methanol, acetonitrile, 2-propanol, and/or water have been reported (Lisa *et al.*, 2011; Bernat *et al.*, 2014a). Different combinations of ammonium formate or acetate (5–10 mM) and formic acid or acetic acid (0.05–0.1%) are highly recommended to improve both the LC separation and lipid detection (Cajka and Fiehn 2014). Other additives used in HPLC systems with UV may harm the ion source of the mass spectrum. Of the other LC systems, the least likely used is hydrophilic interaction chromatography (HILIC). It is a separation alternative to NP chromatography, providing better reproducibility and robustness, and is more compatible with mass spectrometry (Novakova and Vlckova, 2009). Like NP chromatography, HILIC employs traditional polar stationary phases such as silica, amino, or cyano groups, but the mobile phase used is similar to those employed in RP chromatography (Buszewski and Noga, 2012).

6.6 Mass spectrometry analysis of phospholipids

The choice of ionization mode used in LC-MS analyses plays a major role in the lipidome profile that will be obtained (Cajka and Fiehn, 2014). Electrospray ionization (ESI) is the most common mode in LC-MS because it can effectively ionize a wide range of lipids. ESI is one of the gentlest ionization techniques, in which a reproducibility of >95% can be achieved from a prepared lipid sample in the presence of an internal standard after direct infusion (a method known as shotgun lipidomics) (Wang *et al.*, 2015). In positive-ion mode to lipid ions, protons $[M+H]^+$ are added. This mode effectively ionizes a wide range of lipids, including the phospholipid classes PC and PE. However, negative ionization (proton removed, $[M\text{-}H]^-$) provides superior results for certain lipid classes such as PI, PS, and PA (Bernat *et al.*, 2014a). Moreover, in negative-ion mode, fatty acids of all phospholipid classes can be determined. Atmospheric-pressure chemical ionization is preferred for more non-polar lipids (e.g. ergosterol) and has been used in a smaller number of lipidomic investigations (Cajka and Fiehn, 2014). On the other hand, matrix-assisted laser desorption/ionization MS can yield data with high mass accuracy, which cannot be delivered by quadruple instruments.

Owing to the multiplicity of phospholipid molecular species present in samples, some of which could have identical mass but different combinations of acyl chains, it is considerably complicated to accurately identify the phospholipids. In these cases, experiments of tandem mass spectrometry (MS/MS) are needed. In MS/MS experiments, the precursor ion created in the first MS undergoes a further fragmentation either by collision-activated dissociation or by spontaneous dissociation (Frega *et al.*, 2012). A discussion of the product-ion spectra achieved for the phospholipid classes is given below.

All glycerophospholipids form a common fragment at *m/z* 153, with $[\text{glycerophosphate–}H_2O]^-$ in the negative mode. Therefore, precursor ion scanning for *m/z* 153 allows for the detection of all glycerophospholipids that form negative ions. With this scan mode, background signals from a mixture of constituents other than glycerophospholipids are strongly reduced (Brugger *et al.*, 1997). In this type of mode, the fatty acid part of phospholipids is also clearly visible. For example, in eukaryotes, the common fragments at *m/z* 253, 255, 279, 281, and 283 represent the fatty acid anions of palmitoleic acid, palmitic acid, linoleic acid, oleic acid, and stearic acid, respectively.

The product-ion spectra of PAs are suited to be identified in the ESI negative-ion mode. Phosphatidic acid yields a $[M\text{-}H]^-$ ion, and the mass spectrum shows fragments arising from the loss of both the fatty acyl substituents.

PI is another class of phospholipids for which the product-ion spectra are better visualized in negative-ion mode. The head group of this species is inositol, which has an intrinsic negative charge (Frega *et al.*, 2012). Although molecular ions produced by ESI in negative-ion mode correspond to $[M\text{-}H]^-$, there are also fragments that characterize the polar head group of PI; namely, at *m/z* 241, identified as inositol phosphate minus water.

The characterization of PS molecular species by MS/MS in negative-ion mode offers complete structural information about the ions and gives high sensitivity for its structural determination. The polar head group is identified by the loss of a neutral fragment, serine-H_2O of 87D (Fig. 6.4).

PE can form positive and negative ions. Fragmentation of PE anions generates a fragment at *m/z* 196, which is related to the glycerol phosphoethanolamine minus water head group (Fig. 6.5). The product-ion spectra arising from $[M+H]^+$ are often simple and therefore less applicable for the analysis (Frega *et al.*, 2012).

All PC species yield abundant pseudo-molecular ions in positive ionization studies. The product-ion spectra arising from the $[M+H]^+$ ions are dominated by the *m/z* 184 ion, representing a phosphocholine ion. However, the ions related to the structural information provide more data in the negative-ion mode. Under this mode, PC generates $[M\text{-}15]^-$ ions, resulting from the demethylation of the choline moiety and the negative ions created by forming adducts with anions such as chloride

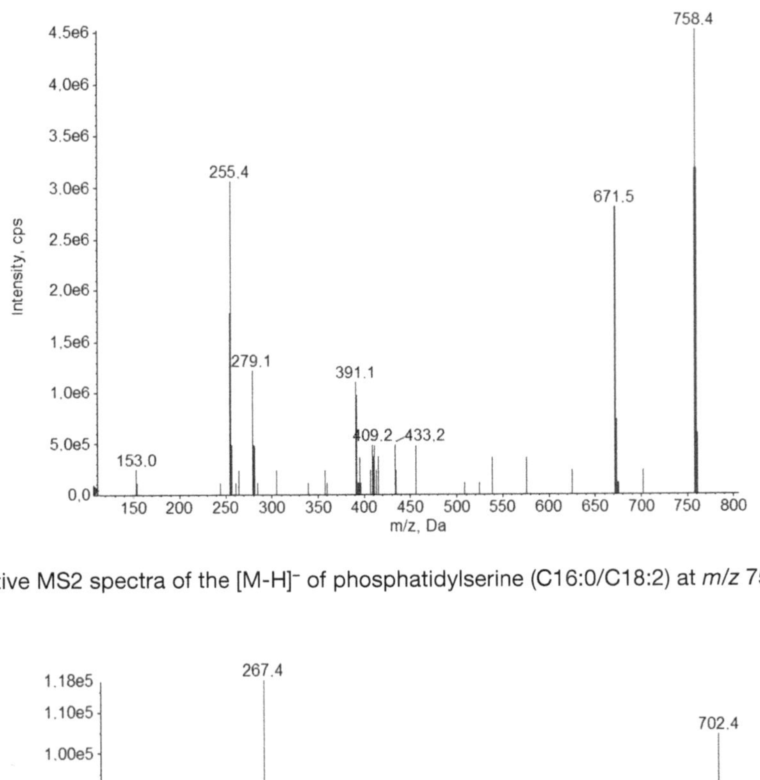

Figure 6.4 Negative MS2 spectra of the [M-H]$^-$ of phosphatidylserine (C16:0/C18:2) at *m/z* 758.5.

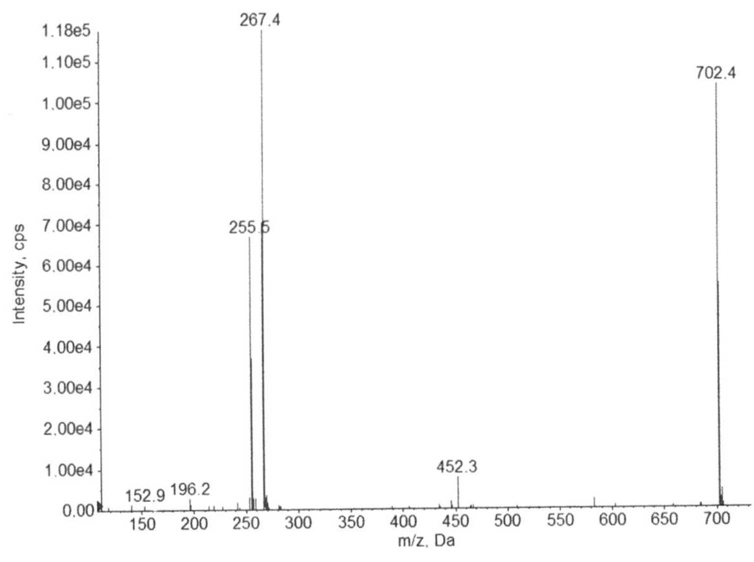

Figure 6.5 Negative MS2 spectra of the [M-H]$^-$ of phosphatidylethanolamine (C16:0/C17:1) at *m/z* 720.5.

[M+35]$^-$, formate [M+45]$^-$, and acetate [M+59]$^-$. The MS2 spectra of the [M-15]$^-$ and adduct ions show anion fragments which are suitable for identification of the carboxylate anions [R1COO]$^-$ and [R2COO]$^-$ (Frega *et al.*, 2012).

In PGs, the tandem mass spectrum of the [M-H]$^-$ ion contains fragment ions at *m/z* 227, 209, and 171, reflecting the glycerol polar head group (Hsu and Turk, 2001).

The general structure of CL includes a unique dimeric phosphatidyl lipid moiety, whereby two phosphatidylglycerols are connected via a glycerol backbone, thus adding up to four acyl (fatty acid) chains and two negative charges of phosphate groups (Tyurina *et al.*, 2014). Because cardiolipin possesses two phosphate charge sites, it forms both [M-H]$^-$ and [M-2H]$^{2-}$ ions when subjected to ESI in the negative-ion mode (Hsu *et al.*, 2005). Structural characterization of cardiolipin can be achieved by multiple-stage (MS^3) MS/MS with ESI in the negative-ion mode to reveal the position of the fatty acyl substituents on the glycerol backbone. Product-ion spectra arising from the [M-2H]$^-$ ions are also useful and provide complementary information to confirm the structure (Hsu *et al.*, 2005).

ESI of sphingomyelin follows closely that of PC. Thus, SM generates more abundant ions in the positive-ion mode than in the negative-ion mode. Moreover, in the negative-ion mode, SM can be analysed only with the addition of an anionic reagent (acetate or formate).

Variations during sample extraction and data acquisition cause analytical errors. To compensate for these errors, internal standards should be applied prior to the extraction by adding a small volume of concentrated stock solution of these standards (Cajka and Fiehn, 2014). For analysis of phospholipids, it is acceptable to use synthetic di-saturated phospholipids (di-C_{14}, di-C_{15} or di-C_{17}), or preferably mixed odd carbon number phospholipids (C_{13}–C_{15} or C_{15}-C_{17}) for each phospholipid class. The alternative is to use isotopically labelled standards (Wolf and Quinn, 2008; Cajka and Fiehn, 2014). To achieve lipid quantification, it is a common practice to normalize the individual molecular ion-peak intensities using an internal standard for each lipid class (Cajka and Fiehn, 2014). The calculated ratio of analyte and internal standard is then multiplied by the concentration of the internal standard to obtain the concentration of a particular analyte. To perform quantitative analysis of lipids by LC-MS, the limit of detection, the standard curves, and their linear ranges are generally determined before sample analysis. In order to deliver high specificity and sensitivity to a major LC-MS technique for quantitative analysis of lipids, selected reaction monitoring (or multiple reaction monitoring) should be used. This approach performs MS/MS and monitors a particular pair (or pairs) of precursor/product ions at a specified elution time for quantification (Yang and Han, 2011). An instrument possessing a high-duty cycle capability is therefore crucial to employ this approach for the quantification of multiple species.

6.7 Comparative lipidomics

Comparative lipidomic analysis uses aligned datasets from samples obtained under at least two biological conditions that differ by biological variables such as growth condition, environmental factors, gene knockout, or cell type (Layre and Moody, 2013). The most frequently reported have been lipidomic analyses of fatty acids. Fatty acid alterations in microbial cells in response to environmental pollution have been described in many papers for the last thirty years (Murinova and Dercova, 2014). In contrast, little information is available about the role of phospholipid head groups.

In many bacterial strains, PE and PG are the most abundant phospholipids, corresponding to 60–95% of total membrane phospholipids (Epand *et al.*, 2007). Murzyn *et al.* (2005) prepared a computer model for the inner bacterial membrane, with the palmitoyloleoyl PE and palmitoyloleoyl PG in a 3:1 proportion. They concluded that the average surface area per PG molecule is larger, and the average vertical location of the PG phosphate group lower, than those of PE. Moreover, the alkyl chains of PG are more ordered and less densely packed than PE chains.

Organic solvents like benzene and toluene destabilize the lamellar structure of a PE bilayer by decreasing the lamellar-to-inverted-hexagonal-phase transition temperature and/or increasing the bilayer permeability (Murzyn *et al.*, 2005). To counter-act these effects, some bacteria have developed several adaptation mechanisms. Cultivation of *Pseudomonas putida* S-12 with toluene decreased its amount of PE and increased its content of PG and CL. This alteration could stabilize the membrane by lowering the fluidity (Segura *et al.*, 1999). Furthermore, because PE–PG interactions are stronger than those of PE–PE, with an increasing PG/PE ratio, the membrane becomes less permeable for lipophilic molecules and thus more stable (Murzyn *et al.*, 2005).

The effect of lipophilic tributyltin (TBT) was assessed in our laboratory using *Pseudomonas proteolytica*. TBT induced a twofold decline in the amounts of many molecular species of PG and an increase in the levels of PA (by 58%) and PE (by 70%) (Bernat *et al.*, 2014b). An increase in the degree of saturation of phospholipid fatty acids in TBT-exposed cells was also observed. Donato *et al.* (1997) described the effect of DDT on the bacterial strain *Bacillus stearothermophilus*. This compound induced an increase of the membrane PE content with a parallel decrease of the PG amount. This alteration was accompanied by an increase of straight chains and a parallel decrease of branched fatty acids in the cytoplasmic membrane, which promoted a more ordered membrane with a shift of the phase transition temperature to higher values. However, an increase in PE and a decrease in PG are not usual responses of the bacteria.

Modification of the PE/PG ratio was also observed in bacterial cells in response to changes in extracellular osmotic pressure. Growth in media of high salinity altered the phospholipid head group and the fatty acid compositions of bacterial cytoplasmic membranes, in many cases increasing the ratio of anionic to zwitterionic lipids (Romantsov *et al.*, 2009). In *Escherichia coli*, the proportion of CL increases as the proportion of PE decreases when osmotic stress is imposed with an electrolyte or a non-electrolyte. Osmotic induction of the gene encoding CL synthase (*cls*) contributes to these changes. The proportion of PG increases at the expense of PE in *cls*$^-$ bacteria, and, in *Bacillus subtilis*, the genes encoding CL and PG synthases (*clsA* and *pgsA*) are both osmotically regulated (Romantsov *et al.*, 2009).

The denitrifying beta-proteobacterial strain *Aromatoleum aromaticum* EbN1 degrades toluene and ethylbenzene under anoxic conditions. Zink and Rabus (2010) observed an increased PG/PE ratio in the strain, at semi-inhibitory concentrations of alkylbenzenes. They concluded that the phenomenon was aimed at preventing maceration of the cell membrane by toluene and ethylbenzene.

PC is the major membrane-forming phospholipid in eukaryotes, with important structural and signalling functions. In contrast, a recent estimate suggests that probably no more than 10% of all bacteria contain PC as a membrane phospholipid (Sohlenkamp *et al.*, 2003). This phospholipid belongs to a bilayer-forming group, similarly to PG (Dercova *et al* 2008). An increase in PC accumulation in the membranes of two polychlorinated biphenyl (PCB)-degrading bacterial strains (*Pseudomonas stutzeri* and *Burkholderia xenovorans* LB400) was observed after addition of non-polar PCBs into the culture media (Zorádová-Murínová *et al.*, 2012).

In contrast to changes in bacterial phospholipids, which are described in some papers, modifications in the phospholipid head groups of fungi are poorly studied. The effect of TBT was assessed in our laboratory using a *Cunninghamella elegans*strain, in which 49 lipid species were identified. TBT caused a decline in the amounts of many PE and PS molecular species and an increase in the levels of PA, Pl, and PC. Because PC can be synthesized from PE, the two are closely related (Fig. 6.6). Moreover, because PE has a strong propensity to form non-bilayer hexagonal phases, and PC is a bilayer-stabilizing lipid, the PC/PE ratio plays a key role in membrane integrity and cell function (Welti *et al.*, 2002). Measurements of membrane competence showed that more ion leakage occurs in leaves, where the PC/PE molar ratio drops after freezing which can lead to cell death (Welti *et al.*, 2002). Owing to the differences in the characteristics of PC and PE, the changes in PC/PE ratios in *C. elegans* from 1.42 for the control to 1.93 for TBT-treated cells (for the stationary phase of growth) are indicative of a significant influence of the organotin on the membrane's composition. In the presence of TBT, it was observed that the overall fatty acid unsaturation was lower than in the control (Bernat *et al.*, 2014a).

Słaba *et al.* (2013) compared the effect of five heavy metals (Zn, Pb, Cd, Ni and Cu) on the phospholipid composition of the soil fungus *Paecilomyces marquandii*, originating from a strongly metal-polluted area and characterized by high tolerance to these elements. Cd, Ni, and Cu caused an increase in PC. Only Pb decreased the PC content,

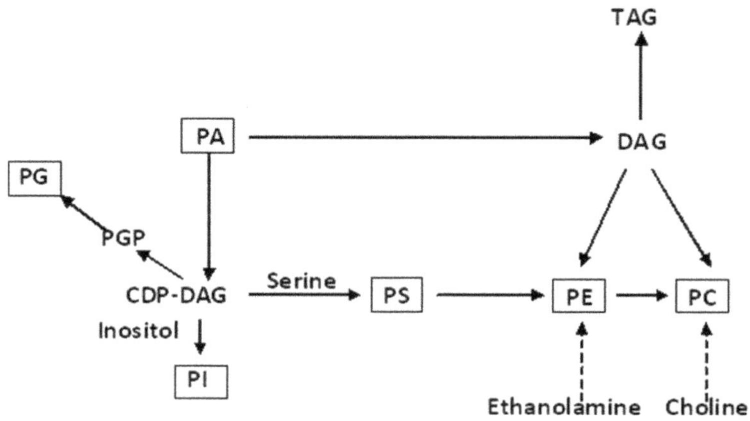

Figure 6.6 Representation of the metabolic network for phospholipids in *Saccharomyces cerevisiae* (Bernat *et al.*, 2014a). CDP, cytidyldiphosphate; DAG, diacylglycerol; TAG, triacylglycerol; PGP, phosphatidylglycerophosphate.

which was accompanied by a significant rise in the PA level, probably due to the activation of phospholipase D that hydrolyses PC to PA.

6.8 The role of cardiolipin in microbial adaptation to stress factors

Cardiolipin (diphosphatidylglycerol) is a unique phospholipid that plays an important role in cell membrane adaptation (Murinova and Dercova, 2014). CL is able to form non-lamellar structures that are required for membrane curvature and lead to the formation of clusters. The advantage of its unique conformation enables the non-lamellar structure to pack tightly, forming micro-domains that are stabilized by membrane proteins (Murinova and Dercova, 2014). In bacteria, CL is regarded as a stress phospholipid (Seydlova *et al.*, 2013). Although CL is only a trace component during exponential growth, it may increase in response to different factors (Schlame, 2008). An increase in the amount of this phospholipid is a known adaptation mechanism under a stress environment. It may reflect a requirement for enhancement of the structural integrity of the cytoplasmic membrane (Murínová and Dercová, 2014). Its enhanced levels were observed in different bacteria in response to various adverse conditions, such as high salinity or alkaline pH (Seydlova *et al.*, 2013). Exposure of *Pseudomonas putida*to stressing agents resulted in an increase in its CL content and enhanced the adaptation ability of the bacterial cell to the presence of organic solvents (Bernal at al., 2007). A possible mechanism of the membrane protection against toxic compounds can be deduced from the rigidizing effect of CL on the membrane (Seydlova *et al.*, 2013). On the other hand, in *Pseudomonas putida* ATCC 12633, a fluidizing effect was observed under chemical stress conditions originated by treatment with the cationic surfactant tetradecyltrimethylammonium bromide (TTAB) (Heredia *et al.*, 2014). An increase in PG and in saturated fatty acids, together with a decrease in CL content, enabled greater membrane resistance, reversing the fluidizing effect of TTAB.

Some studies confirmed that a mutant of *P. putida* P8 that is not able to synthesize CL was more vulnerable to elevated temperatures and to the presence of 4-chlorophenol (Wallbrun *et al.*, 2002).

6.9 Changes in fatty acid composition

Gas chromatography (GC) has become the most popular tool for the separation and analysis of fatty acids. The determination of fatty acids in microbial cells by GC involves transforming the analytes into more volatile and non-polar derivatives after extracting the lipids from biomass before GC analysis. The most commonly used method for the determination of fatty acids is to convert them into their corresponding methyl esters (FAMEs). Many different methylation methods have been established for preparing FAMEs from biomass samples: acid- or base-catalysed transmethylation, boron trifluoride methylation after hydrolysis, methylation with diazomethane, and silylation (Salimon *et al.*, 2014). After GC separation of the FAMEs, two detectors are commonly used to analyse these esters: a flame ionization detector or a mass spectrometer. The former is the most frequently used in quantification of FAMEs. On the other hand, MS offers the abilities to confirm the identity of analytes based on their spectral information in addition to retention time, and to separate peaks from a noisy background or co-eluting peaks if unique ions are available. These features make this technique an increasingly popular one in the lipids laboratory (Dodds *et al.*, 2005).

Aromatic compounds such as benzene, biphenyl, phenol, PCBs, and toluene can accumulate in membrane bilayers between the acyl chains of fatty acids. This phenomenon leads to higher membrane fluidity (Murínová and Dercová, 2014). The major adaptive mechanism of bacteria cells to counter-act that effect is to increase the degree of membrane lipid saturation (Keweloh *et al.*, 1991; Heipieper *et al.*, 1994). Interestingly, similar modifications in the fatty acid profile in response to other lipophilic compounds have been noticed in fungi. In *C. elegans*, it was observed that with an increase in TBT concentration (up to 30 mg/l), the ratio of saturated to unsaturated fatty acids was augmented. It is likely that the increasing toxicity of the substrate was compensated for by the rise in the amount of saturated fatty acids (Bernat and Długoński, 2007). Similar correlations between an increase in the degree of fatty acid saturation and tolerance towards toxic compounds have been observed in *Saccharomyces cerevisiae* (in the presence of 2,4-dichlorophenoxyacetic acid) (Viegas *et al.*, 2005).

Among bacteria, species belonging to the genera *Pseudomonas* and *Vibrio* possess a mechanism of isomerizing cis to trans unsaturated fatty acids. Trans fatty acids are generated by direct isomerization of the respective cis configuration of the double bond (Heipipeper *et al.*, 2003; Murínová and Dercová 2014). The elevated level of trans fatty acids correlates with the increased fluidity caused; that is, by the accumulation of membrane-toxic organic compounds (Heipieper *et al.*, 2003). However, this mechanism cannot fully replace changes in the fatty acid saturation or polar head groups of phospholipids. Von Wallbrunn *et al.* (2002) observed that a mutant of *P. putida* P8 with down-regulated CL production was not able to grow, which proved that cis–trans isomerase was not fully able to replace the adaptation effect of CL.

6.10 Changes in branched fatty acids

Branched-chain fatty acids of the iso and anteiso series occur in many bacteria as the major acyl constituents of membrane lipids (Kaneda, 1991). Iso and anteiso fatty acids show different physico-chemical properties because of the differences in structure and transition temperature (T_M). The T_M of the branched fatty acids is lower for the anteiso fatty acids (e.g. 51.7°C for $C_{15:0}$ iso and 23.0°C for $C_{15:0}$ anteiso) (Kaneda, 1991). This difference causes a change in the fluidity of the membrane. The effect on T_M caused by a change from anteiso to iso branching in gram-positive bacteria is comparable to the isomerization of *cis* to *trans* unsaturated fatty acids in gram-negative bacteria (Murínová and Dercová, 2014). Gram-positive and Gram-negative bacteria that contain branched-chain fatty acids adapt to organic solvents by altering the anteiso/iso ratio in the cell membrane. According to the different physico-chemical properties of those two species of branched fatty acids, the bacteria showed a decrease in the anteiso/iso ratio when grown in the presence of toxic concentrations of phenol, 4-chlorophenol, and 4-nitrophenol, leading to a more rigid membrane and counter-acting the fluidity increase caused by the organic solvents (Unell *et al.*, 2007).

6.11 Conclusion

The lipid composition of the microbial membrane is not static and instead adjusts to changes in environmental conditions. The alterations in the cytoplasmic membrane maintain optimal membrane fluidity. Therefore, increased knowledge of lipidomics is necessary to take advantage of the use of bacterial or fungal strains for remediation of polluted environments. Successful environment decontamination requires strains that are able to survive in and adapt to environmental pollution and then degrade particular (one or more) contaminants.

References

Berg, J.M., Tymoczko, J.L., and Stryer, L. (2002). Biochemistry, 5th edn. New York: W H Freeman.

Bernal, P., Muñoz-Rojas, J., Hurtado, A., Ramos, J.L., and Segura, A. (2007). A *Pseudomonas putida* cardiolipin synthesis mutant exhibits increased sensitivity to drugs related to transport functionality. Environ. Microbiol. *9*, 1135–1145.

Bernat, P., and Długoński, J. (2007). Tributyltin chloride interactions with fatty acids composition and degradation ability of the filamentous fungus *Cunninghamella elegans*. Int. Biodet. Biodeg. *60*, 133–136.

Bernat, P., Gajewska, E., Szewczyk, R., Słaba, M., and Długoński, J. (2014a). Tributyltin (TBT) induces oxidative stress and modifies lipid profile in the filamentous fungus *Cunninghamella elegans*. Environ. Sci. Pollut. Res. Int. *21*, 228–235.

Bernat, P., Siewiera, P., Soboń, A., and Długoński, J. (2014b). Phospholipids and protein adaptation of Pseudomonas sp. to the xenoestrogen tributyltin chloride (TBT). World J. Microbiol. Biotechnol. *30*, 2343–2350.

Bligh, E.G., and Dyer, W.J. (1959). A rapid method of total lipid extraction and purification. Can. J. Biochem. Physiol. *37*, 911–917.

Boeris, P.S., and Lucchesi, G.I. (2012). The phosphatidylcholine synthase of *Pseudomonas putida* A ATCC 12633 is responsible for the synthesis of phosphatidylcholine, which acts as a temporary reservoir for Al3+. Microbiology *158*, 1249–1257.

Brouwers, J.F. (2011). Liquid chromatographic-mass spectrometric analysis of phospholipids. Chromatography, ionization and quantification. Biochim. Biophys. Acta *811*, 763–775.

Brügger, B., Erben, G., Sandhoff, R., Wieland, F.T., and Lehmann, W.D. (1997). Quantitative analysis of biological membrane lipids at the low picomole level by nano-electrospray ionization tandem mass spectrometry. Proc. Natl. Acad. Sci. U.S.A. *94*, 2339–2344.

Buré, C., Ayciriex, S., Testet, E., and Schmitter, J.M. (2013). A single run LC-MS/MS method for phospholipidomics. Anal. Bioanal. Chem. *405*, 203–213.

Buszewski, B., and Noga, S. (2012). Hydrophilic interaction liquid chromatography (HILIC) – a powerful separation technique. Anal. Bioanal. Chem. *402*, 231–247.

Cajka, T., and Fiehn, O. (2014). Comprehensive analysis of lipids in biological systems by liquid chromatography-mass spectrometry. Trends Analyt. Chem. *61*, 192–206.

Chi, Z., Kohlwein, S.D., and Paltauf, F. (1999). Role of phosphatidylinositol (PI) in ethanol production and ethanol tolerance by a high ethanol producing yeast. J. Ind. Microbiol. Biotechnol. *22*, 58–63.

Dercova, Dowhan, W., Bogdanov, M., and Mileykovskaya, E. (2008). Functional roles of lipids in membranes in Biochemistry of Lipids, Lipoproteins and Membranes, D.E. Vance and J.E. Vance, eds. (Elsevier: Amsterdam).

Dodds, E.D., McCoy, M.R., Rea, L.D., and Kennish, J.M. (2005). Gas chromatographic quantification of fatty acid methyl esters: flame ionization detection vs. electron impact mass spectrometry. Lipids. *40*, 419–428.

Donato, M.M., Jurado, A.S., Antunes-Madeira, M.C., and Madeira, V.M.C. (1997). Effects of a lipophilic environmental pollutant (DDT) on the phospholipid and fatty acid contents of *Bacillus stearothermophilus* Arch. Environ. Contam. Toxicol. *33*, 341–349.

Epand, R.F., Savage, P.B., and Epand, R.M. (2007). Bacterial lipid composition and the antimicrobial efficacy of cationic steroid compounds (Ceragenins). Biochim. Biophys. Acta *1768*, 2500–2509.

Folch, J., Lees, M., and Sloane Stanley, G.H. (1957). A simple method for the isolation and purification of total lipides from animal tissues. J. Biol. Chem. *226*, 497–509.

Frega, N.G., Pacetti, D., and Boselli, E. (2012). Characterization of phospholipid molecular species by means of hplc-tandem mass spectrometry. Tandem mass spectrometry – applications and principles. InTech. DOI: 10.5772/31539.

Gibellini, F., and Smith, T.K. (2010). The Kennedy pathway – de novo synthesis of phosphatidylethanolamine and phosphatidylcholine. IUBMB Life *62*, 414–428.

Hawthorne, J.N., and Swilem, A.F. (1982). Phosphatidylinositol metabolism in the adrenal medulla. Cell Calcium. *3*, 351–358.

Heipieper, H.J., Weber, F.J., Sikkema, J., Keweloh, H., and de Bont J.A.M. (1994). Mechanisms of resistance of whole cells to toxic organic solvents. Trends Biotechnol. *12*, 409–415.

Heipieper, H.J., Meinhardt, F., and Segura, A. (2003). The cis–trans isomerase of unsaturated fatty acids in *Pseudomonas* and *Vibrio*: biochemistry, molecular biology and physiological function of a unique stress adaptive mechanism. FEMS Microbiol. Lett. *229*, 1–7.

Heredia, R.M., Boeris, P.S., Biasutti, M.A., López, G.A., Paulucci, N.S., and Lucchesi, G.I. (2014). Coordinated response of phospholipids and acyl components of membrane lipids in *Pseudomonas putida* A (ATCC 12633) under stress caused by cationic surfactants. Microbiology *160*, 2618–2626.

Hsu, F.F., Turk, J., Rhoades, E.R., Russell, D.G., Shi, Y., and Groisman, E.A. (2005). Structural characterization of cardiolipin by tandem quadrupole and multiple-stage quadrupole ion-trap mass spectrometry with electrospray ionization. J. Am. Soc. Mass Spectrom. *16*, 491–504.

Hsu, F.F., and Turk (2001). J. Structural determination of glycosphingolipids as lithiated adducts by electrospray ionization mass spectrometry using low-energy collisional-activated dissociation on a triple stage quadrupole instrument. J. Am. Soc. Mass Spectrom. *12*, 61–79.

Hsu, F.F., Turk, J., Rhoades, E.R., Russell, D.G., Shi, Y., and Groisman. E.A. (2005). Structural characterization of cardiolipin by tandem quadrupole and multiple-stage quadrupole ion-trap mass spectrometry with electrospray ionization. J. Am. Soc. Mass Spectrom. *16*, 491–504.

Iverson, S.J., Lang, S.L., and Cooper, M.H. (2001). Comparison of the Bligh and Dyer and Folch methods for total lipid determination in a broad range of marine tissue. Lipids. *36*, 1283–1287.

Kaneda, T. (1991). Iso- and anteiso-fatty acids in bacteria: biosynthesis, function, and taxonomic significance. Microbiol Rev. *55*, 288–302.

Kay, J.G., and Grinstein, S. (2011). Sensing phosphatidylserine in cellular membranes. Sensors (Basel). *11*, 1744–1755.

Keweloh, H., Diefenbach, R., and Rehm, H.J. (1991). Increase of phenol tolerance of *Escherichia coli* by alterations of the fatty acid composition of the membrane lipids. ArchivMicrobiol *157*, 49–53.
Klose, C., Surma, M.A., and Simons, K. (2013). Organellar lipidomics – background and perspectives. Curr. Opin. Cell Biol. *25*, 406–413.
Köfeler, H.C., Fauland, A., Rechberger, G.N., and Trötzmüller, M. (2012). Mass spectrometry based lipidomics: an overview of technological platforms. Metabolites *5*, 19–38.
Layre, E., and Moody, D.B. (2013). Lipidomic profiling of model organisms and the world's major pathogens. Biochimie *95*, 109–115.
Lísa, M., Cífková, E., and Holčapek, M. (2011). Lipidomic profiling of biological tissues using off-line two-dimensional high-performance liquid chromatography-mass spectrometry. J Chromatogr A. *1218*, 5146–5156.
Loizides-Mangold, U. (2013). On the future of mass-spectrometry-based lipidomics. FEBS J. *280*, 2817–2829.
Lopalco, P., Lobasso, S., Baronio, M., Angelini, R., and Corcelli, A. (2011). Impact of lipidomics on the microbial world of hypersaline environments. Halophiles and Hypersaline Environments (Ed. Ventosa, A.) (Springer-Verlag: Berlin), pp. 123–135.
Matyash, V., Liebisch, G., Kurzchalia, T.V., Shevchenko, A., and Schwudke, D. (2008). Lipid extraction by methyl-tert-butyl ether for high-throughput lipidomics. J. Lipid Res. *49*, 1137–1146.
Mazzella, N., Molinet, J., Syakti, A.D., Dodi, A., Doumenq, P., Artaud, J., and Bertrand, J.C. (2004). Bacterial phospholipid molecular species analysis by ion-pair reversed-phase HPLC/ESI/MS. J. Lipid Res. *45*, 1355–1363.
Minkler, P.E., and Hoppel, C.L. (2010). Separation and characterization of cardiolipin molecular species by reverse-phase ion pair high-performance liquid chromatography-mass spectrometry. J. Lipid Res. *51*, 856–865.
Mrozik, A., Piotrowska-Seget, Z., and Łabuzek, S. (2004). Changes in whole cell-derived fatty acids induced by naphthalene in bacteria from genus *Pseudomonas*. Microbiol. Res. *159*, 87–95.
Murínová, S., and Dercová, K. (2014). Response mechanisms of bacterial degraders to environmental contaminants on the level of cell walls and cytoplasmic membrane. Int. J. Microbiol. Doi:2014:873081.
Murzyn, K., Róg, T., and Pasenkiewicz-Gierula, M. (2005). Phosphatidylethanolamine-phosphatidylglycerol bilayer as a model of the inner bacterial membrane. Biophy.s J. *88*, 1091–1103.
Nováková, L., and Vlcková, H. (2009). A review of current trends and advances in modern bio-analytical methods: chromatography and sample preparation. Anal. Chim. Acta *656*, 8–35.
Oresic, M., Hänninen, V.A., and Vidal-Puig, A. (2008). Lipidomics: a new window to biomedical frontiers. Trends Biotechnol. *26*, 647–652.
Parsons, J.B., and Rock, C.O. (2013). Bacterial lipids: metabolism and membrane homeostasis. Prog. Lipid Res. *52*, 249–276.
Peterson, B.L., and Cummings, B.S. (2006). A review of chromatographic methods for the assessment of phospholipids in biological samples. Biomed Chromatogr. *20*, 227–243.
Rilfors, L., and Lindblom, G. (2002). Regulation of lipid composition in biological membranes – biophysical studies of lipids and lipid synthesizing enzymes. Colloids Surf. B Biointerfaces *26*, 112–124.
Romantsov, T., Guan, Z., and Wood, J.M. (2009). Cardiolipin and the osmotic stress responses of bacteria. Biochim. Biophys. Acta *1788*, 2092–2100.
Rühl, J., Hein, E.M., Hayen, H., Schmid, A., and Blank, L.M. (2012). The glycerophospholipid inventory of *Pseudomonas putida* is conserved between strains and enables growth condition-related alterations. Microb. Biotechnol. *5*, 45–58.
Salimon, J., Omar, T.A., and Salih, N. (2014). Comparison of two derivatization methods for the analysis of fatty acids and trans fatty acids in bakery products using gas chromatography. Sci. World J. *25*, 906407.
Schlame, M. (2008). Cardiolipin synthesis for the assembly of bacterial and mitochondrial membranes. J. Lipid Res. *49*, 1607–1620.
Segura, E., Duque, G., Mosqueda, J.L., Ramos, and Junker, F. (1999). Multiple responses of Gram-negative bacteria to organic solvents. Environ Microbiol*1*, 191–198.
Seydlová, G., Fišer, R., Cabala, R., Kozlík, P., Svobodová, J., and Pátek, M. (2013). Surfactin production enhances the level of cardiolipin in the cytoplasmic membrane of *Bacillus subtilis*. Biochim. Biophys. Acta *1828*, 2370–2378.
Słaba, M., Bernat, P., Różalska, S., Nykiel, J., and Długoński, J. (2013). Comparative study of metal induced phospholipid modifications in the heavy metal tolerant filamentous fungus *Paecilomycesmarquandii* and implications for the fungal membrane integrity. Acta Biochim. Pol. *60*, 695–700.
Sohlenkamp, C., López-Lara, I.M., and Geiger, O. (2003). Biosynthesis of phosphatidylcholine in bacteria. Prog. Lipid Res. *42*, 115–162.
Testerink, C., and Munnik, T. (2005). Phosphatidic acid: a multifunctional stress signaling lipid in plants. Trends Plant Sci. *10*, 368–375.
Tyurina, Y.Y., Domingues, R.M., Tyurin, V.A., Maciel, E., Domingues, P., Amoscato, A.A., Bayir, H., and Kagan, V.E. (2014). Characterization of cardiolipins and their oxidation products by LC-MS analysis. Chem. Phys. Lipids *179*, 3–10.
Unell, M., Kabelitz, N., Jansson, J.K., and Heipieper, H.J. (2007). Adaptation of the psychrotroph *Arthrobacter chlorophenolicus* A6 to growth temperature and the presence of phenols by changes in the anteiso/iso ratio of branched fatty acids. FEMS Microbiol. Lett. *266*, 138–143.
Viegas, C.A., Cabral, M.G., Teixeira, M.C., Neumann, G., Heipieper, H.J., and Sá-Correia, I. (2005). Yeast adaptation to 2,4-dichlorophenoxyacetic acid involves increased membrane fatty acid saturation degree and decreased OLE1 transcription. Biochem. Biophys. Res. Commun. *330*, 271–278.
von Wallbrunn, A., Heipieper, H.J., and Meinhardt, F. (2002). Cis/trans isomerisation of unsaturated fatty acids in a cardiolipin synthase knock-out mutant of

Pseudomonas putida P8. Appl. Microbiol. Biotechnol. *60*, 179–185.

Wang, C., Wang, M., and Han, X. (2015). Applications of mass spectrometry for cellular lipid analysis. Mol. Biosyst. *11*, 698–713.

Welti, R., Li, W., Li, M., Sang, Y., Biesiada, H., Zhou, H.E., Rajashekar, C.B., Williams, T.D., and Wang, X. (2002). Profiling membrane lipids in plant stress responses. Role of phospholipase D alpha in freezing-induced lipid changes in *Arabidopsis*. J. Biol. Chem. *277*, 1994–2002.

Wolf, C., and Quinn, P.J. (2008). Lipidomics: practical aspects and applications, Prog. Lipid Res. *47*, 15–36.

Xia, Y.Q., and Jemal, M. (2009). Phospholipids in liquid chromatography/mass spectrometry bioanalysis: comparison of three tandem mass spectrometric techniques for monitoring plasma phospholipids, the effect of mobile phase composition on phospholipids elution and the association of phospholipids with matrix effects. Rapid Commun Mass Spectrom23, 2125–2138.

Yang, K., and Han, X. (2011). Accurate quantification of lipid species by electrospray ionization mass spectrometry – Meet a key challenge in lipidomics. Metabolites *1*, 21–40.

Yao, J., and Rock, C.O. (2013). Phosphatidic acid synthesis in bacteria. BiochimBiophysActa. *1831*, 495–502.

Zhao, W., Róg, T., Gurtovenko, A.A., Vattulainen, I., and Karttunen, M. (2008). Role of phosphatidylglycerols in the stability of bacterial membranes. Biochimie *90*, 930–938.

Zink, K.G., and Rabus, R. (2010). Stress-induced changes of phospholipids in betaproteobacterium *Aromatoleum aromaticum* strain EbN1 due to alkylbenzene growth substrates. J. Mol. Microbiol. Biotechnol. *18*, 92–101.

Zorádová-Murínová, Dudášová, H., Lukáčová, L., Certík, M., Silharová, K., Vrana, B., and Dercová, K. (2012). Adaptation mechanisms of bacteria during the degradation of polychlorinated biphenyls in the presence of natural and synthetic terpenes as potential degradation inducers. Appl. Microbiol. Biotechnol *94*, 1375–1385.

Microbial Elimination of Endocrine Disrupting Compounds

7

Jerzy Długoński

Contents

Abstract

Endocrine disrupting compounds (EDCs) are chemicals that interfere with the proper functioning of the endocrine system in humans and animals, modulating it in a way that favours the activity of female sex hormones (oestrogens). As well as being slowly degraded by microorganisms, EDCs are insufficiently eliminated in wastewater treatment plants. Some of the compounds are discharged into aquatic environments in quantities of a few to tens of ng/l where they can accumulate in the bodies of different organisms (including humans) and subsequently stimulate numerous deleterious changes including fertility disorders and cancer processes.

In this chapter special attention is focused on the mechanisms of microbial detoxification and elimination of EDCs and the accomplishments of recent lipidomic, proteomic and metabolomic studies in these topics. This review also provides the results of the latest research on the parallel elimination of EDCs and heavy metals in saline environments, which has practical significance for the development of effective strategies for the removal of complex pollutants from contaminated areas.

7.1 Introduction

Increasing amounts of toxic organic chemicals and heavy metals have been contaminating the environment for many years because of human activity. Some particularly deleterious pollutants are named endocrine disrupting compounds (EDCs) or endocrine disruptors (EDs). EDCs are emerging contaminants of concern because they disrupt the normal function of human and animal hormonal systems, modelling their activity in the way characteristic for female sexual hormones at very low concentrations, thereby disrupting fertility and consequently contributing to a decrease in biodiversity. The US Environmental Protection Agency defines EDCs as follows: 'An endocrine disruptor is an exogenous chemical substance or mixture that alters the structure or function(s) of the endocrine system and causes adverse effects at the level of the organism, its progeny, populations or subpopulations of organisms, based on scientific principles, data, weight-of-evidence, and the precautionary principle' (Lintelmann *et al.*, 2003). Chemicals with known significant endocrine modulating activity include synthetic and semi-synthetic steroids as well as a number of other pharmaceuticals, environmental androgens and oestrogens, and some pesticides and industrial pollutants such as dichlorodiphenyltrichloroethane (DDT), alachlor, nonylphenol (NP), pentachlorophenol (PCP), organotins, benzene, styrene and heavy metals (e.g. lead, cadmium, and nickel) (Lintelmann *et al.*, 2003; Choi *et al.*, 2004; Iavicoli *et al.*, 2009; Rahman *et al.*, 2009; European Commission 2011; Barber *et al.*, 2015). These chemicals enter the environment as a result of medical therapies for humans or domestic and breeding animals, and application of oestrogen-like pesticides during agriculture as well as industrial activity (e.g. phenol derivatives with estrogenic properties). Most of them penetrate the water environment either directly or indirectly and then get into the sea areas along with other contaminants through the connected water systems. The entry of EDCs into sea basins that have a limited water exchange with open oceans (e.g. the Baltic Sea or Osaka Bay) is particularly dangerous as evident from the harmful effects documented by several researchers (Albalat 2002 *et al.* 2002; Beck *et al.*, 2006; HELCOM, 2009, 2010a,b; Nurulnadia *et al.*, 2014; Ortiz-Zarragoitia *et al.*, 2014; Rocha *et al.*, 2014; Barber *et al.*, 2015).

Unfortunately, the wealth of information on the threats and the level of environmental contamination by substances from the EDCs group is not accompanied by deep knowledge about the biodegradation course of the compounds and the mechanisms that determine the efficient elimination of EDCs from contaminated environments. Recent developments of new molecular techniques and methods, specially proteomics, metabolomics and lipidomics, have allowed the researchers to track changes in the microbial cell during the elimination of toxic pollutants and to elucidate crucial points in the biodegradation routes (Bernat *et al.*, 2014; Szewczyk *et al.*, 2015). In this chapter the possibilities of '... omic' applications for expanding the knowledge about microbial EDC detoxification and degradation are presented.

7.2 Deleterious effects of EDCs on human and wildlife populations

Although EDCs commonly appear in the environment in very low concentrations, their amounts are nonetheless high enough to have a bad influence on endocrine systems of humans and animals. Negative impacts by EDCs on the reproduction and growth of numerous wildlife populations have been documented in many parts of the world (Dudziak and Lukas-Betlej, 2004; Beck *et al.*, 2005; Rahman *et al.*, 2009; WHO-UNEP, 2013). Aquatic organisms that are frequently endangered by additional environmental stressors are particularly susceptible to toxic contaminants like EDCs (HELCOM, 2010B; WHO-UNEP, 2013). The phenomenon of '*imposex*' (the superimposition of male sexual characteristics on a female) was observed in many species of *Gastropoda* (molluscs) living in sea basins contaminated by organotins (Antizar-Ladislao, 2008; Cruz *et al.* 2011). Exposure to EDCs is also very harmful for humans and is frequently combined with numerous disease outcomes, many of which involve underlying inflammatory and immune dysfunctions, with differential effects based on age (Inadera, 2006; Clayton *et al.*, 2011). Permanent contact with environmental disruptors increases the risk of reduced semen quality as well as adverse pregnancy effects (low birth weight, pre-term birth) or genital malformations in newborn males (WHO-UNEP, 2013). EDCs can also induce neurobehavioural disorders and endocrine – related cancers (ovarian, prostate, endometrial, testicular, thyroid, and brain). The major human endocrine glands and the main EDC-mediated disturbances in activity of the human endocrine system are presented in Fig. 7.1.

Knowledge about the harmful effects of EDCs on wildlife and humans is relatively vast as a consequence of the continuous monitoring of toxic substances for years, especially in coastal and estuarine environments (HELCOM, 2010b; Nurulnadia *et al.*, 2014; Ortiz-Zarragoitia *et al.*, 2014; Rocha

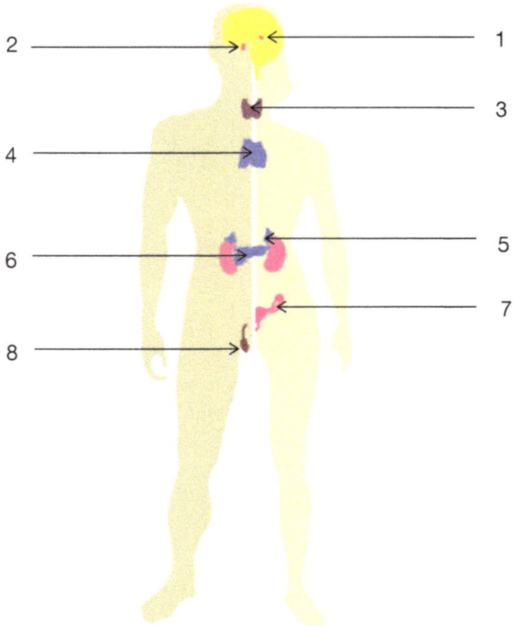

The main ways in which EDCs interfere with the human endocrine system:

- binding to receptors in target tissues or cells and stimulating the normal signal;
- binding to receptors and blocking the normal signal;
- increasing or inhibiting
- the synthesis of natural hormone(s);
- disrupting or modulating hormone gene(s) expression;
- disturbing hormone(s) displacement.

Figure 7.1 Major endocrine glands in the human body, shown in a male (left) and a female (right). 1, Pineal gland; 2, pituitary gland; 3, thyroid gland; 4, thymus gland; 5, adrenal gland; 6, pancreas; 7, ovary; 8, testis.

et al., 2014; Barber *et al.*, 2015). It allowed the leaders and decision-makers in many countries to identify the threats caused by penetration of endocrine system modulators to the environment. Consequently regulations were issued that defined substances particularly dangerous for water environments, and whose removal should be a priority in water protection management (Directive 2008/105/WE of European Parliament and Council). The Directive list consists of 33 positions or clusters of priority substances and their derivatives. However, the levels of known and potential EDCs are still increasing, especially in food and in drinking water. This is a result of the open translocation of a number of EDCs not only though natural processes but also via the worldwide trade market. The global status of scientific knowledge on exposure to and effects of EDCs was recently provided by the World Health Organization (WHO) and the United Nations Environment Programme (UNEP) in the article 'State of the Science of Endocrine Disrupting Chemicals – 2012 (WHO-UNEP, 2013) and was also profoundly discussed by Lamb *et al.* (2014), Bergman *et al.* (2015) and Lamb *et al.* (2015). The sources and properties of dangerous toxic substances established as EDCs and the ones most frequently noticed in contaminated environments are listed in Table 7.1. The above presented facts clearly indicate that, independently of the legal restrictions, it is also important to look for microorganisms and methods that will enable the efficient and productive detoxification and elimination of EDCs.

7.3 Microbial transformation and detoxification of EDCs

7.3.1 Morphological changes and growth inhibition

EDCs appearing in different environments also have strong negative effects on both prokaryotic and eukaryotic microorganisms, especially on reproductive processes and growth of microbial populations (Długoński and Wilmańska, 1998; Rahman *et al.*, 2009; Snyder and Benotti, 2010; Cruz *et al.*, 2011, Krupiński *et al.*, 2013; Krupiński *et al.*, 2014; Różalska *et al.*, 2014).

The study on the reduction of environmental androstenedione (AD) to the sex hormone

Table 7.1 Substances with high endocrine disrupting potentials

Category/use	Chemical name	Major application and/or source of contamination	Induced changes	References
Phenolic compounds	Pentachlorophenol (PCP) 4-Cumylphenol (4-CP 4-*tert*-Octylphenol (4-*t*-OP) *para*-Nonylphenol (*p*NP) Bisphenols (BPs): A, F and S	wood preservative and detergents Supplements; stabilizer for fuels polymers, rubbers; nonyl- and octylphenol ethoxylates degradation	Interference with sexual development and reproductive function in various water organisms; hormone-dependent cancers	Janicki *et al.* (2016) Rochester and Bolden (2015) Stasinakis *et al.* (2008) Szewczyk *et al.* (2009) Zuo and Zhu (2014)
Pesticides	Alachlor Endosulfan (ES) DDT	Extensive applications in agricultural practice	Malformation of sexual features and fertility disruptions in fishes amphibians and mammalians	Jürgens *et al.* (2015) Piazza *et al.* (2015) Słaba *et al.* (2013) Szewczyk *et al.* (2015)
Organotins	Tributyltin (TBT) Octyltins (OT) Tricyclohexyltin (TcyT) Triphenyltin (TPhT)	antifouling paints for ships and fishing nets, wood preservatives	*imposex* (development of male reproductive organs in females)	Cole *at al.* (2015) Cruz *et al.* (2015) Ishihara *et al.* (2012) Soboń. *et al.* (2016)
Heavy metals	Cadmium (II) Nickel (II) Lead (II)	Heavy metals industry	Altering or blocking enzyme activity, inhibiting macromolecule synthesis	Khanna *et al.* 2015 Paraszkiewicz *et al.* (2009) Słaba *et al.* (2013)
Physiological oestrogens	Oestron (E1) 17β-Oestradiol (E2) Oestriol (E3)	The main oestrogen compounds in in the environment and sewage plant effluent which are excreted by all mammals	play a crucial role in number of natural physiological processes in males and females of all mammals	Lucas and Jones (2006) Siewiera *et al.* (2015) Shi *et al.* (2013) Sun *et al.* (2014)
Synthetic oestrogens	17α-ethinyloestradiol (EE2) Mestranol (MeEE2)	Applied in birth control pills and to regulate hormonal disorders; following unmetabolized enters the environment through effluents from	Strong negative effects on aquatic life, disrupting reproductive functions of numerous animals	Kidd *et al.* 2007 Różalska *et al.* 2015 Sun *et al.* (2014) Yu *et al.* (2013)
Pharmaceuticals	Acetaminophen Carbamazepine Diltiazem Ibuprofen Iopromide Naproxen Trimethoprim	Commonly used as active components of numerous medicines; are found in sewage plant effluents as well as in inland and costal waters	Notable sexual abnormalities in fish, birds, reptiles and mammal populations	Hemminger (2005) Komesli *et al.* (2015) Muz *et al.* (2014) Rahman *et al.* (2009

testosterone (TS) and of androstandienedione (ADD) to dehydrotestosterone (DHTS) by the fission yeast *Schizosaccharomyces pombe* (Fig. 7.2) revealed that the substrates, contrary to the products of the transformation, inhibited the yeast growth and heat production rate (Fig. 7.3). Additionally, the formation of aberrant swollen cells with a change in the cell cover structure was observed (Fig. 7.4). It was postulated that the reduction of the substrate keto group to the hydroxyl group by 17β-hydroxysteroid dehydrogenase prevented the harmful effects of environmental androgens on yeast cells viability (Długoński and Wilmańska, 1998; Różalska *et al.*, 2008).

Growth limitation was also observed in nine fungal strains tested for capability of cortexolone 11-hydroxylation to epihydrocortisone and hydrocortisone (Fig. 7.5) (Lisowska and Długoński, 1999). The same filamentous fungi also hydroxylated anthracene and phenanthrene.

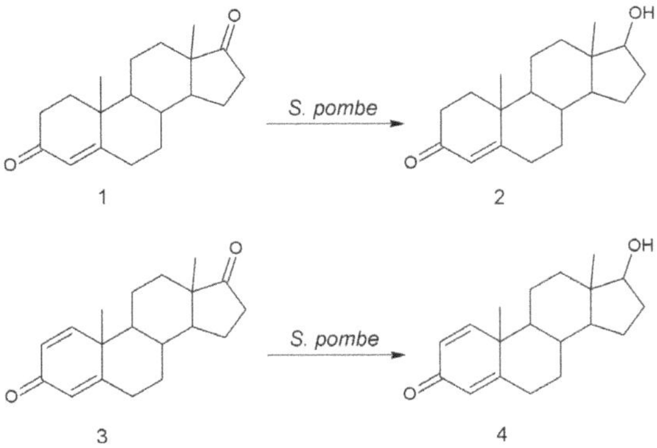

Figure 7.2 Androgens transformations by *Schizosaccharomyces pombe*. 1, AD; 2, TS; 3, ADD; 4, DHTS.

Detailed studies on phenanthrene detoxification by the filamentous fungus *Cunninghamella elegans* revealed that the product of biotransformation – 9-phenanthrol disturbed the metabolic activity and spore germination less than did the substrate (phenanthrene) (Lisowska *et al.*, 2004; Lisowska *et al.*, 2005). *C. elegans* is commonly recommended as a convenient fungal model for mammalian metabolism of toxic xenobiotics (Yadav and Loper, 2000; Asha and Vidyavathi, 2009). It has been documented to catalyse phase I and phase II type biotransformation of a number of chemical compounds, including *polycyclic aromatic hydrocarbons* (PAHs) and steroid drugs, in a way similar to those noticed in mammals (Zhang *et al.*, 1996; Smith and Rosazza, 2004; Asha and Vidyavathi, 2009). A comparative study on the detoxification of phenanthrene (Fig. 7.6) and cortexolone (derivative of the cyclopentane-perhydrophenanthrene system, bearing three condensed

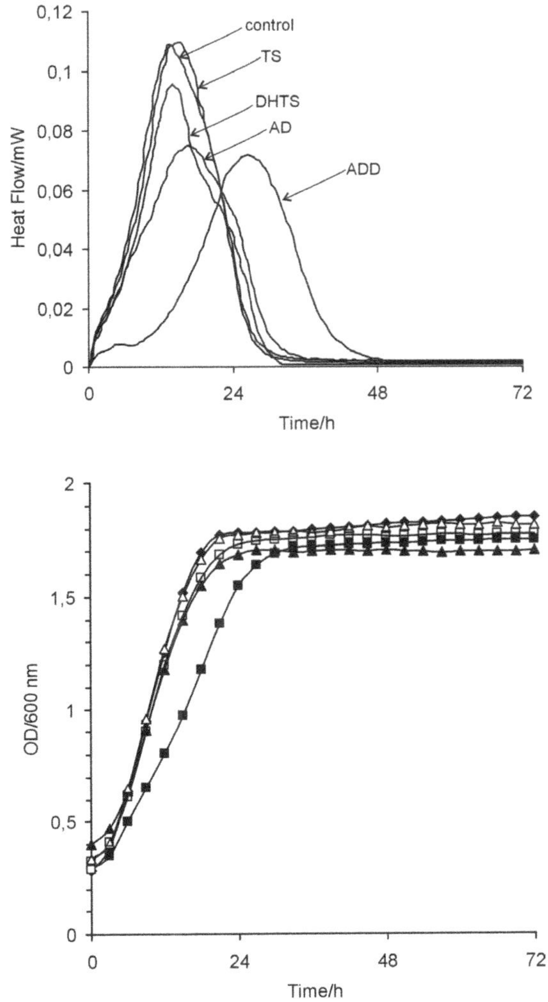

Figure 7.3 Heat production rate (A) and growth (B) of *Schizosaccharomyces pombe* in androgens presence. Control without androgens, black rhombus; AD, black triangles; TS, white triangles; ADD, black squares; DHTS, white squares (Różalska *et al.*, 2008).

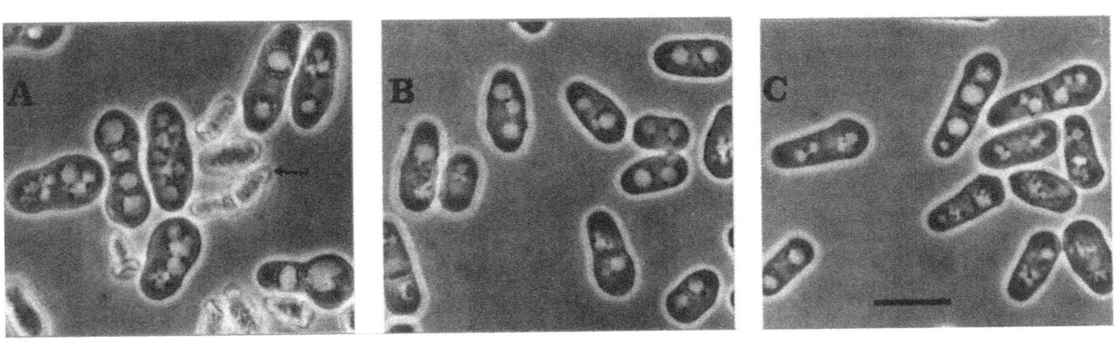

Figure 7.4 Androgens effects on *Schizosaccharomyces pombe* microscopic morphology. (A) Cells cultivated with AD. Note the presence of aberrant swollen cells against background of dead cells (→). (B) Cells cultured with TS. Note that the cells are hardly changed in comparison with the control culture without steroids (C). Bars represent 10 μm (Długoński and Wilmańska, 1998).

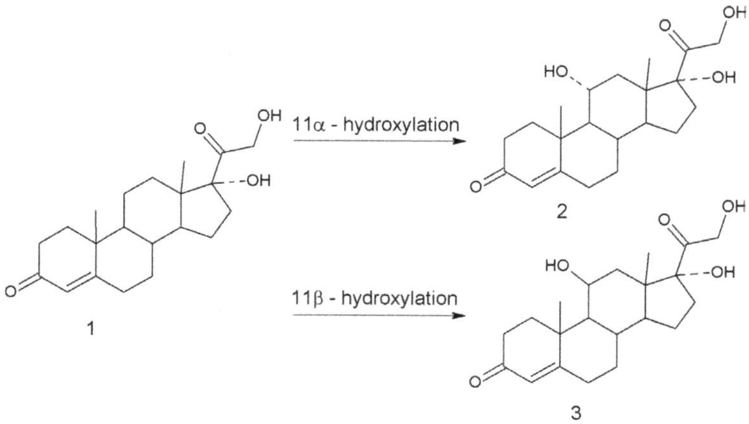

Figure 7.5 11- Hydroxylation of cortexolone by filamentous fungi. 1, Cortexolone; 2, epihydrocortisone; 3, hydrocortisone.

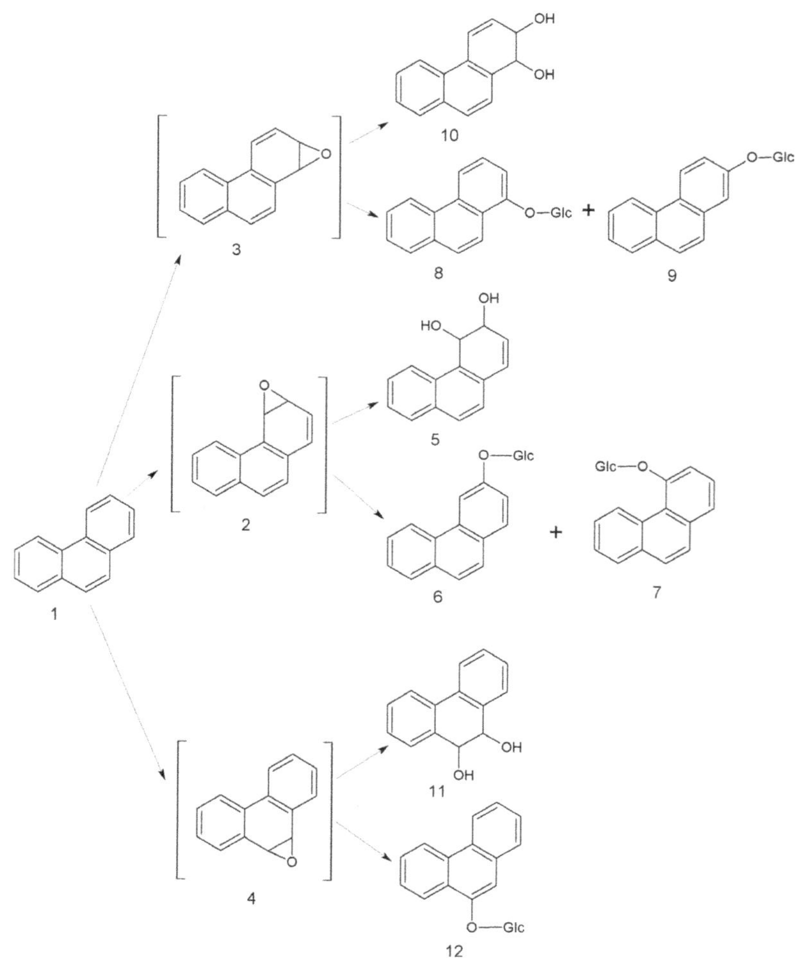

Figure 7.6 Pathway for phenanthrene metabolism in *Cunninghamella elegans.* 1, Phenanthrene; 2, phenanthrene 3,4-epoxide; 3, phenanthrene 1,2-epoxide; 4, phenanthrene 9,10-epoxide; 5, phenanthrene *trans*-3,4-dihydrodiol; 6, 3-phenanthrol glucoside; 7, 4-phenanthrol glucoside; 8, 1-phenanthrol glucoside; 9, 2-phenanthrol glucoside; 10, phenanthrene *trans*-1,2-dihydrodiol; 11, phenanthrene *trans*-9,10-dihydrodiol; 12, 9-phenanthrol glucoside (Lisowska *et al.*, 2006).

rings in the phenanthrene system) by *C. elegans* showed that cortexolone limited the phenanthrene hydroxylation ability of the fungus. At the same time, phenanthrene stimulated the transformation of cortexolone to hydrocortisone and restricted the 11α-hydroxylation of the steroid substrate to epi-hydrocortisone (Lisowska and Długoński, 2003; Lisowska *et al.*, 2006).

C. elegans also exhibits significant potential to metabolize toxic organotin compounds. It converted tributyltin (TBT) to the less-toxic dibutyltin (DBT) and monobutyltin (MBT) at 20 mg/l. Above this concentration its growth was inhibited. Efficient TBT dealkylation was related to the intensity of fungal growth (Bernat and Długoński, 2002, 2006, 2012). A study on simultaneous TBT and cortexolone detoxification by *C. elegans* revealed that the TBT conversion rate in the presence of corticosteroid was lower than that without cortexolone (92% and 70% respectively). Cortexolone 11-hydroxylation was also lower in the presence of the organotin (97% and 80%, respectively). Additionally, both substrates inhibited fungal growth and stimulated changes in the cell morphology (Fig. 7.7).

7.3.2 Lipidomic modifications in microbial cell envelopes

Changes that occur within the surface structures of the microbial cell, especially flexibility in the lipid composition, may not only be a result of harmful effects of EDCs on the microorganism but also a reflection of the intensity of the detoxification processes, which enable the cells to survive in contaminated environments. Lipidomic studies on TBT and cortexolone detoxification in *C. elegans* cultures disclosed significant modifications in the fungal membrane fatty acids (Bernat and Długoński, 2007; Bernat *et al.*, 2009; Bernat and Długoński, 2012). In the presence of TBT, a marked increase in fatty acids saturation was observed. The synthesis of oleic acids ($C_{18:1}$) was reduced suggesting that the organotin had blocked the activity of Δ9 stearoyl-CoA desaturase (SCD).

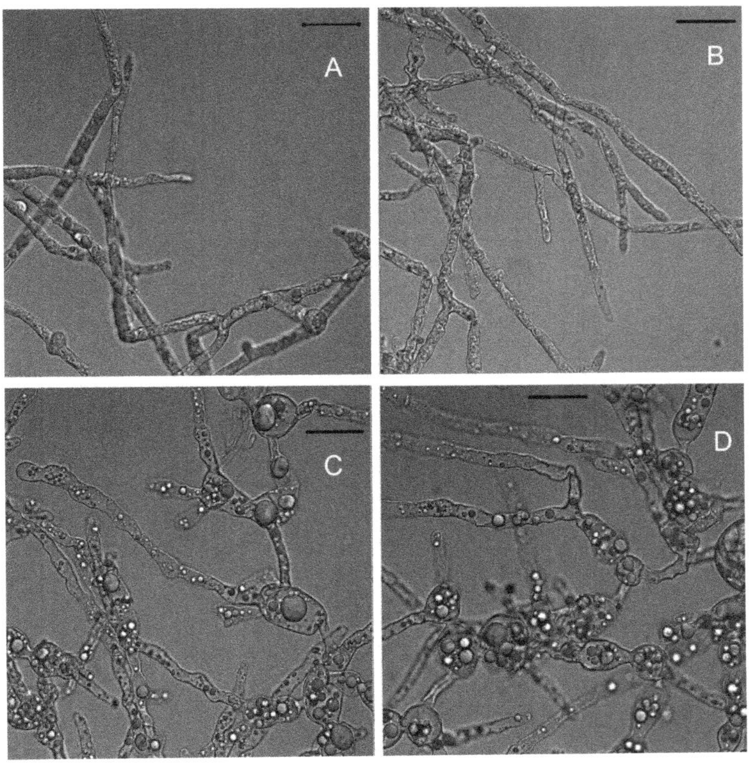

Figure 7.7 Microscopy observation of the morphology of *Cunninghamella elegans* cultured without supplements (A), with cortexolone (B), with TBT (C), and with cortexolone and TBT (D). Bars represent 20 µm (Bernat *et al.*, 2012).

In contrast, the corticosteroid lowered the ratio of fatty acids saturation and enhanced SCD activity. These results indicate a diversity for the detoxification of hydrophobic chemicals within the same fungus (Bernat and Długoński, 2012).

Organotins not only alter lipid homeostasis by disturbing the fatty acids composition but they can also induce oxidative stress, which results from the overproduction of reactive oxygen species (ROS) following cellular oxidative damage processes like protein oxidation, DNA injury, and lipid peroxidation (Inshihara *et al.*, 2012). Apoptosis of *Saccharomyces cerevisiae* was stimulated by the presence of TBT in the yeast culture, a phenomenon that was correlated with significant ROS generation (Chahomechuen *et al.*, 2009). An enhanced level of lipids peroxidation was also noticed in *C. elegans* cultured with TBT. ROS formation in the mycelium was connected with an elevated content of phosphatidic acid (Bernat *et al.*, 2014). It was noticed that the negative effect of lipid peroxidation on mammalian cell survival could be diminished by 17β-oestradiol (E2), which possesses antioxidant features and improves membrane fluidity (Kumar *et al.*, 2011; Inshihara *et al.*, 2012). E2 is also one of the most often detected in water environments and is postulated as an indicator of animal waste contamination (Peterson *et al.*, 2000; HELCOM, 2010a). Additionally, the parallel presence of E2 and TBT was noticed in a number of contaminated aquatic and terrestrial areas (Dubascoux *et al.*, 2008; HELCOM, 2010a,b; Nelson *et al.*, 2011). Our recent studies (Siewiera *et al.*, 2015) with applying of the filamentous fungus *Metarhizium robertsii* as a TBT degrader documented that the TBT induced growth inhibition was reduced in the presence of E2, a lower level of membrane disintegration was observed and the biocide degradation rate was higher. These data indicate that the lowering of ROS generation by E2 improves the elimination of EDCs (TBT).

In addition, it was documented that ascorbic acid (AA) plays a significant role in the protection of fungal biomass against the harmful effects of ROS (Słaba *et al.*, 2013; Słaba *et al.*, 2015). Exogenous ascorbic acid (AA) (1 mM) protected *Paecilomyces marquandii* growth and averted the changes in the fatty acids composition and saturation in the presence of lead ions (Słaba *et al.*, 2013).

7.3.3 Differential protein expression in the presence of EDCs

The microbial adaptation to EDCs contaminants is combined not only with changes in membrane lipid profiles but also with parallel modifications of protein compositions. The expression of proteins is strictly dependent on microbial growth conditions and can change dynamically in response to disadvantageous external factors including the presence of toxic substances like EDCs (Loh and Cao, 2008; Lacerda and Reardon, 2009; Chauhan and Jain, 2010; Kroll *et al.*, 2014). Microscopic fungi and bacteria applied for biodegradation of EDCs biodegradation are relatively weakly characterized in this aspect. Data collected in this field can help in providing a deeper understanding of the processes that occur. The results obtained with the liquid chromatography tandem mass spectrometry (LC MS/MS) technique also allow marking of the key stress proteins (enzymes and structural proteins) involved in EDCs biodegradation processes and selection of potential protein markers of toxicity.

A study on the proteome of *Pseudomonas putida* revealed the expression of a wide variety of proteins in response to the presence of styrene (an EDCs member), involving the catalase/peroxidase subfamily which provides protection from oxidative stress in the tested bacteria (Nikodinovic-Runic *et al.*, 2009). A TBT resistant strain of *Pseudomonas* sp. also reacted to the organotin presence by changing its protein composition (Bernat *et al.*, 2009). The detailed proteomic analyses (in the presence of TBT) disclosed a significant abundance of 16 proteins, including: peroxidases, porins and TonB-dependent outer membrane receptors, which indicates the involvement of these bacterial proteins in the resistance to organic tin compounds (Bernat *et al.*, 2014). TonB-dependent outer membrane receptors are associated with specific uptake of each ferric–siderophore complex in the bacteria (Mirus *et al.*, 2009). It has recently been documented that *Alcaligenes faecalis* is able to use chlorinated TBT as a sole carbon source and transform it into the less-toxic dibutyltin chloride (Khanolkar *et al.*, 2015). The biocide utilization was associated with enhanced siderophores production suggesting their engagement in degradative capability of *A. faecalis*.

A study on the proteomic background of 4-*n*-nonylphenol (4-*n*-NP) degradation by the filamentous fungus *Metarhizum robertsii* disclosed the creation of 14 4-*n*-NP by-products during oxidation of the alkyl chain and the benzene ring of the EDCs substrate (Szewczyk *et al.*, 2014). The main groups of proteins involved in 4-*n*-NP decomposition were oxidation-reduction systems related to nitroreductase-like proteins, ROS defence systems (superoxide dismutase plus peroxiredoxin), the citric acid cycle and energy-related systems. The detailed proteomic analysis concluded that the basic mechanism of xenobiotic removal combines incessant oxidation of $C\text{-}C_{terminal}$ atoms of the aliphatic chain leading to carboxylic acids formation followed by $C_{terminal}$ carbon elimination (Fig. 7.8). Proteomic studies of alachlor degradation by *P. marquandii* disclosed the key protein involved in the elimination of this herbicide to be cyanide hydratase (Szewczyk *et al.*, 2015). Additionally, in the presence of alachlor, there was strong up-regulation of enzymes related to energy, sugar metabolism, and ROS formation, with parallel limitation of fungal growth and glucose consumption. In studies on the deleterious effects of alachlor in mammalian cells it was proved that ROS can induce toxicity of this EDC through oxidative damage to DNA, lipids and proteins as well as increased phosphorylation in an acidic environment (Gizard *et al.*, 2007; Riemann *et al.*, 2011). The studies on the optimization of alachlor degradation with simultaneous oxidative stress limitation in *P. marquandii* cultures showed that the pH of the culture medium could affect ROS generation in mycelia during herbicide removal. A neutral pH was found to be optimal for the degradation of the toxic xenobiotic as well as for oxidative stress inhibition. Additionally, the protection of fungal cells against ROS by AA, glutathione, and anti-oxidative enzymes (dismutase and catalase) was found to be dependent on pH, time of exposure and phase of fungal growth (Słaba *et al.*, 2013, 2015).

7.4 Application aspects of microbial degradation of EDCs

Generally, EDCs are slowly degraded (mineralized) under natural conditions. There are a few basic reasons for this:

- Most of these substrates are barely soluble in water and thus difficult transported into the cells.
- Like in the case of other xenobiotics most microorganisms do not have the adequate enzymes to metabolize EDCs.
- Very often EDCs interfere with the basic physiological processes of microorganisms.
- In EDCs contaminated areas natural organic compounds are available that can easily be metabolized as a source of carbon and energy by microbial cells.

The last factor is especially important in the case of EDC removal in wastewater treatment plants (Lange *et al.*, 2014; Komesli *et al.*, 2015). Microorganisms will take up simple organic compounds first, which are easily transported into the cell and are directly involved in the main metabolic routes (Fig. 7.9).

The majority of EDCs are hardly eliminated in the contaminated environment, thus easily forming complexes with lignins or other organic polymers of natural origin. In sewage treatment systems most of the EDCs (similarly to heavy metals) bind to activated sludge, the excess of which is transferred to and collected on dumping sites or used in compost production. The remainders, usually in the quantity of several dozen ng/l are drained directly into water environments (Pauwels *et al.*, 2008; HELCOM, 2010b; Nelson *et al.*, 2011).

The most common EDCs present in sewage treatment plants are two natural oestrogens: E2 and oestriol (E3) and two xenoestrogens: 17α-oestradiol and NP. Studies on one of the sewage treatment plants in Great Britain (Ifelebuegu, 2011) showed that, depending on the applied conditions of contamination, the removal of the above four EDCs ranged from 41% to 100%, but only part of the toxic contaminants was biodegraded. The majority was bound to the biomass and subsequently eliminated with excess activated sludge into dumps or used for compost production. In anaerobic conditions, EDCs were eliminated in a smaller range from 10% to 48%. Similar results were presented by Komesli *et al.* (2015), who investigated the occurrence, fate, and removal of five different EDCs (the natural hormones – oestrone (E1), progesterone and the pharmaceuticals – carbamazepine, diltiazem, acetaminophen) in Turkish wastewater plants. Only the natural hormones and part of the acetaminophen load were removed with satisfactory yield under

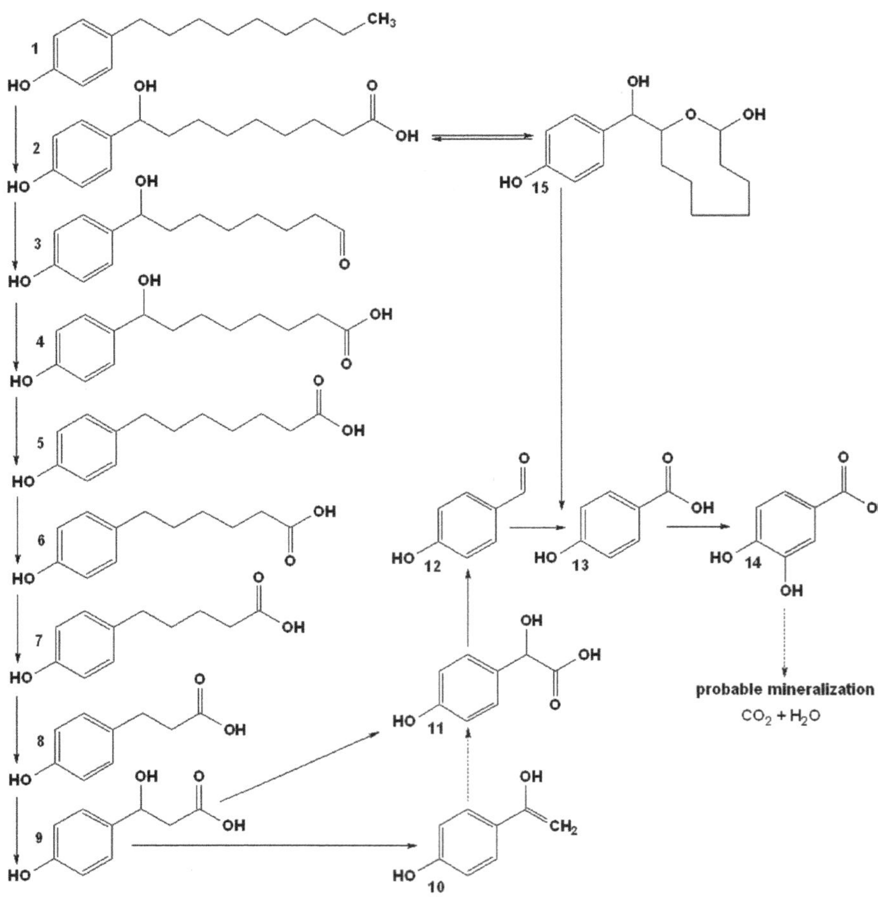

Compound name

1	4-*n*-nonylphenol (4-*n*-NP)
2	9-hydroxy-9-(4-hydroxyphenyl)nonanoic acid
3	8-hydroxy-8-(4-hydroxyphenyl)octanal
4	8-hydroxy-8-(4-hydroxyphenyl)octanoic acid
5	7-(4-hydroxyphenyl)heptanoic acid
6	6-(4-hydroxyphenyl)hexanoic acid
7	5-(4-hydroxyphenyl)pentanoic acid
8	3-(4-hydroxyphenyl)propanoic acid
9	3-hydroxy-3-(4-hydroxyphenyl)propanoic acid
10	4-(1-hydroxyethenyl)phenol (4-HAP)
11	2-(4-hydroxyphenyl)acetic acid
13	4-hydroxybenzoic acid (4-HBA)
14	3,4-dihydroxybenzoic acid
15	9-[hydroxy(4-hydroxyphenyl)methyl]oxonan-2

Figure 7.8 The pathway of the 4-*n*-nonylphenol degradation by *Metarhizium robertsii* (Szewczyk *et al.*, 2014).

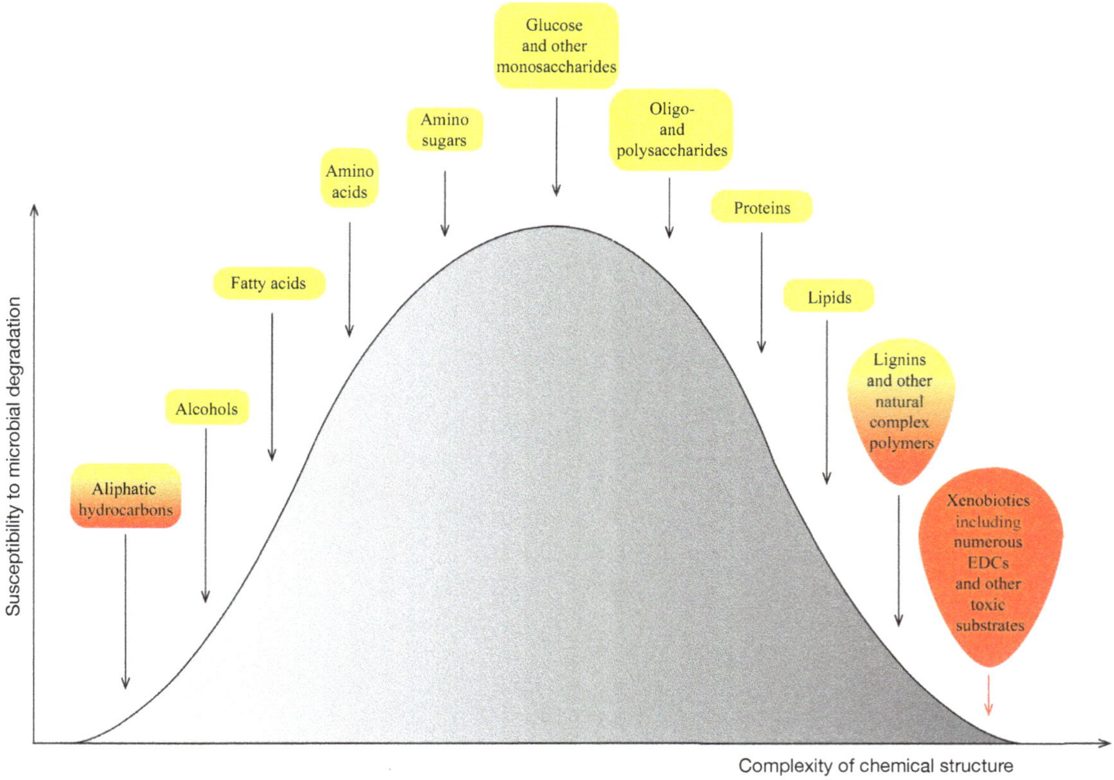

Figure 7.9 The relationship between the complexity of the organic substrate and its susceptibility to microbial degradation. Moreover, compounds especially slowly metabolized by microorganisms are marked in red colour.

optimal treatment conditions. Carbamazepine and diltiazem were resistant to microbial degradation and eliminated mainly by sorption onto sludge, which poses the risk of EDC accumulation in soils if the sludge is used for compost production.

It should also be remembered that compost (including EDCs-laden compost) is used for soil conditioning in human devastated areas such as urban green infrastructure (streets surroundings), recreational regions contaminated with salts and heavy metals, or post-industrial sites (Długoński and Szumański, 2014). This is why it is essential to look for new microorganisms and methods that could be useful for completing removing EDCs and other toxic organic pollutants during biodegradation processes in the presence of salts and heavy metals. It is all the more important because some of the metals like lead, nickel, or cadmium also have endocrine-disrupting properties (European Commission, 2011).

7.4.1 EDCs biodegradation in the presence of NaCl

The deleterious effects of EDCs on microbial cells are frequently enhanced by other physical or chemical constraints like salinity. Microorganisms originating from uncontaminated sites, are usually sensitive or slightly resistant to high salinity. Because of this susceptibility sodium chloride is universally applied for the preservation of food products. On the other hand, salinity is one of the most important environmental factors limiting agricultural productivity because most of the crop plants are sensitive to the salinity caused by high contents of salts (especially NaCl) in the soil (Alexander *et al.*, 2010; Bui, 2013; Długoński and Szumański, 2014; Shrivastava and Kumar, 2015). It was also documented that the presence of different salts, including NaCl, in the microorganism surroundings can stimulate changes in microbial community structures and limit xenobiotic degradation (Wang *et al.*, 2009 a; Wang *et al.*, 2009b). In contrast, some strains of bacteria and fungi selected from highly contaminated areas possess unique and valuable properties, including tolerance to saline conditions and ability to grow in media supplemented with significant amounts of NaCl and/or other inorganic and organic pollutants (Długoński and Słaba, 1997; Fijałkowska *et al.* 1998; Szewczyk *et al.*, 2003; Zheng *et al.*, 2009; Nakbanpote *et al.*, 2014; Różalska *et al.*, 2015). Additionally, such microbes frequently degrade organic pollutants or accumulate heavy metals with similar or even higher efficiency than the appropriate control without the salt.

C. elegans was capable of degrading TBT (10 mg/l) on synthetic medium leading to DBT and MBT formation with yields of 93–96% (after 3 days of degradation). This process was enhanced in the presence of NaCl (14 g/l) and resulted in a smaller concentration of the toxic metabolite DBT, with a simultaneous increase in the practically non-toxic end product of degradation, MBT (1.25 and 2.3 mg/l respectively). It was assumed that the observed changes were caused by a formation of TBT–chlorine adducts, which facilitate the organotin substrate biodegradation. It cannot be excluded that in the presence of TBT and NaCl, changes in the lipid composition of the membrane take place, that may increase the speed of EDCs elimination from the culture (Bernat and Długoński, 2005).

On the contrary, nitrobenzene degradation by bacteria *Micrococcus luteus* in the presence of NaCl presence (30 g/l) was delayed up to 48 hours of incubation and at 120 hours the xenobiotic elimination was on the same level as the control culture (Zheng *et al.*, 2009). Additionally, the negative effect of the salt (range 2–10 g/l) was also observed during anthracene degradation by the filamentous fungus *Aspergillus fumigatus* isolated from the aged crude petroleum-contaminated soil. However, the yields of the PAH elimination after 5 days of incubation were usually maintained at around 60% (Ye *et al.*, 2011).

In the recent study by Różalska *et al.* (2015) on 17α-ethinyloestradiol (EE2) biodegradation, 9 out of 38 tested fungal strains were able to remove synthetic oestrogen (initial content 10 mg/l) during 3 days of incubation. However, only two of the fungal strains (*A. fumigatus* IM 6510 and *A. versicolor* IM 2161) maintained their capability to degrade of EE2 with high efficiency in the presence of NaCl (Fig. 7.10). The detailed LC analyses disclosed that *A. versicolor* IM 2161 transformed EE2 in samples with and without salinity to E2 and nontoxic E1 (Fig. 7.11). NaCl addition slightly inhibited the transformation of E2 to E1 during the first day of experiment, but, by the end of the incubation period (72 h) the amount of E1 dominated in all tested sets.

The above-mentioned examples of biodegradation of toxic xenobiotics in the presence of NaCl clearly indicate that salinity has a significant effect on microbial activity including microorganism's capability of EDCs removal. Subsequently, the widely noticed contamination of sea basins by EDCs, as well as the increasing soil salinity documented in many areas, ought to be taken into considerable

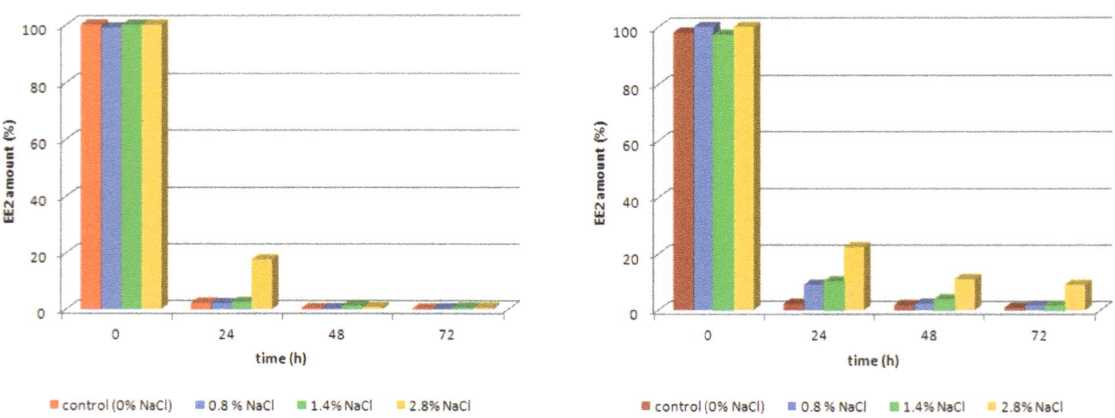

Figure 7.10 EE2 removal from the culture medium supplemented with 0.8%, 1.4%, or 2.8% of NaCl by *Aspergillus versicolor* (A) and *A. fumigatus* (B). (Różalska *et al.*, 2015).

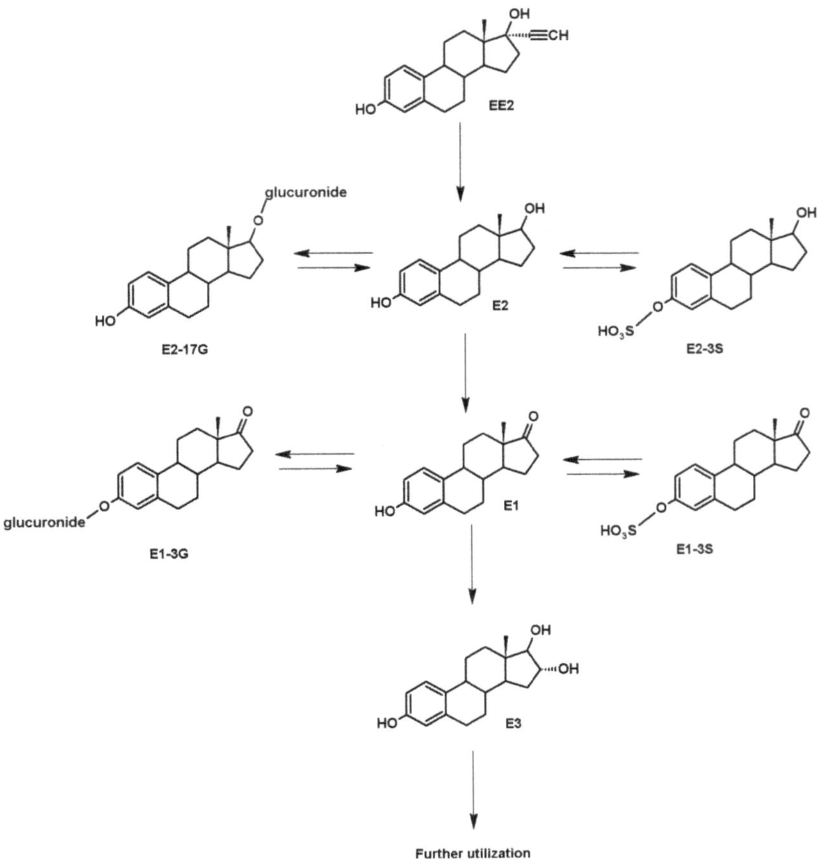

Figure 7.11 EE2 transformation pathway in *Aspergillus* sp. (Różalska *et al.*, 2015).

account during the screening of microorganisms for application in the elimination of hazardous pollutants by microbial EDCs degradation.

7.4.2 Parallel elimination of heavy metals and toxic organic pollutants

As mentioned above, owing to their properties cadmium, lead, zinc and nickel, have been included in the 'List of priority substances in the field of water policy' (Directive 2008/105/WE of European Parliament). Additionally their content in many environments, including the sea basins like the Baltic Sea or the Black Sea, significantly exceeds the allowed norms, which poses a considerable hazard to living organisms and inhibits the decay of organic pollutants (Iavicoli *et al.*, 2009; HELCOM, 2010a,b; Worden *et al.*, 2010; Stancheva *et al.*, 2013; Elekchwi *et al.*, 2014; Dixit *et al.*, 2015). Literature data show that there are significant differences in the resistance of microorganisms to the above- mentioned heavy metals, depending on the environment from which they were isolated, the kind of metal, and the systematic affinity of the microorganisms (Paraszkiewicz *et al.*, 2009; Paraszkiewicz *et al.*, 2010; Worden *et al.*, 2010). Research concerning fungal tolerance to heavy metals showed, that spores germination of the filamentous fungus *P. marquandii* on Sabouraud medium was limited by 50% in the presence of 2.8 mM of lead (Słaba *et al.*, 2005). Cadmium and nickel at a concentration of 5 mM caused strong (over 80%) growth inhibition of *Penicillium pinophilum* IM 6480, whereas zinc and lead introduced up to10 mM did not limit growth of the fungus (Kubiak *et al.*, 2009).

Heavy metals present in the habitat of microorganisms can also inhibit the activity of enzymes involved in organic xenobiotic degradation. This applies particularly to chromium, cadmium, nickel and lead compounds presence in water environments of which (especially in sea water and even in low concentrations) may significantly reduce biodegradation processes (HELCOM, 2010b; Worden *et al.*, 2010; Dixit *et al.*, 2015).

Analyses of the soil bacterial community composition of phenanthrene and phenanthrene- and Cr(VI)-co-contaminated microcosms disclosed that members of the *Sphingomonadaceae* family were the predominant microorganisms (Ibarrolaza *et al.*, 2011). However, the Cr(VI) contamination stimulated a selective change of predominant *Sphingomonas* species, and in co-contaminated soil microcosms, a population closely related to *Sphingomonas paucimobilis* was selected. The isolated *S. paucimobilis* strain showed low Cr(VI) resistance (0.25 mM) in liquid culture and was capable of reducing chromate and degrading phenanthrene simultaneously.

In soil studies on microbial phenanthrene elimination in the presence of zinc, it was documented that the metal at the optimum level of 140 mg/kg in the co-contaminated soil (phenanthrene at 40 mg/kg) resulted in marginal acceleration of the xenobiotic biodegradation rate. However, the Zn^{2+} amount at both the activity level and double the activity level of zinc (720 and 1440 mg/kg) inhibited phenanthrene removal (Wong *et al.*, 2005).

On the other hand, EDCs can stimulate the binding of heavy metals by microbial cells. The filamentous fungus *P. marquandii*, which we isolated from metallurgical waste is capable of binding zinc and lead efficiently (Długoński and Słaba, 1997). It was also shown that this strain can decompose alachlor in the presence of zinc. Biodegradation of the herbicide was not inhibited until the zinc concentration reached 1 mM. Simultaneously, in the presence of xenoestrogens, changes in the lipid content of the membrane and higher ability for zinc binding were observed (Słaba *et al.*, 2009). The increase in the ability of *P. marquandii* to bind zinc in the presence of alachlor correlated with changes within the wall and plasma membrane, even under adverse environmental conditions, such as high salinity or oxygen limitation, which promotes the use of this fungus to eliminate EDCs from contaminated areas.

The filamentous fungus *Penicillium pinophilum* IM 6480 originating from contaminated soil is capable of dehalogenating PCP. It can also bind heavy metals efficiently. We found that in the presence of the xenobiotic the ability of the strain mycelia to bind lead increases more than twice (from 24.5 to 65.7 mg/g d.w.). It seems that the observed phenomena were caused by changes in surface structures occurring in the presence of PCP (Kubiak *et al.*, 2009).

7.4.3 Mixture cultures

In contaminated environments, pollutants are usually not degraded by individual bacterial or fungal strains but simultaneously by microbial consortia with complex ecological interdependencies. Additionally, the biodegradation processes can lead to the formation of by-products that interfere with the metabolic activity of particular members of the microbial consortia. It allows for total contaminant degradation and usually also has an advantageous effect on the biodegradation efficiency.

The filamentous fungus *C. elegans* degrades TBT to its metabolites: DBT and MBT, whereas the fungus *Curvularia tuberculata*, which does not metabolize TBT, has an ability to use DBT and MBT completely (Fig. 7.12). The constructed co-culture of both fungi almost completely removed the organotin compounds (initial TBT content

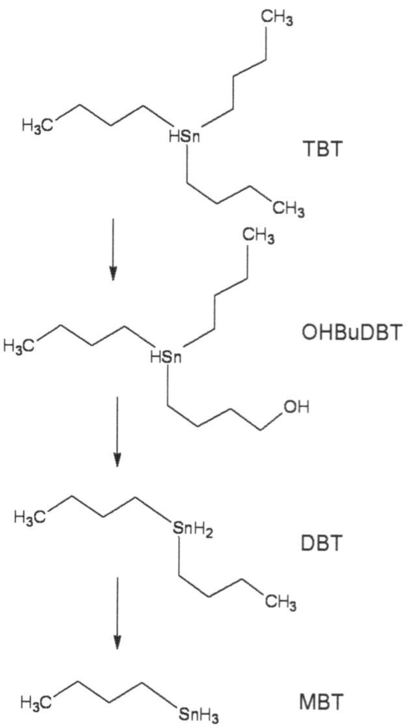

Figure 7.12 TBT degradation pathway in *Cunninghamella elegans* and *Cochliobolus lunatus* co-culture (Bernat *et al.*, 2013).

5 mg/l) during 12 days of culture on mineral medium (Bernat *et al.*, 2013). This microbial co-operation was connected to the fungal growth and glucose utilization, which indicating that butyltins elimination has a co-metabolic character. It opens up the possibility of creating microbial co-cultures with an ability to remove EDCs and their metabolites efficiently.

Zhong *et al.* (2011) applied a mixed bacterial culture of *Mycobacterium* sp. and *Sphingomonas* sp. to phenanthrene, fluoranthene and pyrene degradation. The phenanthrene, was totally degraded by day 3, whereas at day 7 fluoranthene and pyrene were 71.2% and 50% degraded, respectively. The biodegradation of phenanthrene and fluoranthene was decreased, but that of pyrene was increased significantly in relation to the appropriate control monocultures. GC-MS analysis disclosed eight and six additional new metabolites formed from the phenanthrene and fluoranthene degradation, respectively, whereas only two new derivatives were produced from pyrene. The enhancement of pyrene elimination was related to the rapid growth of *Sphingomonas* sp.

More recently, an interesting proposition of eliminating mixture of EDCs from groundwater by crude laccase from the white rot fungus *Pycnoporus sanguineus* was submitted by Garcia-Morales *et al.* (2015). The filtered fungal culture supernatant (laccase cocktail) degraded bisphenol A (BPA), 4-NP, EE2 and triclosan (TCS) with efficiency 89–100%, and an initial analytes concentration of 10 mg/l. The laccase cocktail was also active in real groundwater samples originating from northwestern Mexico, reaching degradation yield between 55% and 93% for all tested EDCs.

7.5 Conclusions

EDCs not only interfere with physiological processes of animals and humans causing harmful changes in the body, but also have deleterious influences on microorganisms, resulting in growth limitation and abnormality in cell morphology. Lipidomic, proteomic, and metabolomic analyses have disclosed significant modifications in the lipid and protein compositions of microbial cells, which help them to survive in unfavourable conditions. Additionally, these modifications can enhance the EDCs detoxification and degradation by microorganisms, which is especially valuable for the efficient removal of EDCs in the presence of substances that have crucial impacts on microbial activity (heavy metals and salinity). The use of modern and very convenient tools such as proteomics, lipidomics and other '... omics' will leads to a better understanding of the mechanisms of EDC detoxification and, consequently, to the development of more efficient techniques for EDCs eliminating from contaminated areas.

Acknowledgement

I am especially thankful to my son, Andrzej Długoński, for his contribution to the elaboration of Figs. 7.1 and 7.9.

References

Albalat, A., Potrykus, J., Pempkowiak, J., and Porte, C. (2002). Assessment of organotin pollution along the Polish coast (Baltic Sea) by using mussels and fish as sentinel organisms. Chemosphere *47*, 1165–1171.

Alexander, J.K., Roberts, A.M., and Pannell, D.J. (2010). Victorian catchment management approaches to salinity: learning from the National Action Plan experience. Australas. J. Environ. Manag. *17*, 45–52.

Antizar-Ladislao B. (2008). Environmental levels, toxicity and human exposure to tributyltin (TBT)-contaminated marine environment. A review. Environment. Internat. *34*, 292–308. Asha S., and Vidyavathi, M. (2009). *Cunninghamella* – a microbial model for drug metabolism studies – a review. Biotechnol. Adv. *27*, 16–29.

Barber, L.B., Loyo-Rosales, J.E., Rice, C.P., Minarik, T.A., and Oskouie, A.K. (2015). Endocrine disrupting alkylphenolic chemicals and other contaminants in wastewater plant effluents, urbans streams, and fish in the Great Lakes and upper Missisippi regions. Sci. Total Environ. *517*, 195–206.

Beck, I.-C., Bruhn, R., and Gandrass, J. (2006). Analysis of estrogenic activity in costal surface waters of the Baltic Sea using the yeast estrogen screen. Chemosphere *63*, 1870–1878.

Bergman, Å., Becher, G., Blumberg, B., Bjerregaard, P., Bornman, R., Brandt, I., Casey, S.C., Frouin, H., Giudice, L.C., Heindel, J.J., *et al.* (2015). Manufacturing doubt about endocrine disrupter science – a rebuttal of industry-sponsored critical comments on the UNEP/WHO report 'State of the Science of Endocrine Disrupting Chemicals 2012'. Regul. Toxicol. Pharmacol.

Bernat, P., and Długoński, J. (2002). Degradation of tributyltin by the filamentous fungus *Cunninghamella elegans*, with involvement of cytochrome P-450. Biotechnol. Let. *24*, 1971–1974.

Bernat, P., and Długoński, J. (2005). Transformation of TBT by the microscopic fungus *Cunninghamella elegans* in NaCl presence. Inz. Apar. Chem. *4*, 7–9 (in Polish with English summary).

Bernat, P., and Długoński, J. (2006), Acceleration of of tributyltin chloride (TBT) degradation in liqiud cultures of the filamentous fungus *Cunninghamella elegans*. Chemosphere *62*, 3-8.

Bernat, P., and Długoński, J. (2007). Tributyltin chloride interactions with fatty acids composition and degradation ability of the filamentous fungus *Cunninghamella elegans*. Int. Biodeter. Biodegr. *60*, 133–136.

Bernat, P., Słaba, M., and Długoński, J. (2009). Action of tributyltin (TBT) on the lipid content and potassium retention in the organotins degradating fungus *Cunninghamella elegans* Curr. Microbiol., *59*, 315–320.

Bernat, P., and Długoński, J. (2012). Comparative study of fatty acids composition during cortexolone hydroxylation and tributyltin chloride (TBT) degradation in the filamentous fungus *Cunninghamella elegans*. Int. Biodeter. Biodegr. *74*, 1–6.

Bernat, P., Szewczyk, R., Krupiński, M., and Długoński, J. (2013). Butyltins degradation by *Cunninghamella elegans* and *Cochliobus lunatus* co-culture. J. Hazard. Mater. *246–247*, 277–282.

Bernat, P., Gajewska, E., Szewczyk, R., Slaba, M., and Długoński, J. (2014). Tributyltin (TBT) induces oxidative stress and modifies liquid profile in the filamentous fungus *Cunninghamella elegans*. Environ. Sci. Pollut. Res. *21*, 4228–4235.

Bui, E.N. (2013). Soil salinity: a neglected factor in plant ecology and biogeography. J. Adrid Environ. *92*, 14–25.

Chahomchuen, T., Akiyama, K., Sekito, T., Sugimoto, N., Okabe, M., Nishimoto, S., and Sugahara, T. (2009). Tributyltin induces Yca1p-dependent cell death of yeast *Saccharomyces cerevisiae*. J. Toxicol. Sci. *34*, 541–545.

Chauhan, A., and Jain, R.K. (2010). Biodegradation: gaining insight through proteomics. Biodegradation *21*, 861–879.

Choi, S.M., Yoo, D.S., and Lee, M.B. (2004). Toxicological characteristics of endocrine-disrupting chemicals: developmental toxicity, carcinogenicity, and mutagenicity. J. Toxicol. Environ. Health, Part B *7*, 1–32.

Clayton, E.M., Todd, M., Dowd, J.B., and Aieto, A.E. (2011). The impact of bisphenol A and triclosan on immune parameters in the U.S. population, NHANES 2003–2006. Environ. Health Persp. *119*, 390–395.

Cole, R.F., Mills, G.A., Parker, R., Bolam, T., Bircheough, A., Kroeger, S., and Fones, G.R. (2015). Trends in the analysis and monitoring of organotins in the aquatic environment. Trends Environ. Anal. Chem. *8*, 1–11.

Cooke, G.M. (2002). Effects of organotins on human aromatase activity *in vitro*. Toxicol. Lett. *126*, 121–130.

Cruz, A., Anselmo, A.M., Suzuki, S., and Mendo, S. (2015). Tributyltin (TBT): a review on microbial resistance and degradation. Crit. Rev. Env. Sci. Tec. *45*, 970–1006.

Directive 2008/105/EC of the European Parliament and of the Council of 16 December 2008. Official Journal of the European Union L 348/84, 24.12.2008.

Dixit, R., Wasiullah, Malaviya, D., Pandiyan, K., Singh, U.B., Sahu, A., Shukla, R., Singh, B.P., Rai, J.P., Sharma, P.K., Lade, H., and Paul, D. (2015). Bioremediation of heavy metals from soil and aquatic environment: An overview of principles and criteria of fundamental processes. Sustainability 7, 2189–2212.

Długoński, A., and Szumański, M. (2014). Analysis of green infrastructure in Lodz, Poland. J. Urban Plann. Dev. *141* (3), A501400-1 – A50140001-9.

Długoński, J., and Słaba, M. (1997). Screening of microorganisms adsorbing zinc and lead from flotation tailings of non ferrous metal works. In International Symposium Proceedings 'Environmental Biotechnology', H. Verachteret and W. Vestraete, eds. (Oostende, Belgium, Technologisch Instituut), pp. 443–446.

Długoński, J., and Wilmańska, D. (1998). Deleterious effects of androstenedione on growth and cell morphology of *Schizosaccharomyces pombe*. Antonie van Leeuwenhoek *73*, 189–194.

Dubascoux, S., Heroult, J., Le Hécho, I., Potin-Gautier, M., and Lespes, G. (2008). Evaluation of a combined fractionation and speciation approach for study of size-based distribution of organotin species on environmental colloids. Anal. Bioanal. Chem. *390*, 1805–1813.

Dudziak, M., and Lukaj-Betlej K. (2004). Occurrence of estrogens – steroid sex hormones – in the riverine water in Poland. Ochr. Srod. *26*, 21–24 (in Polish with English summary).

Elekwachi, C.O., Andresen, J., and Hodgman, T.C. (2014). Global use of bioremediation technologies for decontamination of ecosystems. J. Bioremed. Biodegr. *5*, 225.

European Commission (2011). Commission Staff Working Paper. 4th Report on the implementation of the 'Community Strategy for Endocrine Disrupters' a range of substances suspected of interfering with the hormone systems of humans and wildlife (COM (1999) 706); available on: http://ec.europa.eu/environment// chemicals/endocrine/pdf/sec_2011_1001.pdf

Fijałkowska, S., Lisowska, K., and Długoński, J. (1998). Bacterial elimination of polycyclic aromatic hydrocarbons and heavy metals. J. Basic Microbiol. *38*, 361–369.

Garcia-Morales, R., Rodríguez-Delgado, M., Gomez-Mariscal, K., Orona-Navar, C., Hernandez-Luna, C., Torres, E., Parra, R., Cárdenas-Chávez, D., Mahlknecht, J., and Ornelas- Soto N. (2015). Biotransformation of endocrine-disrupting compounds in groundwater: bisphenol A, nonylphenol, ethynylestradiol and triclosan by a laccase cocktail from *Pycnoporus sanguineus* CS43, Water Air Soil. Pollut. *226*, 251.

Grizard, G., Ouchchane, L., Roddier, H., Artonne, C., Sion, B., Vasson, M.P., and Janny, L. (2007). *In vitro* alachlor effects on reactive oxygen species generation, motility patterns and apoptosis markers in human spermatozoa. Reprod. Toxicol. *23*, 55–62.

HELCOM (2009). Hazardous substances of specific concern to the Baltic Sea – Final report of the Hazardous project. Baltic Sea Environment Proceedings, No. 119, 11–87 (available on http://www.helcom.fi).

HELCOM (2010a). Hazardous substances in the Baltic Sea – An integrated thematic assessment of hazardous substances in the Baltic Sea. Baltic Sea Environment Proceedings, No. 120B, 3–117 (available on http:// www.helcom.fi).

HELCOM (2010b). Towards a tool for quantifying anthropogenic pressures and potential impacts on the Baltic Sea marine environment. A background document on the method, data and testing of the Baltic Sea Pressure and Impact Indices. Baltic Sea Proceedings, No. 125, 7–69 (available on http://www.helcom.fi).

Hemminger, P. (2005). Damming the flow of drugs into drinking water. Environ. Health Persp. *113*, A678-A681.

Iavicoli, I., Fontana L., and Bergamaschi A. 2009. The effects of metals as endocrine disruptors. J. Toxicol. Environ. Health Part B *12*, 206–223.

Ibarrolaza, A., Coppotelli, B.M., Del Panno, M.T., Donati, E.R., and Morelli, I.S. (2011). Application of the knowledge-based approach to strain selection for a bioaugmentation process of phenanthrene-and Cr(VI)-contaminated soil. J. Appl. Microbiol. *111*, 26–35.

Ifelebuegu, A.O. (2011). The fate and behavior of selected endocrine disrupting chemicals in full scale waste water and sludge treatment unit processes. Int. J. Environ. Sci. Technol. *8*, 245–254.

Inadera, H. (2006). The immune system as a target for environmental chemicals: xenoestrogens and other compounds. Toxicol. Lett. *164*, 191–206.

Ishihara, Y., Kawami, T., Ishida, A., and Yamazaki, T. (2012). Tributyltin induces oxidative stress and neuronal injury by inhibiting glutathione S-transferase in rat organotypic hippocampal slice cultures. Neurochem. Int. *60*, 782–790.

Janicki, T., Krupiński, M., and Długoński, M. (2016). Degradation and toxicity reduction of the endocrine disruptors nonylphenol, 4-*tert*-octylphenol and 4-cumylphenol by the non-lininolytic fungus *Umbelopsis isabeliana*. Bioresour. Technol. *200*, 223–229.

Jürgens, M.D., Chaemfa, C., Hughes, D., Johnson, A.C., and Jones, K.C. (2015). PCB and organochlorine pesticide burden in eels in the lower Thames River (UK). Chemosphere *118*, 103–111.

Khanna, I., Khare, M., and Gargava, P. (2015). Health risks associated with heavy metals in fine particulate matter: a case study in Delhi city India J. Geosci. Environ. Protec. *3*, 72-77.

Khanolkar, D., Dubey, S.K., and Naik, M.M. (2015). Tributyltin chloride (TBTCl)-enhanced exopolysaccharide and siderophore production in an estuarine *Alcaligenes faecalis* strain Arch. Environ. Contam. Toxicol. *68*, 612–621.

Kidd, K.A., Blanchfield, P.J., Mills, K.H., Palace, V.P., Evans, R.E., Lazorchak, J.M., and Flick, R.W. (2007). Collapse of a fish population after exposure to a synthetic oestrogen. Proc. Natl. Acad. Sci. USA. *104*, 8897–8901.

Komesli, O.T., Muz, M., Ak, M.S., Bakirdere, S., and Gokcay, C.F. (2015). Occurrence, fate and removal of endocrine disrupting compounds (EDCs) in Turkish wastewater treatment plants. Chem. Eng. J. *277*, 202–208.

Kroll, K., Pähtz, V., and Kniemeyer, O. (2014). Elucidating the fungal stress response by proteomics. J. Proteomics *97*, 151–163.

Krupiński, M., Szewczyk, R., and Długoński, J. (2013). Detoxification and elimination of xenoestrogen nonylphenol by the filamentous fungus *Aspergillus versicolor*. Int. Biodet. Biodegr. *82*, 59–66.

Krupiński, M., Janicki, T., Pałecz, B., and Długoński, J. (2014). Biodegradation and utilization of 4- *n*-nonylphenol by *Aspergillus versicolor* as a sole carbon and energy source. J. Hazard. Mater. *280*, 678–684.

Kubiak, A., Paraszkiewicz K., and Długoński J. 2009. The ability of filamentous fungi isolated from polluted soil to heavy metals uptake and xenobiotics degradation. Ochr. Srod. Zas. Nat. *40*, 321–330.

Kumar, P., Kale, R.K., and Baquer, N.Z. (2011). Estradiol modulates membrane-linked ATPases, antioxidant enzymes, membrane fluidity, lipid peroxidation, and lipofuscin in aged rat liver. J. Aging Res., *2011*, Article ID 580245.

Lacerda, C.M.R., and Reardon, K.F. (2009). Environmental proteomics: applications of proteome profiling in environment al microbiology and biotechnology. Brief. Funct. Genomic. Proteomic. *8*, 75–87.

Lamb IV, J.C., Boffetta, P., Foster, G.W., Goodman, J.E., Hentz, K.L., Rhomberg, L.R., Staveley, J., Swaen, G., Kraak, G. Van Der, and Williams, A.L. (2014). Critical comments on the WHO-UNEP state of the science of endocrine disrupting chemicals – 2012. Regul. Toxicol. Pharmacol. *69*, 22–40.

Lamb IV, J.C., Boffetta, P., Foster, W.G., Goodman, J.E., Hentz, K.L., Rhomberg, L.R., Staveley, J., Swaen, G., Kraak, G. Van Der, and Williams, A.L. (2015). Comments on the opinions published by Bergman *et al.* (2015) on Critical Comments on the WHO-UNEP State of the Science of Endocrine Disrupting Chemicals (Lamb *et al.*, 2014). Regul. Toxicol. Pharm. 73, 754–757.

Lange, C., Kuch, B., and Metzger, J.W. (2014). Determination of the occurrence and elimination of endocrine disrupting compounds (EDCs) in municipal wastewater treatment plants (WWTP). Comp. Water Energy Environ. Eng. *3*, 1–7.

Levesque, H.M., Dorval, J., Hontela, A., Van Der Kraak, G.J., and Campbell, P.G.C. (2003). Hormonal, morphological, and physiological responses of yellow perch (*Perca flavescens*) to chronic environmental exposures. J. Toxicol. Environm. Health – Part A *66*, 657–676.

Lintlemann, J., Katayama, A., Kurihara, N., Shore, L., and Wenzel, A. (2003). Endocrine disruptors in the environment (IUPAC Technical Report). Pure Appl. Chem. 75, 631–681.

Lisowska, K., and Długoński, J. (1999). Removal of anthracene and phenanthrene by filamentous fungus capable of cortexolone 11-hydroxylation. J. Basic Microbiol. 39, 117-125.

Lisowska, K., and Długoński, J. (2003). Concurrent corticosteroid and phenanthrene transformation by filamentous fungus *Cunninghamella elegans*. J. Steroid Biochem. Mol. Biol. 85, 63–69.

Lisowska, K., Pałecz, B., and Długoński, J. (2004). Microcalorimetry as a possible tool for phenanthrene toxicity evaluation to eukaryotic cells. Thermochim. Acta *411*, 181–186.

Lisowska, K., Pałecz, B., and Długoński, J. (2005). Detoxification of phenanthrene by *C. elegans* evaluated by calorimetry. Thermochim. Acta *430*, 43–46.

Lisowska, K., Długoński J., Freeman J.P., and Cerniglia C.E. 2006. The effect of the corticosteroid cortexolone on the metabolites produced during phenanthrene biotransformation in *Cunninghamella elegans*. Chemosphere *64*, 1499–1506.

Loh, K.-C., and Cao, B. (2008). Paradigm in biodegradation using *Pseudomonas putida* – A review of proteomics studies. Enzyme Microb. Tech., *43*, 1–12.

Lucas, S.D., and Jones, D.L. (2006). Biodegradation of estrone and 17 beta-estradiol in grassland soils amended with animal wastes. Soil Biol. Biochem. *38*, 2803–2815.

Mirus, O., Strauss, S., Nicolaisen, K., von Haeseler A., and Schleiff, E. (2009). TonB-dependent transporters and their occurrence in cyanobacteria. BMC Biol. *7*, 68.

Muz, M., Ak, S., Komesli, O.T., and Gokcay, C.F. (2014). Removal of endocrine disrupting compounds in a lab-scale anaerobic/aerobic sequencing batch reactor unit. Environ. Technol. *35*, 1055–1063.

Nakbanpote, W., Panitlurtumpai, N., Sangdee, A., Sakulpone, N., Sirisom, P., and Pimthong, A. (2014). Salt-tolerant and plant growth-promoting bacteria isolated from Zn/Cd contaminated soil: identification and effect on rice under saline conditions. J. Plant Interact. *9*, 379–387.

Nelson, E.D., Do, H., Lewis, R.S., and Carr, S.A. (2011). Diurnal variability of pharmaceutical, personal care product, estrogen and alkylophenol concentrations in effluent from tertiary wastewater treatment facility. Environ. Sci. Technol. *45*, 1228–1234.

Nikodinovic-Runic J., Flanagen, M., Hume, A.R., Cagney, G., and Connor, K.E. (2009). Analysis of the *Pseudomonas putida* CA-3 proteome during on styrene under nitrogen-limiting and non-limiting conditions. Microbiology *155*, 3348–3361.

Nurulnadia, M.Y., Koyama J, Uno, S., Kito, A., Kokushi, E., Bacolod, E.T., Ito K, and Chuman, Y. 2014. Accumulation of endocrine disrupting chemicals (EDCs) in the polychaete *Paraprionospio* sp. from the Yodo River mouth, Osaka Bay, Japan. Environ. Monit. Assess. *186*, 1453–1463.

Ortiz-Zarragoitia, M., Bizarro, C., Rojo-Bartolome I., de Ceiro O.D., Cajaraville, M.P., and Cancio, I. (2014). Mugilid fish are sentinels of exposure to endocrine disrupting compounds in coastal and estuarine environments. Mar. Drugs *12*, 4756–4782.

Paraszkiewicz, K., Bernat, P., and Długoński, J. (2009). Effect of nickel, copper and zinc on emulsifier production and saturation of cellular fatty acids in filamentous fungus *Curvularia lunata*. Int. Biodeter. Biodegr. *63*, 100–105.

Paraszkiewicz, K., Bernat, P., Naliwajski, M., and Długoński, J. (2010). Lipid peroxidation in the fungus *Curvularia lunata* exposed to nickel. Arch. Microbiol., *192*, 135–141.

Paterson, E.T. (2004). Xenoestrogens and brest cancer. J. Orthomol. Med. *19*, 97–103.

Pauwels, B., Noppe, H., De Brabander H., and Verstraete, W. (2008). Comparison of steroid hormone concentrations in domestic and hospital wastewater treatment plants. J. Environ. Eng. *134*, 933–936.

Peterson, E.W., Davis, R.K., and Orndorff, H.A. (2000). 17 β-Estradiol as an indicator of animal waste contamination in mantled karst aquifers. J. Environ. Qual. *29*, 826–833.

Piazza, Y., Pandolfi, M., Da Cuña R., Genovese, G., and Lo Nostro F. (2015). Endosulfan affects GnRH cells in sexually differentiated juveniles of the perciform *Cichlasoma dimerus*. Ecotoxicol. Environ. Saf. *116*, 150–159.

Rahman, M.F., Yanful, E.K., and Jasim, S.Y. (2009). Endocrine disrupting compounds (EDCs) and pharmaceuticals and personal care products (PPCPs) in the aquatic environment: implications for the drinking water industry and global environmental health. J. Water Health *7*, 224–243.

Riemann, A., Schneider, B., Ihling, A., Nowak, M., Sauvant, C., Thews, O., and Gekle, M. (2011). Acidic environment leads to ROS-induced MAPK signalling in cancer cells. PLoS ONE *6*, e22445.

Rocha, M.J., Cruzeiro, C., Peixoto, C., and Rocha, E. (2014). Annual fluctuations of endocrine-disrupting compounds at the lower end of the Lima River, Portugal, and in adjacent coastal waters. Arch. Contam. Toxicol. *67*, 389–401.

Rochester, J.R., and Bolden, A.L. (2015). Bisphenol S and F: A systematic review and comparison of the hormonal activity of bisphenol A substitutes. Environ. Health Perspect. *123*, 643–650.

Różalska, S., Pałecz, B., and Długonski, J. (2008). Calorimetric detection of the toxic effect of androgens on fission yeast. Thermochim. Acta *474*, 91–94.

Różalska, S., Glińska, S., and Długoński, J. (2014). *Metarhizium robertsii* morphological flexibility during nonylphenol removal. Int. Biodeter. Biodegr. *95*, 285–293.

Różalska, S., Bernat, P., Michnicki, P., and Długoński, P. (2015). Fungal transformation of 17α-ethinylestradiol in the presence of various concentrations of sodium chloride. Int. Biodeter. Biodegr. *103*, 77–84.

Shi, J., Chen, Q., Liu, X., Zhan, X., Li, J., and Li, Z. (2013). Sludge/water partition and biochemical transformation of estrone and 17β-estradiol in a pilot-scale step-feed anoxic/oxic wastewater. Biochem. Eng. J. *74*, 107–114.

Shrivastava, P., and Kumar, R. (2015). A serious environmental issue and plant growth promoting bacteria as one of the tools for its alleviation. Saudi J. Biol. Sci. *22*, 123–131.

Siewiera, P., Bernat, P., Różalska, S., and Długoński, J. (2015). Estradiol improves degradation by the filamentous fungus *Metarhizium robertsii*. Int. Bioter. Biodegr. *104*, 258–263.

Słaba, M., Bizukojć, M., Pałecz, B., and Długoński, J. (2005). Kinetic study of toxicity of zinc and lead ions to the heavy metals accumulating fungus *Paecilomyces marquandii*. Bioprocess Biosyst. Eng. *28*, 185–197.

Słaba, M., Szewczyk R., Bernat P., and Długoński J. 2009. Simultaneous toxic action of zinc and alachlor resulted in enhancement of zinc uptake by the filamentous fungus *Paecilomyces marquandii*. Sci. Total Environ. *407*, 4127–4133.

Słaba, M., Gajewska, E., Bernat, P., Fornalska, M., and Długoński, J. (2013). Adaptive alternations in the fatty acids composition under induced oxidative stress in heavy metal-tolerant filamentous fungus *Paecilomyces marquandii* cultured in ascorbic acid presence. Environ. Sci. Pollut. Res. *20*, 3423–3434.

Słaba, M., Różalska, S., Bernat, P., Szewczyk, R., Piątek, M.A., and Długoński, J. (2015). Efficient alachlor degradation by the filamentous fungus *Pacilomyces marquandii* with simultaneous oxidative stress reduction. Bioresour. Technol. *197*, 404–409.

Smith, R.V., and Rosazza, J.P. (2004). Microbial models of mammalian metabolism, aromatic dehalogation. Chem. Bio. Chem. *5*, 87–92.

Snyder, S.A., and Benotti, M.J. (2010). Endocrine disruptors and pharmaceuticals: implications for water sustainability. Water Sci. Technol. *61*, 145–154.

Soboń, A., Szewczyk, R., and Długoński, J. (2016). Tributyltin (TBT) biodegradation induces oxidative stress of *Cunninghamella echinulata*. Int. Biodeter. Biodegr. *107*, 92–101.

Stancheva, M., Makedonski, L., and Petrova, E. (2013). Determination of heavy metals (Pb, Cd, As and Hg) in Black Sea grey mullet *Mugil cephalus*. Bulg. J. Agric. Sci. *19* (Supplement), 30–34.

Stasinakis, A.S., Gatidou, G., Mamais, D., Thomaidis, N.S., and Lekkas, T.D. (2008). Occurrence and fate of endocrine disrupters in Greek sewage treatment plants. water Res. *42*, 1796–1804.

Sun, Y., Huang, H., Sun, Yi., Wang, C., Shi, X., Hu, H., Kameya, T., and Fujie, K. (2014). Occurrence of estrogenic endocrine disrupting chemicals concern in sewage plant effluent. Front. Environ. Sci. Eng. *8*, 18–26.

Szewczyk, R., Bernat, P., Milczarek, K., and Długoński, J. (2003). Application of microscopic fungi isolated from polluted industrial areas for polycyclic aromatic hydrocarbons and pentachlorophenol reduction. Biodegradation *14*, 1–8.

Szewczyk, R., and Długoński, J. (2009). Pentachlorophenol and spent engine oil degradation by *Mucor ramosissimus*. Int. Biodeter. Biodegr. *63*, 123–129.

Szewczyk, R., Soboń., A., Różalska s., Dzitko, K., Waidlich, D., and Długoński, J. (2014). Intracellular proteome expression during 4-*n*-nonylphenol biodegradation by the filamentous fungus *Metharhizium robertsii*. Int. Biodeter. Biogedr., *93*, 44–53.

Szewczyk, R., Soboń, A., Słaba, M., and Długoński, J. (2015). Mechanism study of alachlor biodegradation by *Paecilomyces marquandii* with proteomic and metabolomic methods. J. Hazard. Mater. *291*, 52–64.

Wang, P., Qu, Y., Zhou, J. (2009a). Changes of microbial community structures and functional gens during biodegradation of phenolic compounds under high salt condition. J. Environ. Sci. (China) *21*, 821–826.

Wang, P., Qu, Y., Zhou, J. (2009b). Biodegradation of mixed phenolic compounds under high salt conditions and salinity fluctuations by *Arthrobacter* sp. W1. Appl. Biochem. Biotechnol. *159*, 623–633.

Wong, K.W., Toh, B.A., Ting, Y.P., and Obbard, J.P. (2005). Biodegradation of phenanthrene by the indigenous microbial biomass in a zinc amended soil. Lett. Appl. Microbiol. *40*, 50–55.

Worden, C.R., Kleinheinz, G.T., and Sandrin, T.R. (2010). Effects of metals on microorganisms in environment. In: Handbook of advanced industrial and hazardous wastes treatment, L.K. Wang, Y.-T. Hung, N.K. Smammas, eds. (CRC Press Taylor and Francis Group, New York), pp. 409–428.

WHO (World Health Organization)/UNEP (United Nations Environment Programme) (2013). In The State of Science of Endocrine Disrupting Chemicals – 2012, A. Bergman, J.J. Heindel, S. Jobling, K.A. Kidd and R.T. Zoeller, eds. (Geneva: WHO Press), pp. 4–29, available at: <http://www.who.int/ceh/publications/endocrine/en/index.html> (accessed 12 March 2013).

Yadaw, J.S., and Loper J.C. 2000. Cloning and characterization of the cytochrome P450 oxidoreductase gene from the zygomycete fungus *Cunninghamella*. Biochem. Biophys. Res. Commun. *282*, 345–353.

Ye, J.-S., Yin, H., Qiang j., Qin, H.-M., Zhang, N., and He, B.-Y. (2011).Biodegrdation of anthracene by *Aspergillus fumigatus*. J. Hazard. Mater. *185*, 174–181.

Yu, C.-P., Deeb, R.A., and Chu, K.-H. (2013). Microbial degradation of steroidal estrogens. Chemosphere *91*, 1225–1235.

Zhang, D.L., Yang, Y.F., Leakey, J.E.A., and Cerniglia, C.E. (1996). Phase I and phase II enzymes produced by *Cunninghamella elegans* for the metabolism of xenobiotics. FEMS Microbiol. Lett. *138*, 221–226.

Zhang, X., Li, Q., Li, G., Wang, Z., and Yan, C. (2009). Levels of estrogenic compounds in Xiamen Bay sediment, China. Mar. Pollut. Bull. *58*, 1210–1216.

Zheng, C., Qu, B., Wang, J., Zhou, J., Wang, J., and Lu, H. (2009). Isolation and characterization of a novel nitrobenzene-degrading bacterium with high salinity tolerance: *Micrococcus luteus*. J. Hazard. Mater. *165*, 1152–1158.

Zhong, Y., Luan, T., Lin, L., Liu, H., and Tam, N.F. (2011). Production of metabolites in the biodegradation of phenanthrene, fluoranthene and pyrene by the mixed culture of *Mycobacterium* sp and *Sphingomonas* sp Bioresour. Technol., *102*, 2965–2972.

Zuo, Y., and Zhu, Z. (2014). Simultaneous identification and quantification of 4-cumylphenol, 2,4-*bis*-(dimethylbenzyl)phenol and bisphenol A in prawn *Macrobrachium rosenbergii*. Chemosphere *107*, 447–453.

Dye Decolorization and Degradation by Microorganisms

8

Anna Jasińska, Aleksandra Góralczyk and Jerzy Długoński

Contents

Abstract

Owing to their industrial applications, many synthetic dyes are commonly present in wastewater and cause serious pollution of the aquatic environment. Most synthetic dyes are toxic, mutagenic, and carcinogenic. No universal method is available for the treatment of dye-contaminated wastewater because of the complex and varied chemical structures of these dyes. Of the current methods used for treating wastewater, microbial methods offer considerable advantages such as high efficiency, are environment friendly, and involve low operation costs. This review presents the latest research on microbial decolorization of synthetic dye. Mechanisms involved in the bioremoval of dyes, pathways underlying the biodegradation of most important dye classes, and physicochemical parameters affecting dye decolorization have been summarized in this review. In addition, this review discusses genetic manipulation of microorganisms and enzymes used for dye decolorization.

8.1 Introduction

Over centuries, dyes have been used by humans for dyeing clothes, leather, and daily use objects. Until the mid-nineteenth century, all dyes used by humans were obtained from natural inorganic (e.g. manganese oxide, hematite, or carbon black) and organic (plants or animals) materials. However, low pigment content in biological materials and high production costs made these dyes very expensive and difficult to produce (Clark *et al.*, 1993; Bechtold *et al.*, 2003). This changed radically in the nineteenth century because of the development of organic chemistry. Production of synthetic dyes dates back to 1856 when an 18-year-old student, William Henry Perkin, during his unsuccessful attempts to synthesize quinine, synthesized a blue substance that was patented later as aniline purple. Since then, >100,000 commercial synthetic dyes have been produced, with an annual production of over 7×10^5 tonnes/year (Robinson *et al.*, 2001; Yagub *et al.*, 2014). Production of synthetic dyes in 2008 in different regions worldwide is presented in Fig. 8.1.

Synthetic dyes are used in textile, leather, plastic, and cosmetic industries; paper printing; and colour photography (Forgacs *et al.*, 2004; Rai *et al.* 2005; Saratale *et al.*, 2011). They are also used to determine specific surface area of activated sludge for groundwater tracing and to control the efficiency of sewage and wastewater treatments. Many dyes are used in the fields of life sciences and medicine. Moreover, dyes that are known to be harmlessness to humans are used in food and pharmaceutical industries (Ali, 2010; Rauf and Ashraf, 2012).

8.2 Classification and characteristics of synthetic dyes

Technically, dyes are chemical compounds that associate permanently with materials upon contact, thus giving these materials appropriate colours. Chemically, dyes are compounds that selectively absorb electromagnetic radiation in the visible spectrum of light, i.e. light having wavelength in the range of 400–700 nm (Ejder-Korucu *et al.*, 2015).

Dyes possess at least one chromophore (colour-bearing group that frequently contains heteroatoms such as N, O, and S, with non-bonding electrons) and a conjugated system (a structure with alternating double and single bonds) and show electron resonance, which stabilizes organic compounds. The colour of a dye is lost in the absence of any of these features. In addition, most dyes contain auxochromes (i.e. carboxylic acid, sulfonic acid, and amino and hydroxyl groups) that are responsible for colour shifting and are most often used to modify dye solubility (Lorimer *et al.*, 2001; Sathiya *et al.*, 2007). Dye molecules also contain other substituents that confer specific properties such as increased stability or solubility in water (Khan *et al.*, 2013).

Dyes are mainly classified in 2 ways: based on their chemical structure and based on the type of chromophore present. Approximately 20–30 classes of dyes have been identified to date by using the latter mode of classification (Ramalho, 2005). Most important of these classes are azo (mono-, di-, tri-, and polyazo), triarylmethane, anthraquinone, nitro, nitroso, phthalocyanine, indigoid, thiazine, and sulfur dyes (Table 8.1). Another type of classification is based on the application characteristics of dyes (Table 8.2). This classification method also considers forces responsible for the binding of a dye to a substrate, such as van der Waals forces and hydrogen, ionic, or covalent bonds (O'Neill *et al.*, 2000).

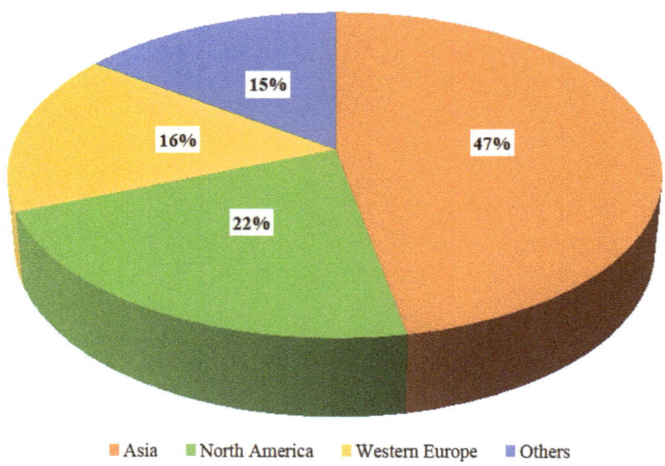

Figure 8.1 Participation of individual regions of the world in the production of synthetic dyes in 2008.

Table 8.1 Colour index classification of dye chemical constituents

Chemical class	CI Constitution numbers	Chemical class	CI Constitution numbers
Nitroso	10000–10299	Indamine	49400–49699
Nitro	10300–10999	Indophenol	49700–49999
Monoazo	11000–19999	Azine	50000–50999
Disazo	20000–29999	Oxazine	51000–51999
Triazo	30000–34999	Thiazine	52000–52999
Polyazo	35000–36999	Sulfur	53000–54999
Azoic	37000–39999	Lactone	55000–55999
Stilbene	40000–40799	Aminoketone	56000–56999
Carotenoid	40800–40999	Hydroxyketone	57000–57999
Diphethylmethane	41000–41999	Anthraquinone	58000–72999
Triarylmethane	42000–44999	Indigoid	73000–73999
Xanthene	45000–45999	Phthalocyanine	74000–74999
Acridine	46000–46999	Natural	75000–75999
Quinoline	47000–47999	Oxidation base	76000–76999
Methine	48000–48999	Inorganic pigment	77000–77999
Thiazole	49000–49399		

Table 8.2 Application categories of dyes

Type of dye	Characteristics	Substrates	Chemical classes
Acid	When in solution are negatively charged; bind to the cationic $NH3^{+}$ – goups present in fibres	Nylon, wool, silk, modified acrylics, paper, leather, inks, food, cosmetics	Azo, anthraquinone, triphenylmethane, azine, xanthene, nitro, nitroso
Basic	Cationic compounds that bind to the acid groups of fibres	Synthetic fibres, paper, inks	Triarylmethane, cyanine, thiazine, oxazine, acridine
Direct	Large molecules bind by Van der Waals forces to the fibres	Cellulose, cotton, viscose, paper, leather, nylon	Polyazo, phthalocyanine, oxazine
Reactive	Form covalent bonds with fibres	Cotton, wool, silk, nylon	Azo, anthraquinone, triarylmethane, phthalocyanine, oxazine
Disperse	Water-insoluble non-ionic dyes used for hydrophobic fibres from aqueous dispersion	Polyester, polyamide, acetate, acrylic, plastics	Azo, anthraquinone, nitro
Solvent	Non ionic dyes that dissolve the substrate to which they bind	Plastics, gasoline, varnish, lacquer, stains, inks, oils, waxes, fats	Xanthene, azine, phthalocyanine, nitro, nitroso, triarylmethane
Vat	Water-insoluble dyes which on reduction give soluble colourless forms with affinity for fibres	Cellulose, cotton, viscose, wool	Anthraquinone, indigoids

Irrespective of its type, each dye has a unique designation called colour index number and is listed in the Colour Index International that is edited systematically since 1924 by the Society of Dyers and Colourists and American Association of Textile Chemists and Colorists (Christie, 2007).

8.3 Methods for treating dye-contaminated wastewater

Not all dyes bind to materials being coloured. Loss of dyes in wastewater varies from 2% for basic dyes, to as high as 50% for reactive dyes (Forgacs *et al.*, 2004). Approximately 280,000 tonnes of dyes produced each year are lost in effluents during application processes (Jin *et al.*, 2007). Dyes should be removed from wastewater because presence of even small quantities of dyes in a water body (<1 ppm) affects its aesthetic value. In addition, dye contamination results in the significant loss of water luminosity, decreases oxygen solubility, and blocks light passage to the depths of aquatic systems. Further, dye contamination leads to eutrophication, disrupts the normal circulation of biogenic elements, and directly or indirectly affects all levels of the trophic chain (Gonçalves *et al.*, 2000; Rodríguez Couto, 2009). Studies have reported that the value of a median lethal dose (LD50/24) in fish for approximately 98% synthetic dyes ranges from 1 to 100 mg/l (Zaharia and Şuteu, 2012). In addition, studies have suggested the accumulation of some dyes in the tissues of fish and other aquatic animals. Dyes reduce the efficiency of wastewater treatment with biological methods by inhibiting the growth and enzymatic activity of microorganisms and by reducing the biodiversity of microorganisms in activated sludge and biological trickling filters (Solecka and Ledakowicz, 2005). Most synthetic dyes have a complex aromatic structure that makes them resistant to light, biological activity, ozone, and other degradative environmental conditions (dos Santos *et al.*, 2007; Srinivasan and Viraraghavan, 2010; Verma *et al.*, 2012). Toxicity and carcinogenicity of many dyes and/or their by-products has increased the interest in controlling environmental pollution caused by these dyes in recent years. The Ecological and Toxicological Association of the Dyestuffs Manufacturing Industry was founded in 1974 to minimize environmental pollution caused by dyes and to protect dye users and consumers (Anliker, 1979; Robinson *et al.*, 2001). Other regulatory agencies such as the US Environmental Protection Agency, Environment Agency and the Scottish Environmental Protection Agency are also working towards achieving cleaner production and restricting the use of toxic dyes (Singh, 2006).

Various methods have been developed for treating dye-contaminated wastewater. However, no universal method is available to date (Crini, 2006; Ali, 2010; Verma *et al.*, 2012). Methods for treating dye-contaminated wastewater can be classified into three major categories, namely physical, chemical and biological methods. Each method alone may not meet the requirements; therefore, a combination of these methods is often used. The choice of treatment method employed depends on factors such as type and concentration of dyes in an effluent, presence of additional pollutants (e.g. heavy metal ions, organic solvents, or mineral), and estimated cost of the method (Robinson *et al.*, 2001; Kandelbauer and Guebitz, 2004; Husseiny, 2008). The most frequently used methods for removing dyes from wastewater are presented in Table 8.3. Although physical and chemical methods are effective in removing colour, they are associated with environmental and economic limitations. Major disadvantages of using physical and chemical methods include high operation costs, low efficiency, limited versatility, interference with other wastewater constituents, formation of hazardous metabolites, and/or intensive energy requirements

Table 8.3 The most frequently used methods for decolorization of dye-loaded wastewaters

Physical	Chemical	Biological
Adsorption	Coagulation	Aerated lagoon
Sedimentation	Precipitation	Anaerobic digestion
Flotation	Neutralization	Stabilization pond
Reverse osmosis	Flocculation	Activated sludge
Nano-and electrofiltration	Ion exchange	Trickling filter
Ultrasonic	Oxidation	Bioaugmentation
Ionizing radiation	Reduction	
	Wet air condition	
	Ozonation	
	Fenton reagent	
	Chlorination	

(Saratale *et al.*, 2011). These disadvantages make these methods economically unsuitable for small enterprises. Therefore, there is a need to develop efficient, eco-friendly methods for decreasing dye concentration in wastewater (dos Santos *et al.*, 2007; Sarayu and Sandhya, 2012). An alternative solution is to use biological methods. Removal of synthetic dyes by using various microorganisms, including bacteria, yeast, filamentous fungi, and algae, is receiving more attention (Fu and Viraraghavan, 2001; Rodríguez Couto, 2009; Kaushik and Malik, 2009).

8.4 Mechanisms underlying microbial removal of dyes

Bioremediation is one of the most promising and intensively developed dye removal technologies. It involves various processes that use natural resources to control xenobiotic pollution (Asgher *et al.*, 2008). Use of biological methods to remove environmental pollutants is associated with significant advantages over other methods because of the adaptability of various microorganisms in degrading different compounds. Biological methods are environmentally friendly and leave the ecosystem intact. Bioremediation can remove lower concentrations of contaminants, which is not feasible with physical or chemical methods. The main advantage of bioremediation is that it is cost-effective compared to the conventional techniques. Moreover, it may lead to the complete mineralization of a pollutant. However, a major drawback of bioremediation is that it is time-consuming and less predictable than that of the conventional methods (McMullan *et al.*, 2001; Bhatnagar and Sillanpaa, 2010; Dubey *et al.*, 2010).

Microorganisms remove dyes through biosorption, bioaccumulation, and biodegradation or a combination of the above mechanisms (Fig. 8.2). Biosorption is a metabolically independent process involving the binding of solutes to biomass; therefore, it can occur in dead or living biomass (Gadd, 2009). Bioaccumulation is an energy- and metabolism-dependent process involving pollutant accumulation within the cytoplasm of actively growing cells (Aksu, 2003). Biodegradation involves the decomposition of pollutants through

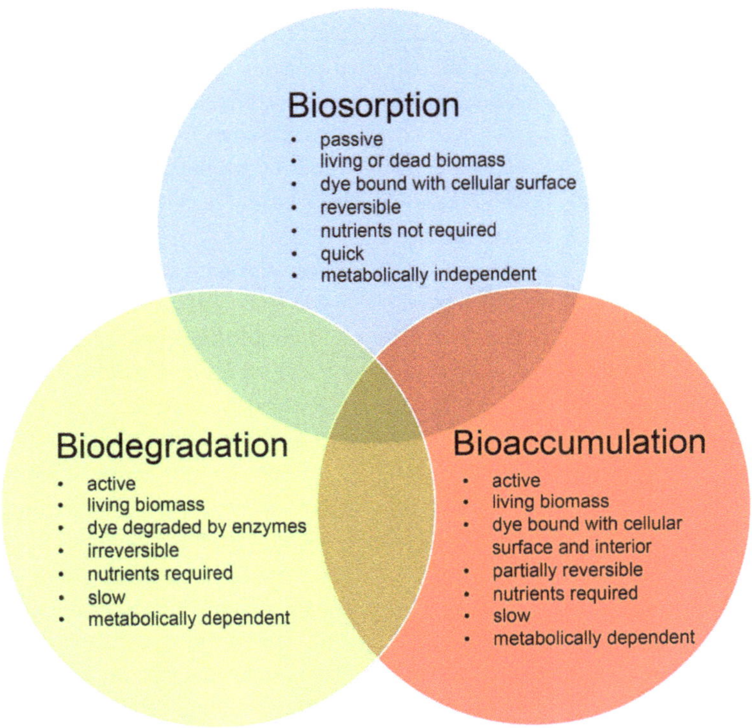

Figure 8.2 Comparison of biosorption, bioaccumulation, and biodegradation.

the action of extra- or intracellular enzymes (after accumulation) (Diaz, 2008). All the above-mentioned mechanisms have been discussed in detail below.

8.4.1 Biosorption

Biosorption is a selective, versatile, and efficient process of removing dye molecules from diluted aqueous solutions by using inactive and dead biomass. Biosorption is a rapid process and can be conducted over a wide range of pH and temperatures. Its efficiency can be significantly increased by appropriately optimizing process parameters (Mou *et al.*, 1991). Biosorption does not require high investment because biological material used for binding to synthetic dyes is often cheap. Often, waste biomass from agricultural and industrial processes is used (Sanghi and Bhattacharya, 2002; Sharma *et al.*, 2011). Biosorption is a physicochemical process involving several metabolism-independent physical and chemical mechanisms such as adsorption, electrostatic interaction, ion exchange, complexation, chelation, and microprecipitation (Crini, 2006; Gupta and Suhas, 2009). Various functional groups present on the microbial cell wall, such as amino, carboxyl, hydroxyl, thiol, and phosphate groups, bind to dye molecules. Structures of individual dye molecules contribute to the adsorption ability of and affinity towards fungal hyphae (Aksu and Dőnmez, 2005; Srinivasan and Viraraghavan, 2010).

Biosorption efficiency can be increased by physically and chemically modifying the biomass used. Physical modifications such as autoclaving, lyophilization, and grinding are simpler and less expensive than chemical modifications (Lim *et al.*, 2010). Of the various chemical pretreatments available, use of acids, alkalis, and organic solvents is preferred because of their simplicity and efficiency (Fu and Viraraghavan, 2001; Won *et al.*, 2006; Vijayaraghavan and Yun, 2007; Ambrósio *et al.*, 2012). Biosorbents can be improved by enhancing their binding ability or by eliminating inhibition groups present in them. Biosorptive capacity of a microbial biomass can be enhanced by optimizing culture conditions or by using genetic engineering techniques at some stage of cultivation (Bakshi *et al.*, 2006; Iqbal and Saeed, 2007; Erden *et al.*, 2011). Adsorbed dyes can be recovered by desorption, and biosorbents (after regeneration) can be used in the next cycle of dye removal. Biosorbents can be recovered using chemicals such as alkalis and chelating agents or by microwaving or heating (Vijayaraghavan *et al.*, 2007; Pavan *et al.*, 2008). The regeneration process is not profitable if the biomass is cheap and widely available. In such cases, dye-associated biomass can be utilized by incineration (Kandelbauer and Guebitz, 2004; Won *et al.*, 2006).

8.4.2 Bioaccumulation

Bioaccumulation involves the uptake of pollutants by living cells through rapid temperature- and metabolism-independent adsorption on the cell wall, followed by metabolism-dependent accumulation of the pollutants into the cytoplasm (Yilmazer and Saracoglu, 2009; Chojnacka, 2010). It is a very complex process that depends on several factors (which are almost identical to factors affecting the cultivation of microorganisms), namely, composition of growth medium or wastewater, pH, temperature, and presence of other pollutants (Kelewou *et al.*, 2014).

Bioaccumulation is usually performed in batch systems to promote dye removal during microbial growth (Aksu and Dönmez, 2000). Bioaccumulation does not require biomass harvesting, drying, processing, and storage and only requires sufficient quantities of easy-to-use carbon and nitrogen sources (Aksu and Dönmez, 2005). Sometimes, it is possible to apply a self-replenishing system, whereby biological material accumulating pollutants becomes generated in a bioaccumulation unit (Chojnacka, 2010). However, bioaccumulation is associated with some limitations such as requirement of a nutrient medium for microbial growth and inhibition of microbial growth in the presence of very high concentrations of dyes and other toxic pollutants in wastewater (Kujan *et al.*, 1995; Yilmazer and Saracoglu, 2009).

Several studies have investigated the bioaccumulation of toxic dyes by microorganisms, especially yeast and filamentous fungi (Table 8.4). Yeasts are good models for studying bioaccumulation because they can be obtained as an industrial by-product from the fermentation industry at almost no cost. Yeasts are efficient in accumulating both metal ions and synthetic dyes (Dönmez, 2002; Aksu and Dönmez, 2005). In addition, many strains of filamentous fungi are useful for dye decolorization,

Table 8.4 Examples of some microorganisms capable for dyes bioaccumulation

Microorganism	Dye	References
Aspergillus lentulus	Acid Navy Blue, Fast Red A, Orange-HF	Kaushik and Malik (2013)
Aspergillus niger	Reactive Black 5	Taskin and Erdal (2010)
Candida tropicalis	Basic violet 3	Das *et al*. (2011), Charumathi and Das (2010)
Candida tropicalis	Remazol Blue, Remazol Black B, Remazol Red RB	Aksu and Donmez (2005), Donmez (2002)
Candida utilis	Remazol Turquoise Blue G (RTBG)	Gonen and Aksu (2009)
Penicillium pinophilum	Basic Green 4	Jasińska *et al*. (2012)
Phormidium sp., *Synechococcus* sp.	Remazol Blue; Reactive Black B; Reactive Red RB	Sadettin and Donmez (2006), Sadettin and Donmez (2007)
Pichia fermentans	Acid Blue 93, DirectRed 28, Basic Violet 3	Das *et al*. (2010)
Rhodotorula glutinis	Reactive Red 11	Saravanan *et al*. (2012)
Rhodotorula mucilaginosa	Remazol Blue	Ertugrul *et al*. (2009)
Saccharomyces cerevisiae	Remazol Blue, Remazol Black B, Remazol Red RB	Aksu (2003)
Saccharomyces cerevisiae	Basic Green 4, Basic Green 2	Kelewou *et al*. (2014)

e.g. bioaccumulation of triphenylmethane dye, malachite green, by a strain of filamentous fungus *Penicillium pinophilum* (Fig. 8.3).

8.4.3 Biodegradation

Biodegradation involves transformation of organic compounds by living organisms. It is an energy-dependent process involving the breakdown of a dye into various by-products through the action of various enzymes (Ali, 2010). Biodegradation results in visible colour removal (achieved by the cleavage of the chromophore; defined as decolorization) and conversion of dyes into simpler compounds with lower toxicities than parent dyes. In some cases, biodegradation leads to the complete mineralization of dyes and their decomposition into carbon dioxide, water, and/or any other inorganic end products (Banat *et al.*, 1996; Rai *et al.*, 2005). Higher mineralization occurs for dyes containing aromatic ring substituents such as amino, acetamido, hydroxyl, or nitro substituents than for dyes with unsubstituted ring substituents. Use of specific

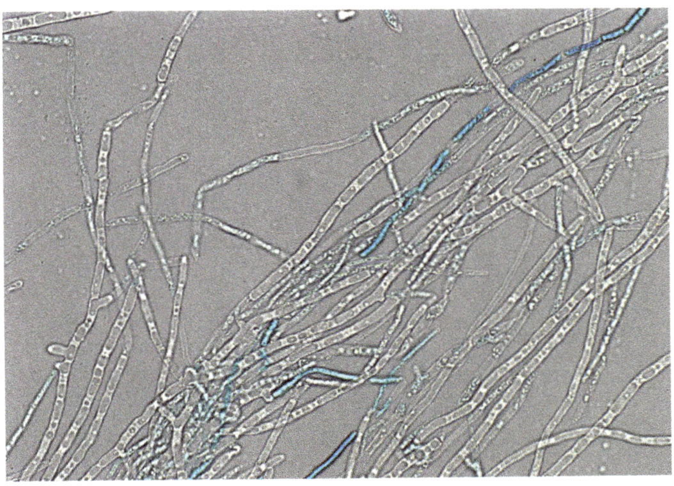

Figure 8.3 Accumulation of malachite green into *Penicillium pinophilum* IM 6480 cells.

microorganisms or a consortium of dye degraders and decolorizers can lead to faster or more complete degradation of dyes (Verma and Madamwar, 2003; Rai *et al.*, 2005; Pandey *et al.*, 2007; Kaushik and Malik, 2009). The relationship among biodegradation, biotransformation, mineralization, and decolorization is presented in Fig. 8.4.

8.4.3.1 Enzymatic dye degradation

Various microorganisms, including fungi, bacteria, yeast, and microscopic algae, have been used for degrading synthetic dyes. However, some groups of microorganisms offer specific advantages over others. Degradation of dyes is best studied with white rot fungi (Gadd, 2001; McMullan *et al.*, 2001; Christian *et al.*, 2005). White rot fungi produce one or more extracellular lignin-modifying enzymes (LMEs) that non-specifically mediate the transformation of various recalcitrant compounds, including synthetic dyes (Kandelbauer and Gűebitz, 2004; Saratale *et al.*, 2007). The main LMEs include oxidoreductases such as lignin peroxidase (LiP, EC 1.11.1.14), manganese peroxidase (MnP, EC 1.11.1.13), and laccase (EC 1.10.3.2) (Wasenberg *et al.*, 2003).

Lacasses show great potential for bioremediation because of their non-specific oxidation capacity, lack of requirement for cofactors, and use of readily available oxygen as an electron acceptor. Lacasses contain 4 copper atoms in their active site and catalyse the oxidation of substrates by inducing simultaneous reduction of 1 dioxygen molecule to 2 water molecules (Morozova *et al.*, 2007; Sharma *et al.*, 2007; Sharma and Kuhad, 2008). To date, approximately 100 different compounds such as polyphenols, methoxy-substituted monophenols, and aromatic amines have been identified as substrates of fungal laccases (Bourbonnais and Paice, 1990; Bourbonnais *et al.*, 1995; Baldrian, 2006). Laccases induce the oxidation of their substrates to corresponding radicals through a direct interaction between their catalytic centres and the substrates. However, most laccases do not directly oxidize their substrates and require chemical mediators. Most common synthetic mediators include 1-hydroxybenzotriazole, 2,2′-azino-bis(3-ethylbenzothiazoline-6-sulfonic acid) (ABTS), and 3-hydroxyanthranilic acid (HAA) (Riva, 2006; Arora and Sharma, 2010). Laccase-mediated dye degradation is presented in Fig. 8.5. Most dyes

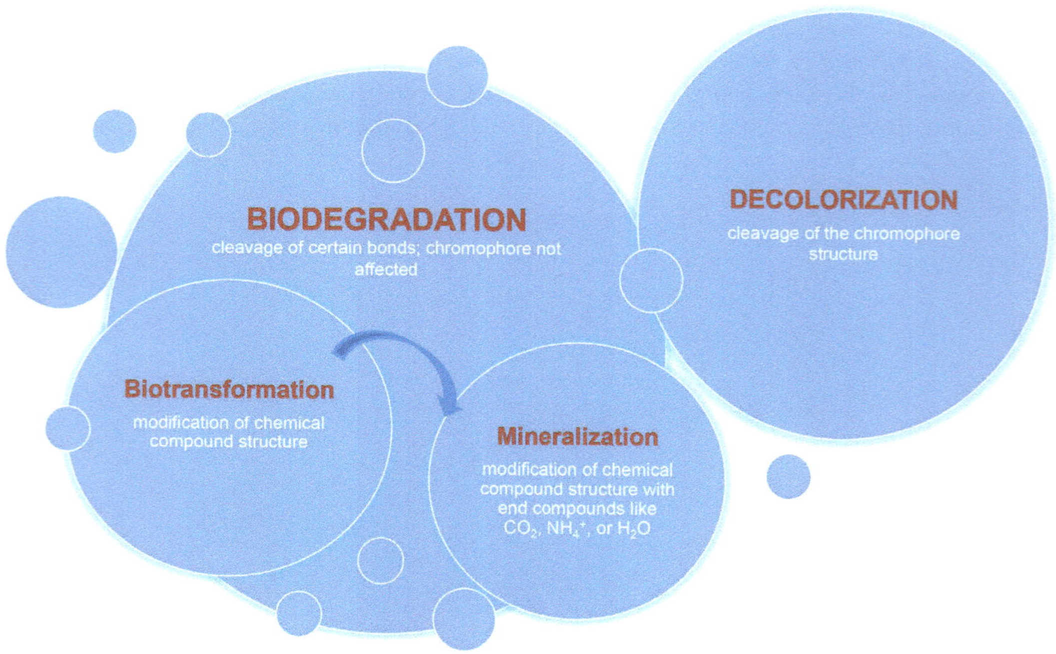

Figure 8.4 The relationship between the phenomena of biodegradation, biotransformation, mineralization and decolorization.

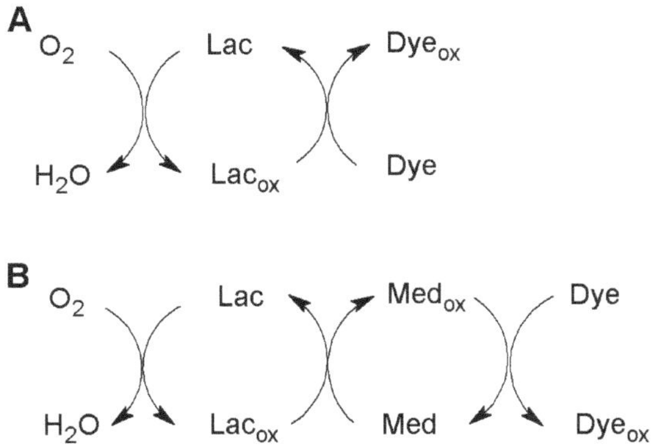

Figure 8.5 A laccase-involved dye degradation mechanism (modified from Desai and Nityanand, 2011).

(e.g. anthraquinone or triphenylmethane dyes) are good laccase substrates, and their decolorization is proportional to laccase activity. However, some dyes (e.g. azo and indigo dyes) are not suitable substrates for laccase-mediated degradation, and their degradation depends on the presence of mediators. Sometimes, substrate dyes present in industrial effluents mediate the interaction between laccases and non-substrate dyes, thus improving the rate of decolorization (Singh, 2006).

MnP and LiP from different sources transform various recalcitrant synthetic dyes. Peroxidases are haemoproteins that catalyse reactions in the presence of hydrogen peroxide (Duran *et al.*, 2000). MnP catalyses the transformation of phenolic compounds, with simultaneous Mn(II) to Mn(III) oxidation (Shin *et al.*, 2005). LiP catalyses one-electron oxidation of phenolic and non-phenolic aromatic compounds (Husain, 2006). The decolorization activity of LiP is enhanced in the presence of compounds such as veratryl alcohol, which acts as a redox mediator. In this, veratryl alcohol is first oxidized by a single electron to form a cation, which oxidizes other substrates. This mechanism was proposed based on the observation that some non-substrate compounds are oxidized by LiP in the presence of veratryl alcohol (Alam *et al.*, 2009). Decolorization of synthetic dyes by fungal and bacterial peroxidases has been studied widely (Boer *et al.*, 2004; Shedbalkar *et al.*, 2008; Parshetti *et al.*, 2009; Yang *et al.*, 2011).

Several studies have been performed on the degradation of industrial dyes by using microbial reductases (Bhosale *et al.*, 2006; Saratale *et al.*, 2007). The most frequently used microbial reductases include azoreductases, triphenylmethane reductase (TMR), and NADH-DCIP reductases. These enzymes require reducing cofactors such as nicotinamide adenine dinucleotide (NAD+) and nicotinamide adenine dinucleotide phosphate (NADP+) for reducing dyes (Chen *et al.*, 2005a,b; Wang *et al.*, 2007).

TMR catalyses the reduction of triphenylmethane dyes to their leuco derivatives. This enzyme was first purified and biochemically characterized from cultures of *Citrobacter* sp. KCTC 18061P (Jang *et al.*, 2005). TMR is also produced by *Pseudomonas aeruginosa* NCIM 2074, *Exiguobacterium* sp. MG2, and *Mucor mucedo* (Moturi and Charya, 2009; Wang *et al.*, 2012). In addition, food-borne pathogens *Listeria monocytogenes* (Nelson *et al.*, 2004) and *Aeromonas hydrophila* (Ren *et al.*, 2006) express the gene encoding TMR (*tmr*). The decolorization activity of TMR depends on the chemical structure of its substrate dyes. Jang *et al.* (2005) reported that malachite green is the most favourable substrate of TMR because of its appropriate structure and that crystal violet is a less favourable substrate because it contains an additional dimethylamino group. According to Kim *et al.* (2008), structures of triphenylmethane dyes such as brilliant green, bromophenol blue, methyl red and Congo red make them incompatible with the substrate-binding pocket of TMR, thus affecting its enzymatic activity. Further, Kim *et al.* (2008) proposed the mechanism underlying the decolorization of malachite green by TMR based on structural inspection and modelled ternary protein/cofactor/substrate complex structure and by performing mutagenesis studies.

Azoreductases use NADH and/or NADPH as electron donors to catalyse the cleavage of the azo linkage. There are 2 types of azoreductases, namely, oxygen-sensitive azoreductases secreted by anaerobic or microaerophillic bacteria and oxygen-insensitive azoreductases secreted by obligate aerobic bacteria.. Biochemical characteristics and protein structures of many bacterial azoreductases have been determined recently (Chen *et al.*, 2005; Ito *et al.*, 2006; Liu *et al.*, 2007; Wang *et al.*, 2007). Azoreductases reduce azo compounds through a ping-pong-bi-bi mechanism (Liu *et al.*, 2007). The currently accepted mechanism involves two cycles of reduction, with the first cycle involving the reduction of an azo substrate to hydrazine and the second cycle involving the reduction of the hydrazine to two amines. Recently, Ryan *et al.* (2010) proposed a novel mechanism of substrate reduction by azoreductases, in which the second reduction step occurred in an enzyme-independent manner. Because human activities are mainly responsible for the dissemination of azo compounds into the environment, reduction of azo dyes may not be the primary role of these enzymes. The physiological role of azoreductases has recently been a subject of debate. Liu *et al.* (2009) showed that *Escherichia coli*-encoded azoreductase AzoR is a quinone reductase that provides resistance against thiol-specific stress induced by electrophilic quinones.

Intracellular enzymes such as cytochrome P450 monooxygenases or other hitherto uncharacterized enzymes may also be involved in the biodegradation of synthetic dyes. Cha *et al.* (2001) suggested the role of cytochrome P-450 monooxygenase of *Cunninghamella elegans* ATCC 36112 in the demethylation and reduction of malachite green to leucomalachite green. Jones and Falkinham (2003) and Wang *et al.* (2012) proposed a similar mechanism for the decolorization of malachite green by *Mycobacterium chelonae*, *Mycobacterium avium*, and *Exiguobacterium* sp. MG2.

8.4.3.2 Non-enzymatic degradation of dyes

Recent findings suggest that high-temperature-resistant, non-enzymatic, and low-molecular-weight compounds induce the decolorization of synthetic dyes (Yan *et al.*, 2009; Gomaa *et al*, 2008; Wang *et al*, 2011; Gomaa, 2012). Hydroxyl radicals produced by white and brown rot fungi by Fenton-like reaction ($Fe^{2+} + H_2O_2 \rightarrow Fe^{3+} + {}^{-}OH + {}^{-}OH$) are suggested to be involved in dye decolorization (Karimi *et al*, 2012; Moldes *et al*, 2012). Peroxides directly oxidize some chemical structures or promote their oxidation by using hydroxyl radicals formed during the Fenton reaction. Advantages of using reactive oxygen species such as hydroxyl radicals or superoxide anions are their high reactivity and low specificity towards a broad spectrum of organic substrates (Nerud *et al*, 2006; Gomaa *et al*, 2008). Hu *et al.* (2006) isolated and purified a new low-molecular-weight peptide with phenol oxidase activity (Pc factor) from a liquid culture of *Phanerochaete chrysosporium*. This Pc factor had a molecular weight is approximately 600 Da and high thermostability and remained active at a weakly alkalescent pH range.

8.5 Pathways underlying the biodegradation of dyes

In the following section, we will discuss mechanisms underlying the microbial decomposition of some important dyes. In addition, we will discuss the characterization and application of azo, triphenylmethane, and anthraquinone dyes.

8.5.1 Azo dyes

Azo dyes are a large class of very effective and versatile synthetic dyes that account for approximately 70% of all dyestuffs used worldwide by weight. These dyes can be synthesized easily, are available in all colours, are easily taken up by fabrics, and have good fastness. Because of their high chemical, biological, and photocatalytic stability and resistance to breakdown induced by time, water, soap, and sunlight exposure, azo dyes are widely used in textile dyeing, paper printing, leather, photography, food, pharmaceuticals, and cosmetics (Chang *et al.*, 2000, 2004; Telke *et al.*, 2008). Azo dyes are the most common class of dyes released into aquatic and terrestrial environments. However, these dyes are resistant to biological degradation. Only those microorganisms that are exposed to the toxic activity of these dyes over a long period can produce enzymes for degrading these dyes (Ghodake *et al.*, 2009; Chacko and Subramaniam, 2011).

Decolorization of azo dyes by different bacteria occurs under anaerobic, facultatively anaerobic, and aerobic conditions. Microbial degradation of azo dyes involves two main mechanisms. The first step in the bacterial decolorization of azo dyes under both anaerobic and aerobic conditions is the reduction of the azo bond in the chromophore. This reduction may involve various intra- and extracellular mechanisms such as enzymatic reduction, resulting in the formation of colourless hazardous aromatic amines (Chang and Kuo, 2000; Van der Zee and Villawerde, 2005). Some non-enzymatic pathways that reduce azo dyes involve low-molecular-weight redox mediators, biogenic reductants like sulfide, or a combination of these. Aromatic amines formed in the first step are then mineralized through oxidation (Joshi *et al.*, 2008).

Bacterial reduction of azo dyes is usually non-specific and is faster than fungal reduction. Various aerobic and anaerobic bacteria have been extensively used for degrading azo dyes. Many recent studies have investigated dye removal by using single bacterial cultures of *Bacillus subtilis, Aeromonas hydrophila, B. cereus, Proteus mirabilis, Pseudomonas luteola* and *Pseudomonas* sp. (Table 8.5). Decolorization of azo dyes by using pure cultures of microorganisms is efficient and fast. However, individual bacterial strains usually cannot degrade azo dyes completely and produce carcinogenic aromatic amines as intermediates, which need to be decomposed further (Joshi *et al.*, 2008). Therefore, use of treatment systems containing mixed microbial populations for the degradation of synthetic azo dyes helps in achieving a higher degree of biodegradation and mineralization because of the synergistic activities of the microbes and provides considerable advantages over the use of pure cultures (Khehra *et al.*, 2005; Chen and Chang, 2007). In a treatment system containing mixed microbial populations, individual strains may attack a dye molecule at different positions or may use metabolites produced by co-existing strains for further decomposition (Chang *et al.*, 2004, Forgacs *et al.*, 2004, Jadhav *et al.*, 2008, Lade *et al.*, 2012). In contrast, isolation of a single bacterial strain from dye-contaminated wastewater samples is time-consuming and difficult and requires long-term adaptation procedures for the effective decolorization and degradation of azo dyes (Cui *et al.* 2012; Zabłocka-Godlewska *et al.*, 2012; Sathish and Amarnath, 2015).

8.5.2 Anthraquinone dyes

Anthraquinone dyes, which are the most important class of dyes after azo dyes, are bright and show good fastness. However, their inherent low tinctorial strength makes them uneconomical compared with azo dyes and is the prime reason for their gradual replacement by azo dyes (Banat *et al.*, 1996; Robinson *et al.*, 2001). Nevertheless, anthraquinone dyes, particularly red and blue anthraquinone dyes, are still used to some extent because other dyes do not provide the combination of properties offered by these dyes but at a price. Other major applications of anthraquinone dyes include use as a pesticide, bird repellent, and additive in chemical alkaline pulp process in paper and pulp industry. Anthraquinone dyes are also used as a seed dressing or in seed treatments (Griffiths, 1990; Hunger, 2007).

Under anaerobic conditions, transformation of anthraquinone dyes is presumably limited to the reduction of quinone to hydroquinone, a reaction that is reversed when the molecule is exposed to oxygen. Deng *et al.* (2008) reported the decolorization of acid blue 25 under anaerobic conditions by *B. cereus* DC11 through an NADH-/NADPH-dependent mechanism. A similar mechanism was described for the decolorization of reactive blue 19 by Giwa and Ifu (2012). Under anaerobic conditions, some anthraquinone dyes are removed from wastewater through the formation of an 'insoluble pigment'. This is in line with the observation that electrochemical reduction of anthraquinone dyes increases their adsorptive properties. Thongachai and Worrawit (2000) reported that reactive blue 5 and reactive blue 19 were not degraded under anaerobic conditions but were decolorized, which was thought to have occurred because of their adsorption.

Limited studies have been performed on the biodegradation of anthraquinone dyes under aerobic conditions. Aerobic degradation of anthraquinone dyes may be similar to that of anthraquinone or anthraquinone-2-sulfonate.

8.5.3 Triphenylmethane dyes

Triphenylmethane dyes are relatively inexpensive and suitable for dyeing a wide range of materials. These dyes are majorly used in photography, foods, cosmetics, and stationeries (Thetford, 2000). In addition, triphenylmethane dyes are extensively used in medicine and experimental biology, e.g. in cytology, histology, and microbiology to stain cells (Kiernan, 2001). Because of the antibacterial, antifungal, and antiprotozoal properties of many triphenylmethane dyes (e.g. malachite green and crystal violet), they are used as medical disinfectants.

However, use of triphenylmethane dyes is highly controversial because of their associated risks. Malachite green and its metabolites induce carcinogenesis and mutagenesis in humans (Sudova *et al.*, 2007), and crystal violet binds to DNA and causes replication errors (Ayed *et al.*, 2009).

Table 8.5 Summary of reports from the last five years on azo dyes decolorization by pure bacteria cultures

Bacteria	Dye concentration	Colour removal (%)	Time required (h)	References
Alishewanella sp.	Mixture of textile dyes (Raspberry red, Orange 2R, Orange M2R, Golden Yellow HER each at 0.5 g/l) Reactive Blue 59 (2.5 g/l)	95–98 95	8 6	Kolekar *et al.* (2013) Kolekar and Kodam (2012)
Bacillus sp.	Metanil Yellow (200 mg/l)	100	12	Anjaneya *et al.* (2011)
Bacillus cereus	Reactive Black B (20 mg/l)	NA	NA	Liao *et al.* (2013)
Bacillus lentus	Reactive Red 141 (500 mg/l), Reactive Red 2 (500 mg/l)	98.22–99.11	6	Oturkar *et al.* (2013)
Bacillus thuringiensis	Congo Red (100 mg/l)	72,84 ± 3,25	12	Olukanni *et al.* (2013)
Brevibacillus laterosporus	Remazol Red (50 mg/l)	100	30	Kurade *et al.* (2013)
Enterococcus faecalis	Acid Orange 7 (100 mg/l)	98	5	Lim *et al.* (2013)
Geobacter metallireducens	Amaranth (25–800 μM)	93,7	40	Liu *et al.* (2013a)
Klebsiella sp.	Evans blue (0.05 g/l)	98,8		Zabłocka-Godlewska *et al.* (2015)
Klebsiella oxytoca	Disperse Orange 3 (24.2 mg/l) Methyl Red (269 mg/l)	100 100	7 7	Yu *et al.* (2012)
Lysinibacillus sp.	Metanil Yellow (200 mg/l)	100	24	Anjaneya *et al.* (2011)
Pseudomonas aeruginosa	Amaranth (50 mg/l) Remazol Orange (400 mg/l)	97 92	6 6.5	Jadhav *et al.* (2013) Jadhav *et al.* (2012)
Psychrobacter alimentarius	Reactive Black 5 (100 mg/l) Reactive Golden Ovifix (100 mg/l) Reactive Blue BRS (100 mg/l)	100 100 100	8 8 3	Khalid *et al.* (2012)
Shewanella algae	Amaranth (200 μM) Acid Orange 52 (200 μM)	97,5 49,1	6 6	Liu *et al.* (2013b)
Shewanella marisflavi	Amaranth (200 μM) Acid Orange 52 (200 μM)	53.4 32.7	6 6	Liu *et al.* (2013b)
Shewanella oneidensis	Dye mixture (Methyl Orange and Naphthol Green B each at 0.1 mM)	NA	NA	Cao *et al.* (2013)
Shewanella oneidensis	Methyl Orange (100 mg/l)	NA	NA	Cui *et al.* (2012)
Shewanella oneidensis	Acid Red 27 (200 μM)	78.4	6	Liu *et al.* (2011)
Shewanella xiamenensis	Methyl Orange (100 mg/l) Reactive Red 198 (200 mg/l) Congo Red (200 mg/l)	96.2 93.0 87.5	6 6 6	Ng *et al.* (2013)
Staphylococcus equorum	Reactive Black 5 (100 mg/l) Reactive Golden Ovifix (100 mg/l) Reactive Blue BRS (100 mg/l)	100 100 100	8 8 3	Khalid *et al.* (2012)

NA, not available.

Fungi degrade a broad spectrum of triphenylmethane dyes (Table 8.6). Fungal degradation of these dyes was first reported by Bumps and Brock (1988) who described the biodegradation of crystal violet in ligninolytic (nitrogen-limited) cultures of *Phanerochaete chrysosporium* based on its disappearance and formation of 3 metabolites, namely, *N,N,N′,N′,N″*-pentamethylpararosaniline, *N,N,N′,N″*-tetramethylpararosaniline, and *N,N′,N″*-trimethylpararosaniline, through the sequential *N*-demethylation of the parent compound. In addition, six triphenylmethane dyes, namely pararosaniline, cresol red, bromophenol blue, ethyl violet, malachite green and brilliant green, were removed by the lignin-degrading system of this fungus. Substantial degradation of crystal violet was also observed in non-ligninolytic (nitrogen-sufficient) cultures of *Phanerochaete chrysosporium*, suggesting that this fungus used other mechanisms in addition to lignin degradation for degrading crystal violet.

The most widely studied for triphenylmethane dyes decolorization have been white rot fungi belonging to *Basidiomycota* (Pointing, 2001; Shah and Nerud, 2002; Levin *et al.*, 2004). Strains of *Phanerochaete chrysosporium, Lentinula edodes, Ischnoderma resinosum, Dichomitus squalens, Trametes versicolor, T. trogii, Phlebia fascicularia, Phlebia floridensis, Grammothele subargentea, Pleurotus ostreatus, Polyporus picipes,* and *Gloeophyllum odoratum* are good decolorizers of various triphenylmethane dyes such as crystal violet, pararosaniline, cresol red, bromophenol blue, ethyl violet, malachite green, and brilliant green. Many microscopic filamentous fungi belonging to Ascomycota can degrade complex triphenylmethane dyes (Pointing, 2001; Brar *et al.*, 2006; Haritash and Kaushik, 2009; Ali, 2010). Because of their adaptability, various microbial strains efficient in removing triphenylmethane dyes are usually isolated from dye-contaminated environments (soil, effluents, and sludges). Removal of triphenylmethane dyes has been observed in cultures of *Phialophora* sp., *Penicillium* sp., *Cladosporium* sp., *Aspergillus* sp., *Fusarium solani, Acremonium kiliense, Penicillium pinophilum, Myrothecium roridum, Penicillium ochrochloron, Cunninghamella elegans,* and *Mucor mucedo.* Yeasts degrade triphenylmethane dyes through rapid biosorption and bioaccumulation (Šafařik *et al.* 2002, 2007; Šafaříková *et al.* 2005; Yu and Wen, 2005; Jadhav and Govindwar, 2006; Jadhav *et al.*, 2008; Deivasigamani and Das, 2011).

Several studies have been performed on the decolorization of triphenylmethane dyes by bacteria. Yatome *et al.* (1981) used thin-layer chromatography to show the degradation of crystal violet by *Pseudomonas pseudomallei.* Development of analytical methods enables the identification of biodegradation products. The main degradation product of crystal violet, i.e. Michler's ketone [*N,N′*-bis(dimethylamino) benzophenone], has been identified in cultures of *B. subtilis* IFO 13719 (Yatome *et al.*, 1991) and *Nocardia corallina* (Yatome *et al.*, 1993). Several studies have shown that crystal violet and malachite green are converted to their colourless leuco derivatives after reduction because of the loss of double bonds (Chen *et al.*, 2008). Leucocrystal violet and 5 other intermediates, namely, *N,N′*-bis(dimethylamino) benzophenone, (*N,N*-dimethylaminophenyl) (*N*-methylaminophenyl) benzophenone, *N,N*-dimethylaminobenzaldehyde, *N,N*-dimethylaminophenol, and 4-methylaminophenol, were detected in the cultures of *Shewanella* sp. NTOU1 after crystal violet degradation (Chen *et al.*, 2008). Michler's ketone was also detected after methyl violet B degradation by this strain (Chen *et al.*, 2010) and after malachite green degradation by *B. cereus.* Degradation of these dyes involved sequential demethylation, resulting in the removal of penta-, tetra-, and trimethyl groups by extracellular enzymes, followed by their biosorption.

8.6 Optimization of biodegradation conditions

The ability of microorganisms to degrade synthetic dyes is primarily regulated by their metabolic activities. However, external factors such as structure and concentration of dyes; pH of the environment; temperature; agitation and aeration; availability of nutrients and oxygen; and presence of other contaminants, including heavy metal ions, affect the efficiency of dye decolorization. Synthetic dyes are rarely used by microorganisms as the sole energy source. Therefore, presence of nutrients that serve as carbon and nitrogen sources exert a significant effect on decolorization. Presence of sucrose and glucose favours the microbial degradation of synthetic dyes (Jadhav

Table 8.6 Summary of reports from the last five years on TPM dyes decolorization by fungi

Fungal strain	Decolorized dye	Initial dye concentration	Effectivity and time of dye removal	Mechanism of decolorization	References
Aspergillus oryzae	Methylene blue	100 mg/l	53; 30 min	Lac (in absence of HBT)	Forootanfar *et al.* (2012)
Aspergillus sp.	Brilliant green Methyl violet	10 mg/l 20 mg/l	100; 3 days 100; 1 day	ND	Kumar *et al.* (2012)
Candida crusei	Basic violet 3	10 mg/l	100; 1 day	Lac, NADH-DCIP reductase	Deivasigamani and Das (2011)
Fusarium sp.	Basic violet 2 Rhodamine B	30 ppm	45; 4 days 79; 4 days	ND	Ansari *et al.* (2011)
Fusarium oxysporum	Brilliant green	60 mg/l	75; 5 days	ND	Przystaś *et al.* (2012)
Ganoderma sp.	Bromophenol blue	50 mg/l	98.3; 3 days		Zhou *et al.* (2011)
Gloephyllum odoratum	Brilliant green	60 mg/l	74; 5 days	ND	Przystaś *et al.* (2012)
Myrothecium roridum	Malachite green	10 mg/l	97; 4 days	Lac	Jasińska *et al.* (2012, 2015)
Paraconiothyrium variabile	Bromophenol blue	120 mg/l	100, 3h in presence of HBT 5 mM	Lac	Forootanfar *et al.* (2011)
Penicillium ochrochloron	Malachite green	50 mg/l	93; 14 h	Lip	Shadbalkar and Jadhav (2011)
Penicillium pinophilum	Malachite green	10 mg/l	87; 4 days	Biosorption and MG reductase	Jasińska *et al.* (2012)
Penicillium sp.	Crystal violet	0.01%	91; 28 days	Enzymes and biosorption	Torres *et al.* (2011)
Penicillium sp.	Malachite green	25 mgl	98; 6 days	LiP, NADH-DCIP reductase	Yang *et al.* (2011)
Pleurotus ostreatus *Pleurotus ostreatus* *Polyporus picipes*	Brilliant green	60 mg/l	95; 5 days 83; 5 days 82; 5 days	ND	Przystaś *et al.* (2012)
Trametes versicolor	Coomassie brilliant blue	120 mg/l	30.3; 3h	Lac (in presence of HBT 5 mM)	Forootanfar *et al.* (2012)

and Govindwar, 2006; Deivasigamani and Das, 2011). These compounds are rapidly degraded by microorganisms, which increases the production of microbial biomass and improves decolorization. However, studies are investigating the use of cheaper nutrients (e.g. starch and molasses) to accelerate dye decolorization (Parshetti *et al.*, 2006; Abedin, 2008; Kaushik and Malik, 2009). Nitrogen sources affect the ability of microbes to decolorize synthetic dyes, mainly by modulating the synthesis and activity of oxidoreductive enzymes. Presence of organic nitrogen sources such as yeast extract and peptone accelerates the decolorization of synthetic dyes (Cheriaa and Bakhrouf, 2009; Parshetti *et al.*, 2011). Yeast extract affects the regeneration of NADH, which serves as an electron donor during dye reduction (Jang *et al.*, 2005; Ogugbue and Sawidis, 2011). Another factor that significantly affects the efficiency of decolorization is environmental pH. Environmental pH affects dye decolorization in two ways: (1) by inducing changes in the synthesis and activity of ligninolytic enzymes and (2) by affecting the transport of dye molecules through cell membranes (Papinutti *et al.*, 2006; Sarnthima *et al.*, 2009; Rajeshkannan *et al.*, 2010). Activity of most ligninolytic enzymes produced by fungi belonging to class Basidiomycota is higher at acidic pH. However, dye decolorization occurs at neutral or slightly alkaline pH (Shedbalkar *et al.*,

2008; Deivasigamani and Das, 2011; Du *et al.*, 2011; Shedbalkar and Jadhav, 2011; Yang *et al.*, 2011; Jadhav *et al.*, 2012). Temperature strongly affects microbial metabolism. The rate of dye degradation usually increases at a temperature that favours the growth of the microorganism and decreases at a temperature that inhibits its growth (Panswad and Luangdilok, 2000). However, in most cases, the rate of dye degradation decreases at temperatures >40°C because of the loss of cell viability and inactivation of enzymes involved in degradation (Cetin and Donmez, 2006). However, some enzymes, e.g. TMR produced by *Citrobacter* sp. KCTC and laccase produced by *B. vallismortis* FMB-103, remain active at high temperatures and decolorize synthetic dyes at temperatures ranging from 60°C to 70°C (Jang *et al.*, 2005; Zhang *et al.*, 2012). Most fungi require oxygen for growth. Therefore, maintenance of appropriate gas content by shaking a fungal culture usually increases the efficiency of fungal decolorization (Kaushik and Malik, 2009). Under shaking conditions, fungal cells have better contact with dye molecules or with essential nutrients. Moreover, most enzymes involved in microbial decolorization of dyes work most effectively under adequate oxygen availability. However, presence of oxygen may adversely affect the degradation of some dyes. Particularly, degradation of azo dyes is reduced under aerobic conditions (Moosvi *et al.*, 2005; Deng *et al.*, 2008). Some studies indicate that dye decolorization depends on dye type and concentration. For example, decolorization of azo dyes and molecules containing $-SO_2NH_2$ and $-SO_3H$ groups is difficult (Hsueh *et al.*, 2009). Efficiency of dye removal generally decreases with an increase in its concentration in the environment. This is because of the inhibitory effect of most dyes on microbial growth and/or enzymes involved in decolorization (Ali, 2010). Microbial removal of dyes can be stimulated by using compounds that support the production and/or activity of ligninolytic enzymes. Most extensively described compounds that stimulate microbial removal of dyes include heavy metal ions (Cu^{2+}, Mn^{2+}, and Co^{2+}) and organic compounds (xylidine, ferulic acid, some synthetic dyes, and alcohols) (Pointing and Vrijmoed, 2000; Kaushik and Malik, 2009).

8.7 Methods for evaluating the toxicity of products of dye biodegradation

The main aim of treating synthetic dye-contaminated wastewater is to reduce the associated toxicity. However, metabolites produced after dye degradation are in many cases more toxic than parent dyes. For example, several azo dyes and amines produced after their degradation exert mutagenic effect. Therefore, it is very important to assess the toxicity of a dye and its metabolites produced after degradation (Jadhav *et al.*, 2011). Many *in vivo* and *in vitro* assays are available for assessing the toxicity of dyes and their metabolites. Of these assays, phytotoxicity assay has gained attention because of its easy performance, low cost, and simplicity of analysis. Plants most frequently used for toxicity assessment are *Sorghum vulgare* (Parshetti *et al.*, 2010; Jadhav *et al.*, 2011; Waghamode *et al.*, 2011), *Phaseolus mungo* (Parshetti *et al.*, 2010; Jadhav *et al.*, 2011, Waghmode *et al.*, 2011; Kumar *et al.*, 2012), *Triticum aestivum* (Parshetti *et al.*, 2010; Ayed *et al.*, 2011; Kumar *et al.*, 2012), and *Oryza sativa* (Sharma *et al.*, 2009; Zhuo *et al.*, 2011). Determination of the toxicity of dyes and their metabolites against standard microorganisms is a more sensitive method with high reproducibility. For example, *Salmonella* mutagenicity assay is widely used to detect the mutagenic and carcinogenic potential of synthetic dyes and their metabolites. Microtox uses marine gram-negative bacteria (*Vibrio fischeri*) to determine the toxicity of specific substances (Kunz *et al.*, 2002; Gottlieb *et al.*, 2003; Anastasi *et al.*, 2011). These bacteria naturally emit light as a part of cellular respiration, which is measured as luminescence. Exposure of these bacteria to toxic substances decreases their luminescence, and the percentage change in luminescence is directly correlated to toxicity. Besides phytotoxicity and microbial toxicity assays, cytotoxicity and genotoxicity assays by using *Allium cepa* are widely used to determine the toxicity of synthetic dyes and their metabolites. *Allium cepa* is an excellent genetic model for evaluating the mutagenicity of toxic chemicals. Moreover, it is cheap and easy to use. Chromosome aberrations in the root cells of *Allium cepa* help in detecting genotoxicity while mitotic index and nuclear abnormalities in these cells help in evaluating the cytotoxicity of dyes. Oxidative stress response has recently been studied for assessing the toxicity of dyes. Analysis of antioxidant enzymes such as ascorbate peroxidase, superoxide dismutase, catalase, glutathione reductase, and peroxiredoxins and analysis of lipid peroxidation and protein oxidation are performed to assess oxidative stress response to toxic dyes (Solis *et al.* 2012). Animal assays have been mainly conducted in crustaceans, fish, rat and mice. The most frequently used, well-established, and standardized acute lethality test uses *Daphnia magna* (Elisangela *et al.*, 2009; Porri *et al.*, 2011; Rizzo, 2011) to determine the toxicity of dyes and their metabolites.

8.8 Genomics and proteomics in decolorization studies

Different microbial enzymes are directly involved in dye decolorization. Enzymes inducing efficient dye decolorization should be determined using technologies like proteomics and genomics. In addition, already known degradative enzymes should be cloned into a single microorganism by using genetic engineering to enhance the decolorization efficiency that microorganism for removing synthetic dyes.

Jang *et al.* (2005) were the first to clone *tmr* of *Citrobacter* sp. KCTC 18061P. The sequence of TMR encoded by this gene was not similar to that of proteins with known functions. Therefore, they proposed that this enzyme was a novel member of short-chain dehydrogenase reductase family. Ren *et al.* (2006) determined the properties of an enzyme from *Aeromonas hydrophila* DN322 that decolorized triphenylmethane dyes. Purification of this enzyme to homogeneity indicated that it was an NADH-/NADPH-dependent, haem-containing oxygenase with a molecular weight of 87 kDa. The gene encoding this enzyme was also cloned and sequenced. *Tmr2*, encoding TPM reductase, of TPM dye-degrading *Pseudomonas* sp. MDB-1 was cloned, sequenced, and effectively expressed in *Escherichia coli* (Li *et al.*, 2009). This gene was 99% similar to the reductase gene of *Citrobacter* sp. KCTC 18061P and *Aeromonas hydrophila* DN322. Sequence alignment showed that this gene was considerably conserved in these strains. Lu *et al.* (2009) expressed laccase from *Pycnoporus sanguineus* in *Pichia pastoris* SMD1168H under the control of an alcohol oxidase promoter. The native signal peptide of laccase efficiently directed the secretion of active recombinant laccase. The behaviour of the purified recombinant enzyme was similar to that of native laccase produced by *Pycnoporus sanguineus*, and it efficiently decolorized crystal violet. Chengalroyen and Dabbs (2013) showed that *Amycolatopsis* genes encoding 3-deoxy-7-phospho-heptulosonate synthase, N^5,N^{10}-methylenetetrahydromethanopterin reductase, polycystic kidney domain I, and glucose/sorbosone dehydrogenase were involved in the degradation of triphenylmethane dyes. Synergistic action of enzymes encoded by these genes in *Streptomyces lividans* led to the complete decolorization of crystal violet. The activity of these genes was also tested in *Mycobacterium* sp. and *Rhodococcus* sp. The range of dye classes decolorized by enzymes encoded by these genes in both the species indicated that these genes adopted novel functions in their hosts. A previous study developed a phytoremediation system involving *Arabidopsis* plants by overexpressing TMR of *Citrobacter* sp. The transgenic *Arabidopsis* plants expressing TMR showed significantly enhanced tolerance towards and ability to decolorize crystal violet and malachite green (Fu *et al.*, 2013).

8.9 Conclusion

To date, several strategies have been developed for treating dye-contaminated wastewaters. However, cost-effective and ecologically suitable bioremediation seems to be the most promising process for removing dyes. Studies indicate that several synthetic dyes can be efficiently decolorized by bacteria and fungi, mainly by the action of extra- and/or intracellular enzymes such as laccases, peroxidases, and reductases. Many microbial strains, especially those isolated from dye-contaminated environments, metabolize dyes to non-toxic intermediates or completely mineralize dyes by converting them to CO_2, H_2O, and/or other inorganic end products. Biodegradation of synthetic dyes is strongly affected by various process parameters such as nitrogen and carbon source, pH of medium, temperature, dye concentration, and agitation. Therefore, process parameters should be appropriately selected while assessing the biodegradation ability of microorganisms. In summary, microbial decolorization is a promising method for removing dyes from the environment. However, efforts should also be taken to scale up the process and to apply microbial decolorization techniques on real industrial effluents.

Acknowledgements

The authors appreciate the support of the National Science Centre of Poland (project No. UMO 2013/11/D/NZ9/02776) in this work.

References

Abedin, R.M.A. (2008). Decolorization and biodegradation of crystal violet and malachite green by *Fusarium solani* (martius) saccardo. A comparative study on biosorption of dyes by the dead fungal biomass. American-Eurasian J. Bot. *1*, 17–31.

Aksu, Z., and Dönmez, G. (2000). The use of molasses in copper(II) containing wastewaters: effects on growth and copper(II) bioaccumulation properties of *Kluyveromyces marxianus*. Process Biochem. *36*, 451–458.

Aksu, Z. (2003). Reactive dye bioaccumulation by *Saccharomyces cerevisiae*. Process Biochem. *38*, 1437–1444.

Aksu, Z., and Dönmez, G. (2005). Combined effects of molasses sucrose and reactive dye on the growth and dye bioaccumulation properties of *Candida tropicalis*. Process Biochem. *40*, 2443–2454.

Alam, M.Z., Mansor, M.F., and Jalal, K.C.A. (2009). Optimization of lignin peroxidase production and stability by *Phanerochaete chrysosporium* using sewage-treatment-plant sludge as substrate in a stirred-tank bioreactor. J. Ind. Microbiol. Biotechnol. *36*, 757–764.

Ali, H. (2010). Biodegradation of synthetic dyes – a review. Water Air Soil Pollut. *213*, 251–273.

Ambrósio, S.T., Vilar Júnior, J.C., da Silva, C.A.A., Okada, K., Nascimento, A.E., Longo, R.L., and Campos-Takaki, G.M. (2012). A biosorption isotherm model for the removal of reactive azo dyes by inactivated mycelia of *Cunninghamella elegant* UCP542. Molecules *17*, 452–462.

Anastasi, A., Parato, B., Spina, F., Tigini, V., Prigione, V., and Varese, G.C. (2011). Decolourisation and detoxification in the fungal treatment of textile wastewaters from dyeing processes. Nat. Biotechnol. *29*, 38–45.

Anjaneya, O., Souche, S.Y., Santoshkumar, M., and Karegoudar, T.B. (2011). Decolorization of sulfonated azo dye Metanil Yellow by newly isolated bacterial strains: *Bacillus* sp. strain AK1 and *Lysinibacillus* sp. strain AK2. J. Hazard. Mater. *190*, 351–358.

Anliker, R. (1979). Ecotoxicology of dyestuffs – a joint effort by industry. Ecotoxicol. Environ. Saf. *3*, 59–74.

Arora, D.S., and Sharma, R.K. (2010). Ligninolytic fungal laccases and their biotechnological applications. Appl. Biochem. Biotechnol. *160*, 1760–1788.

Asgher, M., Bhatti, H.N., Ashraf, M., and Legge, R.L. (2008). Recent developments in biodegradation of industrial pollutants by white rot fungi and their enzyme system. Biodegradation *19*, 771–783.

Ayed, L., Chaieb, K., Cheref, A., and Bakhrouf, A. (2009). Biodegradation of triphenylmethane dye Malachite Green by *Sphingomonas paucimobilis*. World J. Microbiol. Biotechnol. *25*, 705–711.

Ayed, L., Chaieb, K., Cheref, A., and Bakhrouf, A. (2011). Biodegradation and decolorization of triphenylmethane dyes by *Staphylococcus epidermidis*. Desalination *260*, 137–146.

Bakshi, D.K., Saha, S., Sindhu, I., and Sharma, P. (2006). Use of *Phanerochaete chrysosporium* biomass for the removal of textile dyes from a synthetic effluent World J. Microbiol. Biotechnol. *22*, 835–839.

Baldrian, P. (2006). Fungal laccases occurrence and properties. FEMS Microbiol Rev. *30*, 215–242.

Banat, I.M., Nigam, P., Singh, D., and Marchant, R. (1996). Microbial decolorization of textile – dye-containing effluents: A review. Bioresour. Technol. *58*, 217–227.

Bechtold, T., Turcanu, A., Ganglberger, E., and Geissler, S. (2003). Natural dyes in modern textile dyehouses – how to combine experiences of two centuries to meet the demands of the future? J. Clean Prod. *11*, 499–509.

Bhatnagar, A., and Sillanpaa, M. (2010). Utilization of agro-industrial and municipal waste materials as potential adsorbents for water treatment – A review. Chem. Eng. J. *157*, 277–296.

Bhosale, S., Saratale, G., and Govindwar, S. (2006). Biotransformation enzymes in *Cunninghamella blakesleeana* (NCIM-687). J. Basic Microbiol. *46*, 444–448.

Boer, C.G., Obici, L., de Souza, C.G., and Peralta, R.M. (2004). Decolorization of synthetic dyes by solid state cultures of *Lentinula* (Lentinus) edodes producing manganese peroxidase as the main ligninolytic enzyme. Bioresour. Technol. *94*, 107–112.

Bourbonnais, R., and Paice, M.G. (1990). Oxidation of non-phenolic substrates. An expanded role for laccase in lignin biodegradation. FEBS Lett. *267*, 99–102.

Bourbonnais, R., Paice, M.G., Reid, I.D., Lanthier, P., and Yaguchi, M. (1995). Lignin oxidation by laccase isozymes from *Trametes versicolor* and role of the mediator 2,2′-azinobis(3-ethylbenzthiazoline-6-sulfonate) in kraft lignin depolymerization. Appl. Environ. Microbiol. *61*, 876–1880.

Brar, S., Verma, M., Surampalli, R., Misra, K., Tyagi, R., Meunier, N., and Blais, J. (2006). Bioremediation of hazardous wastes – a review. Practice Periodical of Hazardous, Toxic, and Radioactive Waste Management *10*, 59–72.

Bumpus, J.A., and Brock, B.J. (1988). Biodegradation of Crystal Violet by the white rot fungus *Phanerochaete chrysosporium*. Appl. Environ. Microbiol. *54*, 1143–1150.

Cao, D.M., Xiao, X., Wu, Y.M., Ma, X.B., Wang, M.N., Wu, Y.Y., and Du, D.L. (2013). Role of electricity production in the anaerobic decolorization of dye mixture by exoelectrogenic bacterium *Shewanella oneidensis* MR-1. Bioresour. Technol. *136*, 176–181.

Cetin, D., and Donmez, G. (2006). Decolorization of reactive dyes by mixed cultures isolated from textile effluent under anaerobic conditions. Enzyme Microb. Technol. *38*, 926–930.

Cha, Ch.-J., Doerge, D.R., and Cerniglia, C.E. (2001). Biotransformation of Malachite Green by the fungus *Cunninghamella elegans*. Appl. Environ. Microbiol. *67*, 4358–4360.

Chacko, J.T., and Subramaniam, K. (2011). Enzymatic degradation of azo dyes: a review. Int. J. Environ. Sci. Technol. (Tehran) *1*, 1250–1260.

Chang, J.S., and Kuo, T.S. (2000). Kinetics of bacterial decolorization of azo dye with *Escherichia coli* NO3. Bioresour. Technol. 75, 107–111.

Chang, J.S., Kuo, T.S., Chao, Y.P., Ho, J.Y., and Lin, P.J. (2000). Azo dye decolorization with a mutant *Escherichia coli* Strain, Biotechnol Lett 22, 807–812.

Chang, J.S., Chen, B.Y., Lin, and Y.S. (2004). Stimulation of bacterial decolorization of an azo dye by extracellular metabolites from *Escherichia coli* strain NO3. Bioresour. Technol. *91*, 243–248.

Charumathi, D., and Das, N. (2010). Biotechnological approach to assess the performance of dried biomass of *Candida tropicalis* for removal of Basic Violet 3 from aqueous solutions. Int. J. Sci. *1*, 47–52.

Chen, B.Y., Chen, S.Y., and Chang, J.S. (2005). Immobilized cell fixed-bed bioreactor for wastewater decolorization. Process Biochem. *40*, 3434–3440.

Chen, B.Y., and Chang, J.S. (2007). Assessment upon species evolution of mixed consortia for azo dye decolorization. J Chinese Inst Chem Eng *38*, 259–266.

Chen, Ch., Chang, C.F., and Liu, S.M. (2010). Partial degradation mechanisms of Malachite Green and Methyl Violet B by *Shewanella decolorationis* NTOU1 under anaerobic conditions. J. Hazard. Mater. *177*, 281–289.

Chen, Ch.-H., Chang, Ch.-F., Ho, Ch.-H., Tsai, T.-L., and Liu, S.-M. (2008). Biodegradation of crystal violet by a *Shewanella* sp. NTOU1. Chemosphere 72, 1712–1720.

Chen, H., Hopper, S.L., and Cerniglia, C.E. (2005). Biochemical and molecular characterization of an azoreductase from *Staphylococcus aureus*, a tetrameric NADPH-dependent flavoprotein. Microbiology *151*, 1433–1441.

Chengalroyen, M.D., and Dabbs, E.R. (2013). Identification of genetic elements from A. orientalis contributing to triphenylmethane decolorization. Int. J. Curr. Microbiol. Appl. Sci. 2, 41–56.

Cheriaa, J., and Bakhrouf, A. (2009). Triphenylmethanes, malachite green and crystal violet dyes decolourisation by *Sphingomonas paucimobilis*. Ann. Microbiol. *59*, 57–61.

Chojnacka, K. (2010). Biosorption and bioaccumulation – the prospects for practical applications. Environ. Int. *36*, 299–307.

Christian, V., Shrivastava, R., Shukla, D., Modi, H.A., and Vyas, B.R. (2005). Degradation of xenobiotic compounds by lignin-degrading white-rot fungi: enzymology and mechanism involved. Ind. J. Exp. Biol. *43*, 301–312.

Christie, R.M. (2007). Colour Chemistry (The Royal Society of Chemistry, Cambridge).

Clark, R.J.H., Cooksey, C.J., Daniels, M.A.M., and Withnall, R. (1993). Indigo, woad, and Tyrian Purple: important vat dyes from antiquity to the present. Endeavour *17*, 191–199.

Crini, G. (2006). Non-conventional low-cost adsorbents for dye removal: a review. Bioresour. Technol. *97*, 1061–1085.

Cui, D., Li, G., Zhao, D., Gu, X., Wang, C., and Zhao, M. (2012). Microbial community structures in mixed bacterial consortia for azo dye treatment under aerobic and anaerobic conditions. J. Hazard. Mater. *221–222*, 185–192.

Das, D., Charumathi, D., and Das, N. (2010). Combined effects of sugarcane bagasse extract and synthetic dyes on the growth and bioaccumulation properties of *Pichia fermentans* MTCC 189. J. Hazard. Mater. *183*, 497–505.

Das, D., Charumathi, D., and Das, N. (2011). Bioaccumulation of synthetic dye Basic Violet 3 and heavy metals in single and binary systems by *Candida tropicalisgrown* in sugarcane bagasse extract medium: Use of response surface methodology (RSM) and inhibition kinetics. J. Hazard. Mater. *186*, 1541–1552.

Deivasigamani, Ch., and Das, N. (2011). Biodegradation of Basic violet 3 by *Candida krusei* isolated from textile wastewater. Biodegradation 22, 1169–1180.

Deng, D., Guo, J., Zeng, G., and Sun, G. (2008). Decolorization of anthraquinone, triphenylmethane and azo dyes by a new isolated *Bacillus cereus* strain DC11. Int Biodeterior Biodegradation *62*, 263–269.

Desai, S.S., and Nityanand, C. (2011). Microbial laccases and their applications: a review. Asian J. Biotechnol. *3*, 98–124.

Diaz, E. (2008). Microbial Biodegradation: Genomics and Molecular Biology (Caister Academic Press: Norwich).

Donmez, G. (2002). Bioaccumulation of the reactive textile dyes by *Candida tropicalis* growing in molasses medium. Enzyme Microb. Technol. *30*, 363–366.

Dos Santos, A.B., Cervantes, F.J., and van Lier, J.B. (2007). Review paper on current technologies for decolourisation of textile wastewaters: perspectives for anaerobic biotechnology. Bioresour. Technol. *98*, 2369–2385.

Du, L.-N., Wang, S., Li, G., Wang, B., Jia, X.-M., Zhao, Y.-H., and Chen, Y.-L. (2011). Biodegradation of malachite green by *Pseudomonas* sp. strain DY1 under aerobic condition: characteristics, degradation products, enzyme analysis and phytotoxicity. Ecotoxicology *20*, 438–446.

Dubey, A., Mishra, N., and Deb, A. (2010). Isolation of dye degrading microorganism. Electronic J Environ Agri Food Chem 9, 1534–1539.

Duran, N., and Esposito, E. (2000). Potential applications of oxidative enzymes and phenoloxidase-like compounds in wastewater and soil treatment: a review. Appl Catal B *28*, 83–99.

Ejder-Korucu, M., Gürses, A., Doğar, Ç., Sharma, S.K., and Açıkyıldız, M. (2015). Removal of organic dyes from industrial effluents: An overview of physical and biotechnological applications. In *Green Chemistry for Dyes Removal from Wastewater: Research Trends and Applications*, S.K. Sharma, ed. (John Wiley and Sons, Hoboken, NJ). pp. 1–22.

Elisangela, F., Andrea, Z., Fabio, D.G., de Menezes Christiano, R., Regina, D.L., and Artur, C.P. (2009). Biodegradation of textile azo dyes by facultative *Staphylococcus arletta* strain VN-11 using a sequential microaerophilic/aerobic process. Int Biodeterior Biodegradation *63*, 280–288.

Erden, E., Kaymaz, Y., and Pazarlioglu, N.K. (2011). Biosorption kinetics of a direct azo dye Sirius Blue K-CFN by *Trametes versicolor*. Electron J. Biotechnol. *14*, 1–10.

Ertugrul, S., San, N.O., and Donmez, G. (2009). Treatment of dye (Remazol Blue) and heavy metals using yeast cells with the purpose of managing polluted textile wastewaters. Ecol Eng 35, 128–134.

Forgacs, E., Cserha´ti, T., and Oros, G. (2004). Removal of synthetic dyes from wastewaters: a review. Environ Int *30*, 953–971.

Franciscon, E., Zille, A., Dias, G.F., Ragagnin de Menezes C., Durrant, L.R., and Cavaco, P.A. (2009). Biodegradation of textile azo dyes by a facultative *Staphylococcus arlettae* strain VN-11 using a sequential microaerophilic/aerobic process. Int Biodeterior Biodegradation *63*, 280–288.

Fu, X.Y., Zhao, W., Xiong, A.S., Tian, Y.S., Zhu, B., Peng, R.H., and Yao, Q.H. (2013). Phytoremediation of triphenylmethane dyes by overexpressing a *Citrobacter* sp. triphenylmethane reductase in transgenic *Arabidopsis*. Appl. Microbiol. Biotechnol. *97*, 1799–1806.

Fu, Y., and Viraraghavan, T. (2000). Removal dye from an aqueous solution by fungus *Aspergillus niger*. Water Qual Res J Can *35*, 95–111.

Fu, Y., and Viraraghavan, T. (2001). Fungal decolorization of dye wastewaters. A review. Bioresour. Technol. *79*, 251–262.

Gadd, G.M. (2001). Fungi in Bioremediation (The British Mycological Society, Cambridge, UK).

Gadd, G.M. (2009). Biosorption: critical review of scientific rationale, environmental importance and significance for pollution treatment. J. Chem. Technol. Biotechnol. *84*, 13–28.

Ghodake, G., Jadhav, S., Dawkar, V., and Govindwar, S. (2009). Biodegradation of Diazo Dye Direct Brown MR by *Acinetobacter calcoaceticus* NCIM 2890. Int. Biodeterior. Biodegrad. *63*, 433–439.

Giwa, A., and Ifu, B.J. (2012). Microbial decolorization of an anthraquinone dye C.I. reactive blue 19 using *Bacillus cereus*. Am. Chem. Sci. J. *2*, 60–68.

Gomaa, O.M. (2012). Ethanol induced response in *Phanerochaete chrysosporium* and its role in the decolorization of triarylmethane dye. Ann. Microbiol. *62*, 1403–1409.

Gomaa, O.M., Linz, J.E., and Reddy, C.A. (2008). Decolorization of Victoria blue by the white rot fungus, *Phanerochaete chrysosporium*. World J Microbiol Biotechnol *24*, 2349–2356.

Gonçalves, I.M.C., Gomes, A., Brás, R., Ferra, M.I.A., Amorim, M.T.P., and Porter, R.S. (2000). Biological treatment of effluent containing textile dyes. Color Technol. *116*, 393–397.

Gonen, F., and Aksu, Z. (2009). Single and binary dye and heavy metal bioaccumulation properties of *Candida tropicalis*: Use of response surface methodology (RSM) for the estimation of removal yields. J. Hazard. Mater. *172*, 1512–1519.

Gottlieb, A., Shaw, C., Smith, A., Wheatley, A., and Forsythe, S. (2003). The toxicity of textile reactive azo dyes after hydrolysis and decolourisation. J. Biotechnol. *101*, 49–56.

Griffiths, J. (1990). Introduction: The Evolution of Present-Day Dye Technology. In *The Chemistry and Application of Dyes*. Topics in Applied Chemistry, J. Griffiths, ed. (US, Springer), pp. 1–16.

Gupta, V.K., and Suhas, I.A. (2009). Application of low-cost adsorbent for dye removal – A review. J Environ Manage *90*, 2313–2342.

Haritash, A.K., and Kaushik, C.P. (2009). Biodegradation aspects of polycyclic aromatic hydrocarbons (PAHs): a review. J. Hazard. Mater. *169*, 1–15.

Hsueh, C.C., Chen, B.Y., and Yen, C.Y. (2009). Understanding effects of chemical structure on azo dye decolorization characteristics by *Aeromonas hydrophila* J. Hazard. Mater. *167*, 995–1001.

Hu, M., Zhang, W., Lu, X., and Gao, P. (2006). Purification and characteristics of a low-molecular-weight peptide possessing oxidative capacity for phenol from *Phanerochaete chrysosporium*. Sci. China C. Life Sci. *49*, 243–250.

Hunger, K. (2007). Industrial Dye: Chemistry, Properties, Applications (Weinheim: Wiley-VCH).

Husain, Q. (2006). Potential applications of the oxidoreductive enzymes in the decolorization and detoxification of textile and other synthetic dyes from polluted water: a review. Crit. Rev. Biotechnol. *26*, 201–221.

Husseiny, S.M. (2008). Biodegradation of the reactive and direct dyes using egyptian isolates. J Appl Sci Res *4*, 599–606.

Iqbal, M., and Saeed, A. (2007). Biosorption of reactive dye by loofa sponge-immobilized fungal biomass of *Phanerochaete chrysosporium* Process Biochem. *42*, 1160–1164.

Ito, K., Nakanishi, M., Lee, W.C., Sasaki, H., Zenno, S., Saigo, K., Kitade, Y., and Tanokura, M. (2006). Three-dimensional structure of AzoR from *Escherichia coli*. An oxidereductase conserved in microorganisms. J. Biol. Chem. *281*, 20567–20576.

Jadhav, J.P., and Govindwar, S.P. (2006). Biotransformation of malachite green by *Saccharomyces cerevisiae* MTCC 463. Yeast *23*, 315–323.

Jadhav, S.U., Kalme, S.D., and Govindwar, S.P. (2008). Biodegradation of Metyl red by *Galactomyces geotrichum* MTCC 1360. Int. Biodeterior. Biodegrad. *62*, 135–142.

Jadhav, S.B., Phugare, S.S., Patil, P.S., and Jadhav, J.P. (2011). Biochemical degradation pathway of textile dye Remazol red and subsequent toxicological evaluation by cytotoxicity, genotoxicity and oxidative stress studiem. Int. Biodeterior. Biodegrad.*6*, 733–743.

Jadhav, S.B., Yedurkar, S.M., Phugare, S.S., and Jadhav, J.P. (2012). Biodegradation studies on acid violet 19, a triphenylmethane dye, by *Pseudomonas aeruginosa* BCH. Clean (Weinh) *40*, 551–558.

Jadhav, S.B., Patil, N.S., Watharkar, A.D., Apine, O.A., and Jadhav, J.P. (2013). Batch and continuous biodegradation of Amaranth in plain distilled water by *P. aeruginosa* BCH and toxicological scrutiny using oxidative stress studies. Environ. Sci. Pollut. Res. Int. *20*, 2854–2866.

Jang, M.S., Lee, Y.M., Kim, C.H., Lee, J.H., Kang, D.W., Kim, S.-J., and Lee, Y.-Ch. (2005). Triphenylmethane reductase from *Citrobacter* sp. strain KCTC 18061P: purification, characterization, gene cloning and overexpression of a functional protein in *Escherichia coli*. Appl. Environ. Microbiol. *71*, 7955–7960.

Jasińska, A., Różalska, S., Bernat, P., Paraszkiewicz, K., and Długoński, J. (2012). Malachite green decolorization by non-basidiomycete filamentous fungi of *Penicillium pinophilum* and *Myrothecium roridum*. Int. Biodeterior. Biodegrad. 73, 33–40.

Jin, X.C., Liu, G.Q., Xu, Z.H., and Tao, W.Y. (2007). Decolorization of a dye industry effluent by *Aspergillus fumigatus* XC6. Appl. Microbiol. Biotechnol. *74*, 239–224.

Jones, J.J., and Falkinham, J.Q. (2003). Decolorization of malachite green and crystal violet by waterborne pathogenic mycobacteria. Antimicrob Agents Chemother (Bethesda) 47, 2323–2326.

Joshi, T., Iyengar, L., Singh, K., and Garg, S. (2008). Isolation, identification and application of novel bacterial consortium TJ-1 for the decolorization of structurally different azo dyes. Bioresour. Technol. 99, 7115–7121.

Kandelbauer, A., and Guebitz, G.M. (2004). Bioremediation for the decolorization of textile dyes – a review. In Environmental Chemistry, Lichtfouse, E., Schwarzbauer, J., Robert, D., eds. (Springer, Berlin, Heidelberg New York), pp. 269–288.

Karimi, A., Aghbolaghy, M., Khataee, A., and Bargh, S.S. (2012). Use of enzymatic bio-Fenton as a new approach in decolorization of malachite green. Scientific World J, Article ID 691569, 1–5.

Kaushik, P., and Malik, A. (2009). Fungal dye decolorization: recent advances and future potential. Environ Int 35, 127–141.

Kaushik, P., and Malik, A. (2013). Comparative performance evaluation of *Aspergillus lentulus* for dye removal through bioaccumulation and biosorption. Environ. Sci. Pollut. Res. Int. *20*, 2882–2892.

Kelewou, H., Merzouki, M., and Hassani, A.L. (2014). Biosorption of textile dyes Basic Yellow 2 (BY2) and Basic Green 4 (BG4) by the live yeast *Saccharomyces cerevisiae*. J Mater Environ Sci 5, 633–640.

Khalid, A., Kausar, F., Arshad, M., Mahmood, T., and Ahmed, I. (2012). Accelerated decolorization of reactive azo dyes under saline conditions by bacteria isolated from Arabian seawater sedyment. Appl. Microbiol. Biotechnol. *96*, 1599–1606.

Khan, R., Bhawana, P., and Fulekar, M.H. (2013). Microbial decolorization and degradation of synthetic dyes: a review. Rev. Environ. Sci. Biotechnol. *12*, 75–97.

Khehra, M.S., Saini, H.S., Sharma, D.K., Chadha, B.S., and Chimni, S.S. (2005). Comparative studies on potential of consortium and constituent pure bacterial isolates to decolorize azo dyes. Water Res. *39*, 5135–5141.

Kiernan, J.A. (2001). Classification and naming of dyes, stains and fluorochromes. Biotech. Histochem. *76*, 261–278.

Kim, M.H., Kim, Y., Park, H.J., Lee, J.S., Kwak, S.N., Jung, W.H., Lee, S.G., Kim, D., Lee, Y.C., and Oh, T.K. (2008). Structural insight into bioremediation of triphenylmethane dyes by *Citrobacter* sp. triphenylmethane reductase. J. Biol. Chem. *283*, 31981–31990.

Kolekar, Y.M., and Kodam, K.M. (2012). Decolorization of textile dyes by *Alishewanella* sp. KMK6. Appl. Microbiol. Biotechnol. 95, 521–529.

Kolekar, Y.M., Pawar, S.P., Adav, S.S., Zheng, L.Q., Li, W.J., Shouche, Y.S., Dastager, S.G., and Kodam, K.M. (2013). Alishewanella solinquanti sp. nov., isolated from soli contaminated with textile dyes. Curr. Microbiol. *67*, 454–459.

Kujan, P., Votruba, J., and Kamenik, V. (1995). Substrate-dependent bioaccumulation of cadmium by growing yeast *Candida utilis*. Folia Microbiol. *40, 288–292.*

Kumar, C.G., Mongolla, P., Joseph, J., and Sarma, V.U.M. (2012). Decolorization and biodegradation of triphenylmethane dye, brilliant green, by *Aspergillus* sp. isolated from Ladakh, India. Process Biochem. *47*, 1388–1394.

Kunz, A., Mansilla, H., and Durán, N. (2002). A degradation and toxicity study of three textile reactive dyes by ozone. Environ Technol 23, 911–918.

Kurade, M.B., Waghmode, T.R., Tamboli, D.P., and Govindwar, S.P. (2013). Differential catalytic action of *Brevibacillus laterosporus* on two dissimilar azo dyes Remazol red and Rubine GFL. J. Basic Microbiol. 53, 136–146.

Lade, H.S., Waghmode, T.R., Kadam, A.A., and Govindwar, S.P. (2012). Enhanced biodegradation and detoxification of disperse azo dye Rubine GFL and textile industry effluent by defined fungal-bacterial consortium. Int. Biodeterior. Biodegrad.72, 94–107.

Levin, L., Papinutti, L., and Forchiassin, F. (2004). Evaluation of Argentinean white rot fungi for their ability to produce lignin-modifying enzymes and decolorize industrial dyes. Bioresour. Technol. *94*, 169–176.

Li, L.T., Hong, Q., Yan, X., Fang, G.H., Ali, S.W., and Li, S.P. (2009). Isolation of a malachite green-degrading *Pseudomonas* sp. MDB-1 strain and cloning of the *tmr2* gene. Biodegradation *20*, 769–776.

Liao, C.S., Hung, C.H., and Chao, S.L. (2013). Decolorization of azo dye reactive black B by *Bacillus cereus* strain HJ-1. Chemosphere *90*, 2109–2114.

Lim, S.L., Chu, W.L., and Phang, S.M. (2010). Use of *Chlorella vulgaris* for bioremediation of textile wastewater. Bioresour. Technol. *101*, 7314–7322.

Lim, Ch.K., Bay, H.H., Aris, A., Majid, Z.A., and Ibrahim, Z. (2013). Biosorption and biodegradation of Acid Orange 7 by *Enterococcus faecalis* strain ZL: optimization by response surface methodological approach. Environ. Sci. Pollut. Res. Int. *20*, 5056–5066.

Liu, G., Zhou, J., Lv, H., Xiang, X., Wang, J., Zhou, M., and Qv, Y. (2007). Azoreductase from *Rhodobacter sphaeroides* AS1.1737 is a flavodoxin that also functions as nitroreductase and flavin mononucleotide reductase. Appl. Microbiol. Biotechnol. 76, 1271–1279.

Liu, G., Zhou, J., Fu, Q.S., and Wang, J. (2009). The *Escherichia coli* azoreductase AzoR is involved in resistance to thiol-specific stress caused by electrophilic quinones. J. Bacteriol. *191*, 6394–6400.

Liu, G., Zhou, J., Chen, C., Wang, J., Jin, R., and Lv, H. (2011). Decolorization of azo dyes by *Shewanella oneidensis* MR-1 in the presence of humic acids. Appl. Microbiol. Biotechnol. *91*, 417–424.

Liu, G., Zhou, J., Chen, C., Wang, J., Jin, R., and Lv, H. (2013a). Decolorization of azo dyes by *Geobacter metallireducens*. Appl. Microbiol. Biotechnol. 97, 7935–7942.

Liu, G., Zhou, J., Chen, C., Wang, J., Jin, R., and Lv, H. (2013b). Decolorization of azo dyes by marine *Shewanella* strains under saline conditions. Appl. Microbiol. Biotechnol. *97*, 4187–4197.

Lorimer, J.P., Mason, T.J., Plattes, M., Phull, S.S., and Walton, D.J. (2001). Degradation of dye effluent. Pure Appl Chem 73, 1957–1968.

Lu, L., Zhao, M., Liang, S.-C., Zhao, L.-Y., Li, D.-B., and Zhang, B.-B. (2009). Production and synthetic dyes decolorization capacity of a recombinant laccase from *Pichia pastoris*. J. Appl. Microbiol. *107*, 1149–1156.

McMullan, G., Meehan, C., Conneely, A., Kirby, N., Robinson, T., Nigam, P., Banat, I.M., Marchant, R., and Smyth, W.F. (2001). Microbial decolourisation and degradation of textile dyes. Appl. Microbiol. Biotechnol. *56*, 81–87.

Moldes, D., Fernández-Fernández, M., and Ángeles Sanromán, M. (2012). Role of laccase and low molecular weight metabolites from *Trametes versicolor* in dye decolorization. Scientific Word J Article ID 398725, 1–9.

Moosvi, S., Keharia, H., and Madamwar, D. (2005). Decolorization of textile dye Reactive Violet 5 by a newly isolated bacterial consortium RVM 11.1. World J. Microbiol. Biotechnol. *21*, 667–672.

Morozova, O.V., Shumakovich, G.P., Gorbacheva, M.A., Shleev, S.V., and Yaropolov, A.I. (2007). 'Blue' Laccases. Biochemistry 72, 1136–1150.

Moturi, B., and Singara Charya, M.A. (2009). Decolourisation of crystal violet and malachite green by fungi. Science World J. *4*, 28–33.

Mou, D.G., Lim, K.K., and Shen, H.P. (1991). Microbial agents for decolorization of dye wastewater. Biotechnology Advance *9*, 613–622.

Nelson, K.E., Fouts, D.E., Mongodin, E.F., Ravel, J., DeBoy, R.T., Kolonay, J.F., Rasko, D.A., Angiuoli, S.V., Gill, S.R., Paulsen, I.T., *et al.* (2004). Whole genome comparisons of serotype 4b and 1/2a strains of the food-borne pathogen *Listeria monocytogenes* reveal new insights into the core genome components of this species. Nucleic Acids Res. *32*, 2386–2395.

Nerud, N., Baldrian, P., Eichlerova, I., Merhautova, V., Homolka, L., Gabriel, J., Benes, M.J., and Hruby, M. (2006). Fungal and chemical decolorization of synthetic dyes. In Conference materials Protection 2006.

Ng, I.S., Chen, T., Lin, R., Zhang, X., Ni, C., and Sun, D. (2014). Decolorization of textile azo dye and Congo red by an isolated strain of the dissimilatory manganese-reducing bacterium *Shewanella xiamenensis* BC01. Appl. Microbiol. Biotechnol. *98*, 2297–2308.

O'Neill, C., Lopez, A., Esteves, S., Hawkes, F.R., Hawkes, D.L., and Wilcom, S. (2000). Azo-dye degradation in an anaerobic-aerobic treatment system operating on simulated textile effluent. Appl. Microbiol. Biotechnol. *53*, 249–254.

Ogugbue, Ch.J., and Sawidis, T. (2011). Bioremediation and detoxification of synthetic wastewater containing triarylmethane dyes by *Aeromonas hydrophila* isolated from industrial effluent. Biotechnol. Res. Int. 2011, Article ID 967925, 1–12.

Olukanni, O.D., Adenopo, A., Awotula, A.O., and Osuntoki, A.A. (2013). Biodegradation of malachite green by extracellular laccase producing *Bacillus thuringiensis* RUN1. Int. J. Basic Appl. Sci. *9*, 543–549.

Oturkar, C.C., Patole, M.S., Gawai, K., and Madamwar, D. (2013). Enzyme based cleavage strategy of *Bacillus lentus* BI377 in response to metabolism of azoic recalcitrant. Bioresour. Technol. *130*, 360–365.

Pandey, A., Singh, P., and Iyengar, L. (2007). Bacterial Decolorization and Degradation of Azo Dyes, Int Biodeterior Biodegradation 59, 73–84.

Panswad, T., and Luangdilok, W. (2000). Decolorization of reactive dyes with different molecular structures under different environmental conditions. Water Res. *34*, 4177–4184.

Papinutti, L., Mouso, N., and Forchiassin, F. (2006). Removal and degradation of the fungicide dye malachite green from aqueous solution using the system wheat bran–*Fomes sclerodermeus*. Enzyme Microb. Technol. *39*, 848–853.

Parshetti, G., Kalme, S., Saratale, G., and Govindwar, S. (2006). Biodegradation of Malachite Green by *Kocuria rosea* MTCC 1532. Acta Chimica Slovenica *53*, 492–498.

Parshetti, G., Saratale, G., Telke, A., and Govindwar, S. (2009). Biodegradation of hazardous triphenylmethane dye methyl violet by *Rhizobium radiobacter* (MTCC 8161). J. Basic Microbiol. *49*, 36–42.

Parshetti, G.K., Telke, A.A., Kalyani, D.C., and Govindwar, S.P. (2010). Decolorization and detoxification of sulfonated azo dye methyl orange by *Kocuria rosea* MTCC 1532. J. Hazard. Mater. *176*, 503–509.

Parshetti, G.K., Parshetti, S.G., Telke, A.A., Kalyani, D.C., Dong, R.A., and Govindwar, S.P. (2011). Biodegradation of Crystal Violet by *Agrobacterium radiobacter*. J Environ Sci 23, 1384–1393.

Pavan, F.A., Lima, E.C., Dias, S.L.P., and Mazzocato, A.C. (2008). Methylene blue biosorption from aqueous solutions by yellow passion fruit waste. J. Hazard. Mater. *150*, 703–712.

Pointing, S.B., and Vrijmoed, L.L.P. (2000). Decolorization of azo and triphenylmethane dyes by *Pycnoporus sanguineus* producing laccase as the sole phenoloxidase. World J Microbiol Biotechnol *16*, 317–318.

Pointing, S.B. (2001). Feasibility of bioremediation by white-rot fungi. Appl. Microbiol. Biotechnol. *57*, 220–233.

Porri, A., Baroncelli, R., Guglielminetti, L., Sarrocco, S., Guazzelli, L., Forti, M., Catelani, Valentini, G., Bazzichi, A., Franceschi, M., and Vannacci, G. (2011). *Fusarium oxysporum* degradation and detoxification of a new textile-glycoconjugate azo dye (GAD). Fungal Biol. *115*, 30–37.

Rai, H.S., Bhattacharyya, M.S., Singh, J., Bansal, T.K., Vats, P., and Banerjee, U.C. (2005). Removal of dyes from the effluent of textile and dyestuff manufacturing industry: A review of emerging techniques with reference to biological treatment. Crit Rev Environ Sci Technol 35, 219–238.

Rajeshkannan, R., Rajasimman, M., and Rajamohan, N. (2010). Removal of Malachite Green from aqueous solution using *Hydrilla verticillata* – optimization, equilibrium and kinetic studies. Int. J. Civil Environ. Eng. *2*, 222–229.

Ramalho, P.A. (2005). Degradation of dyes with microorganisms – studies with ascomycete yeasts. PhD thesis, Biology Department, University of Minho.

Rauf, M.A., and Ashraf, S.S. (2012). Survey of recent trends in biochemically assisted degradation of dyes. Chem. Eng. J. *209*, 520–530.

Ren, S., Guo, J., Zeng, G., and Sun, G. (2006). Decolorization of tripehnylmethane, azo, and anthraquinone dyes by a newly isolated *Aeromonas hydrophila* strain. Appl. Microbiol. Biotechnol. 72, 1316–1321.

Riva, S. (2006). Laccases: blue enzymes for green chemistry. Trends Biotechnol. *24*, 219–226.

Rizzo, L. (2011). Bioassays as a tool for evaluating advanced oxidation processes in water and wastewater treatment. Water Res. *48*, 4311–4340.

Robinson, T., McMullan, G., Marchant, R., and Nigam, P. (2001). Remediation of dyes in textile effluent: a critical review on current treatment technologies with a proposed alternative. Bioresour. Technol. *77*, 247–255.

Rodriguez Couto, S.R. (2009). Dye removal by immobilised fungi. Biotechnol Adv *27*, 227–235.

Ryan, A., Laurieri, N., Westwood, I., Wang, C.J., Lowe, E., and Sim, E. (2010). A novel mechanism for azoreduction. J Mol Biol *400*, 24–37.

Sadettin, S., and Dönmez, G. (2006). Bioaccumulation of reactive dyes by thermophilic cyanobacteria. Process Biochem. *41*, 836–841.

Sadettin, S., and Dönmez, G. (2007). Simultaneous bioaccumulation of reactive dye and chromium(VI) by using thermophil *Phormidium* sp. Enzyme Microb. Technol. *41*, 175–180.

Šafařík, I., Ptáčková, L., and Šafaříková, M. (2002). Adsorption of dyes on magnetically labeled baker's yeast cells. Eur. Cell Mater. *3*, 52–55.

Šafařík, I., Rego, L.F.T., Borovská, M., Mosiniewicz-Szablewska, E., Weyda, F., and Šafaříková, M. (2007). New magnetically responsive yeast-based biosorbent for the efficient removal of water-soluble dyes. Enzyme Microb. Technol. *40*, 1551–1556.

Šafaříková, M., Ptáčková, L., Kibriková, I., and Šafařík, I. (2005). Biosorption of water-soluble dyes on magnetically modified *Saccharomyces cerevisiae* subsp. *uvarum* cells. Chemosphere *59*, 831–835.

Sanghi, R., and Bhattacharya, B. (2002). Review on decolorisation of aqueous dye solutions by low cost adsorbents. Color Technol. *118*, 256–269.

Saratale, G.D., Bhosale, S.K., Kalme, S.D., and Govindwar, S.P. (2007). Biodegradation of Kerosene in *Aspergillus ochraceus* (NCIM 1146). J. Basic Microbiol. *47*, 400–405.

Saratale, R.G., Saratale, G.D., Chang, J.S., and Govindwar, S.P. (2011). Bacterial decolorization and degradation of azo dyes: A review. J. Taiwan Inst. Chem. Eng. *42*, 138–157.

Saravanan, P., Sivakumar, P., Suganya, T., and Nagendra, N. (2012). Bioaccumulation of reactive Red 11 using live yeast *Rhodotorula glutinis*. Ind. J. Environ. Protect. *32*, 249–255.

Sarayu, K., and Sandhya, S. (2012). Current technologies for biological treatment of textile wastewater–a review. Appl Biochem Biotechnol *167*, 645–661.

Sarnthima, R., Khammuang, S., and Svasti, J. (2009). Extracellular ligninolytic enzymes by *Lentinus polychrous* Lev. under solid-state fermentation of potential agro-industrial wastes and their effectiveness in decolorization of synthetic dyes. Biotechnol. Bioprocess. Eng. *14*, 513–522.

Sathish, S., and Amarnath, D.J. (2015). Study on decolorization of dye stuff (azo dye–Congo red) by using bacterial consortia. Int. J. Pharm. Pharm. Sci. *7*, 143–146.

Sathiya, M., P., Periyar, S., Sasikalaveni, A., Murugesan, K., and Kalaichelvan, P.T. (2007). Decolorization of textile dyes and their effluents using white rot fungi. African J. Biotechnol. *6*, 424–429.

Shah, V., and Narud, F. (2002). Lignin degrading system of white-rot fungi and its exploitation for dye decolorization. Can J Microbiol *48*, 857–870.

Sharma, K.K., and Kuhad, R.C. (2008). Laccase: enzyme revisited and function redefined. Ind. J. Microbiol. *48*, 309–316.

Sharma, P., Goel, R., and Caplash, N. (2007). Bacterial laccases. World J. Microbiol Biotechnol. *23*, 823–832.

Sharma, P., Kaur, H., Sharma, M., and Sahore, V. (2011). A review on applicability of naturally available adsorbents for the removal of hazardous dyes from aqueous waste. Environ. Monit. Assess. *183*, 151–195.

Sharma, S., Sharma, S., Upreti, N., and Sharma, K.P. (2009). Monitoring toxicity of an azo dye methyl red and a heavy metal Cu, using plant and animal bioassays. Toxicol. Environ. Chem. *91*, 109–120.

Shedbalkar, U., Dhanve, R., and Jadhav, J. (2008). Biodegradation of triphenylmethane dye cotton blue by *Penicillium ochrochloron* MTCC 517. J. Hazard. Mater. *157*, 472–479.

Shedbalkar, U., and Jadhav, J. (2011). Detoxification of malachite green and textile industrial effluent by *Penicillium ochrochloron*. Biotechnol. Bioprocess Eng. *16*, 196–204.

Shin, K.S., Kim, Y.H., and Lim, J.S. (2005). Purification and characterization of manganese peroxidase of the white-rot fungus *Irpex lacteus*. J. Microbiol. *43*, 503–509.

Singh., H. (2006). Fungal Decolorization and Degradation of Dyes. In Mycoremediation: Fungal Bioremediation, H. Singh, ed. (John Wiley and Sons, Inc., Hoboken, NJ), pp. 420–483.

Solecka, M., and Ledakowicz, S. (2005). Biologiczne procesy oczyszczania barwnych ścieków włókienniczych. Biotechnologia *2*, 103–124.

Solís, M., Solís, A., Pérez, H.I., Manjarrez, N., and Flores, M. (2012). Microbial decolouration of azo dyes: A review. Process Biotechnol 47, 1723–1748.

Srinivasan, A., and Viraraghavan, T. (2010). Decolorization of dye wastewaters by biosorbents: a review. J Environ Manage *91*, 1915–1929.

Sudova, E., Machova, J., Svobodova, Z., and Vesely, T. (2007). Negative effects of malachite green and possibilities of its replacement in the treatment of fish eggs and fish: a review. Vet. Med. *52*, 527–539.

Taskin, M., and Erdal, S. (2010). Reactive dye bioaccumulation by fungus *Aspergillus niger* isolated from the effluent of sugar fabric-contaminated soil. Toxicol. Ind. Health *26*, 239–247.

Telke, A., Kalyani, D., Jadhav, J., and Govindwar, S. (2008). Kinetics and Mechanism of Reactive Red 141 Degradation by a Bacterial Isolate *Rhizobium radiobacter* MTCC 8161. Acta Chim. Slovenica *55*, 320–329.

Thetford, D. (2000). Triphenylmethane and related dyes. In: Kirk-Othmer Encyclopedia of Chemical Technology (John Wiley and Sons, New York).

Thongchai, P., and Worrawit, L. (2000). Decolorization of reactive dyes with different molecular structures under different environmental conditions. Water Res. *34*, 4177–4184.

Van der Zee, F.P., and Villaverde, S. (2005). Combined Anaerobic–Aerobic Treatment of Azo Dyes – A Short Review of Bioreactor Studies. Water Res. *39*, 1425–1440.

Verma, A.K., Dash, R.R., and Bhunia, P. (2012). A review on chemical coagulation/flocculation technologies for removal of colour from textile wastewaters. J. Environ. Manage. *93*, 154–168.

Verma, P., and Madamwar, D. (2003). Decolorization of synthetic dyes by a newly isolated strain of *Serratia maerascens*. World J Microbiol.. Biotechnol. *19*, 615–618.

Vijayaraghavan, K., Han, M.H., Choi, S.B., and Yun, Y.S. (2007). Biosorption of Reactive black 5 by *Corynebacterium glutamicum* biomass immobilized in alginate and polysulfone matrices. Chemosphere *68*, 1838–1845.

Vijayaraghavan, K., and Yun, Y.-S. (2007). Chemical modification and immobilization of *Corynebacterium glutamicum* for biosorption of Reactive black 5 from aqueous solution. Ind. Eng. Chem. Res. *46*, 608–617.

Waghmode, T.R., Kurade, M.B., andd Govindwar, S.P. (2011). Time dependent degradation of mixture of structurally different azo and non azo dyes by using *Galactomyces geotrichum* MTCC 1360. Int. Biodeterior. Biodegrad. *65*, 479–486.

Wang, C.J., Hagemeier, C., Rahman, N., Lowe, E., Noble, M., Coughtrie, M., Sim, E., and Westwood, I. (2007). Molecular Cloning, Characterisation and Ligand-Bound Structure of an Azoreductase from *Pseudomonas aeruginosa*. J. Mol. Biol. *373*, 1213–1228.

Wang, J., Qiao, M., Wei, K., Ding, J., Liu, Z., Zhang, K.Q., and Huang, X. (2011). Decolorizing activity of malachite green and its mechanisms involved in dye biodegradation by *Achromobacter xylosoxidans* MG1. J. Mol. Microbiol. Biotechnol. *20*, 220–227.

Wang, J., Gao, F., Liu, Z., Qiao, M., Niu, X., Zhang, K.Q., and Huang, X. (2012). Pathway and molecular mechanisms for Malachite Green biodegradation in *Exiguobacterium* sp. MG2. PLoS ONE 7, e51808.

Wasenberg, D., Kyriakides, I., and Agathos, S.N. (2003). White-rot fungi and their enzymes for the treatment of industrial dye effluents. Biotechnol. Adv. *22*, 161–187.

Won, S.W., Kim, H.-J., Choi, S.-H., Chung, B.-W., Kim, K.-J., and Yun, Y.-S. (2006). Performance, kinetics and equilibrium in biosorption of anionic dye Reactive Black 5 by the waste biomass of *Corynebacterium glutamicum* as a low-cost biosorbent. Chem Eng J *121*, 37–43.

Yagub, M.T., Sen, T.K., Afroze, S., and Ang, H.M. (2014). Dye and Its Removal from Aqueous Solution by Adsorption: A Review. Adv Colloid Interface Sci *209*, 172–184.

Yan, K., Wang, H., and Zhang, X. (2009). Biodegradation of crystal violet by low molecular mass fraction secreted by fungus. J Biosci Bioeng *108*, 421–424.

Yang, Y., Wang, G., Wang, B., Du, L., Jia, X., and Zhao, Y. (2011). Decolorization of malachite green by a newly isolated *Penicillium* sp. YW 01 and optimization of decolorization parameters. Environ. Eng. Sci. *28*, 555–562.

Yatome, C., Ogawa, T., Koga, D., and Idaka, E. (1981). Biodegradability of azo and triphenyl methane dyes by *Pseudomonas pseudomallei* 13NA. Journal of The Society of Dyers and Colourists *97*, 166–169.

Yatome, C., Ogawa, T., and Matsui, M. (1991). Degradation of crystal violet by *Bacillus subtilis*, Journal of Environmental Science and Health. Part A: Environ. Sci. Eng.Toxicol. *26*, 75–87.

Yatome, C., Yamada, S., Ogawa, T., and Matsui, M. (1993). Degradation of crystal violet by *Nocardia coralline*. Appl. Microbiol. Biotechnol. *38*, 565–569.

Yilmazer, P., and Saracoglu, N. (2009). Bioaccumulation and biosorption of copper(II) and chromium(III) from aqueous solutons by *Pichia stiptis* yeast. J. Chem. Technol. Biotechnol. *84*, 604–610.

Yu, Z., and Wen, X. (2005). Screening and identification of yeasts for decolorizing synthetic dyes in industrial wastewater. Int. Biodeterior. Biodegrad.*56*, 109–114.

Yu, L., Zhang, X., Xie, T., Hu, J., Wang, S., and Li, W. (2014). Intracellular azo decolorization is coupled with aerobic respiration by a *Klebsiella oxytoca* strain. Appl. Microbiol. Biotechnol. *99*, 2431–2439.

Zabłocka-Godlewska, E., Przystaś, W., and Grabińska-Sota, E. (2012). Decolorization of diazo Evans blue by two strains of *Pseudomonas fluorescens* isolated from different wastewater treatment plants. Water Air Soil Pollut. *223*, 5259–5266.

Zabłocka-Godlewska, E., Przystaś, W., and Grabińska-Sota, E. (2015). Dye decolourisation using two *Klebsiella* strains. Water Air Soil Pollut. *226*, 2249–2263.

Zaharia, C., and Suteu, D. (2012). Textile organic dyes – characteristics, polluting effects and separation/elimination procedures from industrial effluents – a critical overview. In Organic Pollutants Ten Years After the Stockholm Convention – Environmental and Analytical Update, T. Puzyn, ed. (InTech Publisher, Rijeka, Croatia), pp. 55–86.

Zhang, Ch., Diao, H., Lu, F., Bie, X., Wang, Y., and Lu, Z. (2012). Degradation of triphenylmethane dyes using a temperature and pH stable spore laccase from a novel strain of *Bacillus vallismortis*. Bioresour. Technol. *126*, 80–86.

Zhuo, R., Ma, L., Fan, F., Gong, Y., Wan, X., Jiang, M., Zhang, X., and Yang, Y. (2011). Decolorization of different dyes by a newly isolated white-rot fungi strain *Ganoderma* sp.En3 and cloning and functional analysis of its laccase gene. J. Hazard. Mater. *192*, 855–873.

Novel Insights into Polycyclic Aromatic Hydrocarbon Biodegradation Pathways Using Systems Biology and Bioinformatics Approaches

9

Ohgew Kweon, Seong-Jae Kim, John B. Sutherland and Carl E. Cerniglia

Contents

Abstract

Biodegradation of polycyclic aromatic hydrocarbons (PAHs) entails a complex and diverse set of biological reactions. Although there has been a massive effort over the years, understanding of the mechanism of PAH biodegradation has been limited when using the traditional approaches of genetics and biochemistry. The application of systems biology approaches, with advanced high-throughput analytical technologies, provides new global insights into not only the direct molecular mechanisms but also the genome-wide cellular ecophysiological responses involved in PAH degradation. This review describes research accomplishments from earlier traditional genetic and biochemical studies as well as the recent achievements of a combination of genomic, proteomic, and bioinformatics approaches to elucidate pathways for the degradation of high-molecular-weight (HMW) PAHs.

9.1 Introduction

As a class of more than 100 chemicals, PAHs are characterized by two or more fused benzene rings arranged in various configurations that do not typically carry other functional groups or branched groups (Fig. 9.1). PAHs are ubiquitous in the natural environment and are continually introduced petrogenically, such as from seeps and spills of crude oil; pyrogenically, such as from forest fires and car exhaust; and biogenically, such as from vegetation (Blumer, 1976; Hites *et al.*, 1977; IARC, 1983). Due to their ecotoxic, mutagenic, and in some cases, carcinogenic effects, PAHs pose a significant risk to humans and the environment, having been listed as priority chemicals with many other hazardous substances (ATSDR, 2011). Differences in the structure and size of individual PAHs result in substantial variability in the physicochemical properties, water solubility, and bioavailability. Their genotoxicity increases with molecular weight (IARC, 1983). PAHs are classified into low-molecular-weight (LMW) PAHs, containing two or three aromatic rings, and HMW PAHs, containing four or more aromatic rings (Fig. 9.1).

Microbial degradation is a major route for the environmental elimination of PAHs in addition to atmospheric photolysis, sorption, chemical oxidation, and volatilization (Atlas, 1988; Leahy and Colwell, 1990; Van Hamme *et al.*, 2003). For this reason, studies of PAH biodegradation have been one of the major research topics in environmental microbiology and considerable attention has been paid to the practical application of biological systems to remediate PAH contaminated environments (Cerniglia, 1992; Van Hamme *et al.*, 2003). In recent years, because of the higher persistency and toxicity of HMW PAHs, more research efforts have been focused on microorganisms capable of degrading these compounds (Kanaly and Harayama, 2011; Fuentes *et al.*, 2014).

Landmarks in the research of HMW PAH biodegradation include a series of studies that began with the isolation of a *Mycobacterium* strain that degraded HMW PAHs, including pyrene, and their ring fission products (Heitkamp and Cerniglia, 1988; Heitkamp *et al.*, 1988a,b). The isolate, which was also shown to degrade other HMW PAHs, including fluoranthene, 1-nitropyrene, benz[*a*]anthracene, benzo[*a*]pyrene, and 7,12-dimethylbenz[*a*]anthracene, was described later as *M. vanbaalenii* strain PYR-1 (Khan *et al.*, 2002; Kim *et al.*, 2005). Since then, this bacterium has been the subject of extensive metabolic, genetic, and biochemical studies and, recently, genomic and proteomic studies. Other HMW PAH-degrading bacterial strains, mostly in the genus *Mycobacterium*, have also been isolated (Grosser *et al.*, 1991; Boldrin *et al.*, 1993; Kleespies *et al.*, 1996; Cheung and Kinkle, 2001). This indicates that the physiology and enzyme systems of this group of bacteria have probably evolved, and are evolving, to survive under various conditions in PAH-contaminated environments (Kweon *et al.*, 2015).

During the last three decades, studies on *M. vanbaalenii* PYR-1 have significantly expanded our understanding of prokaryotic HMW PAH biodegradation. Over 50 scientific papers in refereed journals tell a research story from the initial isolation of the bacterium to the elucidation of individual metabolic steps and then to pathways for each PAH. A PAH–metabolic network (MN) has been proposed for a set of 10 PAHs, and the PAH degradation phenotype in the genus *Mycobacterium* has been analysed from an evolutionary standpoint (Kim *et al.*, 2010; Kweon *et al.*, 2010a; Kweon *et al.*, 2011; Kweon *et al.*, 2015). The HMW

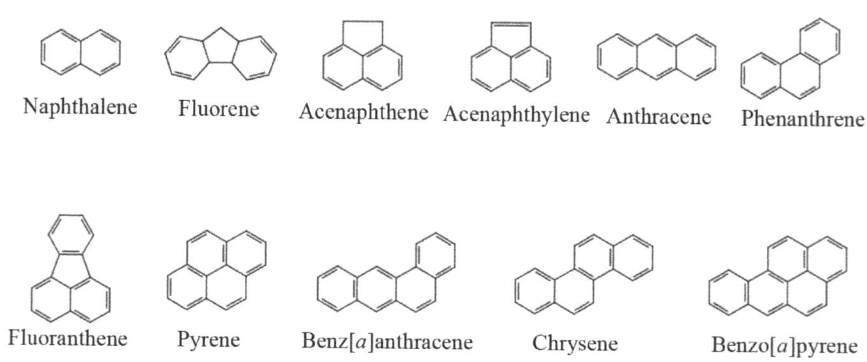

Figure 9.1 Structures of typical PAHs, including LMW PAHs, with two or three rings, and HMW PAHs, with four or more rings.

PAH biodegradation studies show an apparent shift in emphasis from classical microbiology to systems biology with the development of analytical techniques (Fig. 9.2). In this chapter, we describe research endeavours to understand the mycobacterial PAH metabolism, highlighting novel aspects in HMW PAH biodegradation that the recent systems biology-based studies have provided (Cerniglia, 1992, 2003; Kim *et al.*, 2009, 2010; Kweon *et al.*, 2010a).

9.2 Characterization of PAH degrading mycobacteria

9.2.1 Isolation and taxonomic classification of PAH-degrading mycobacteria

Mycobacteria capable of degrading PAHs and many other recalcitrant hydrocarbons are widely distributed in diverse natural habitats. More than 70 aromatic hydrocarbon-degrading mycobacterial strains, belonging to at least 12 valid species, have been isolated from geographically different regions (Kim *et al.*, 2010). Although they are found in pristine or non-polluted soils, ground water, and even such harsh environments as soils with acidic conditions below pH 2, isolation of PAH-degrading mycobacteria is often associated with contamination by various hydrocarbon pollutants (Kim *et al.*, 2010). It includes, for example, tar/PAH-contaminated soils, estuarine or marine sediments exposed to petrogenic chemicals, activated sewage sludge, drainage ditches, and insecticide-polluted soils (Kim *et al.*, 2010). Culture-independent molecular ecological studies have shown the presence of more diverse strains of mycobacteria (Cheung and Kinkle, 2005; Debruyn *et al.*, 2007; Ding *et al.*, 2010; Marcos *et al.*, 2009; Atlas *et al.*, 2015). The molecular ecophysiology of this group of bacteria

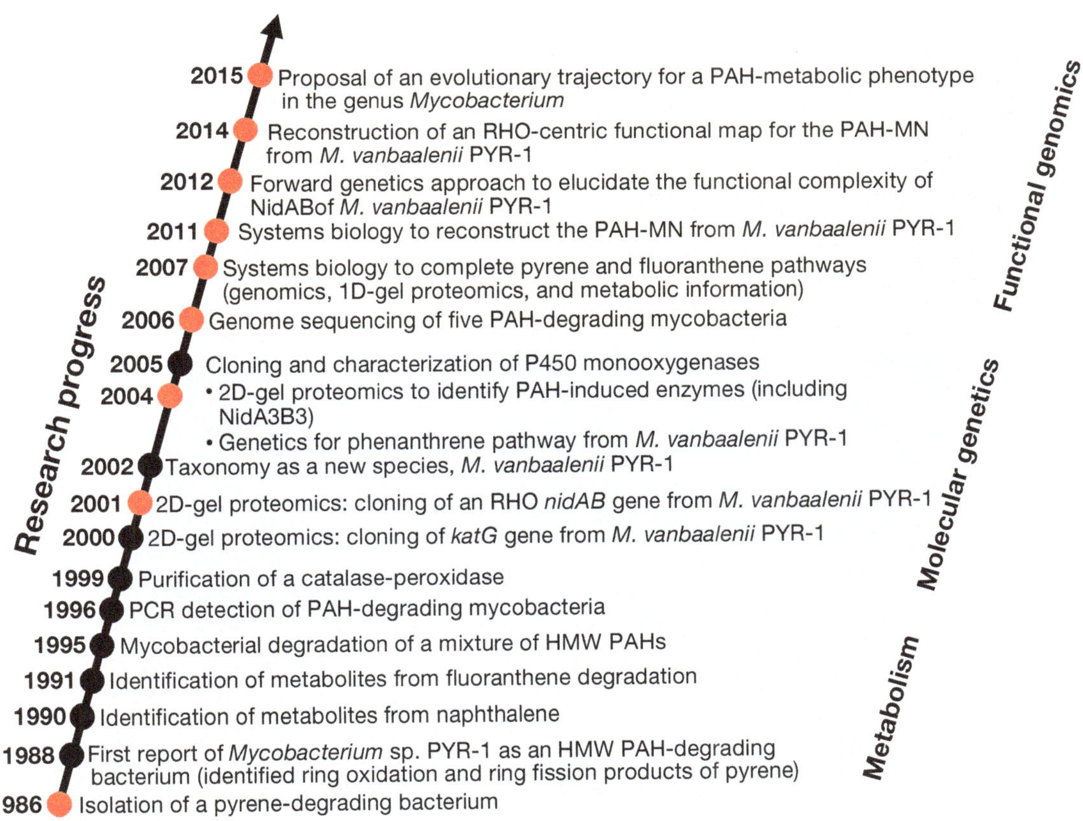

Figure 9.2 Studies on the degradation of HMW PAHs by *M. vanbaalenii* PYR-1 shows an apparent systematic shift in terms of research progress from traditional microbiology to omics-based systems biology.

appears to be well suited to survive, multiply, compete, and degrade PAHs in various environmental niches.

Mycobacteria are aerobic, Gram-positive, GC-rich, catalase-positive, acid-fast, mycolate-containing actinobacteria (Hartmans *et al.*, 2006; Saviola and Bishai, 2006). Although the mycobacteria degrading PAHs are comprised of diverse species, they fall into the fast-growing group (colony formation requires less than 7 days), instead of those clinical mycobacteria that grow slowly (colony formation requires more than 7 days) (Stahl and Urbance, 1990). The genus *Mycobacterium* is classified in the order Actinomycetales, which also encompasses other well-known metabolically versatile genera, including *Corynebacterium, Rhodococcus, Gordonia, Dietzia* and *Nocardia* (Stackebrandt *et al.*, 1997).

9.2.2 Biochemistry of PAH degradation pathways

Most of the biochemical pathways for the degradation of HMW PAHs have been elucidated from studies of mycobacteria. Bacterial isolates were cultured with HMW PAHs and the metabolic intermediates were identified by HPLC, UV spectra, and thin-layer chromatography, often compared with authentic compounds, as well as by GC-MS, LC-MS, and proton NMR spectroscopy. *M. vanbaalenii* PYR-1, *M. flavescens* PYR-GCK, *M. aromaticivorans* JS19b1 and *Mycobacterium* spp. strains RJGII-135, AP1, KMS, and KR2 have been extensively investigated, significantly expanding our understanding of HMW PAH biodegradation pathways (Schneider *et al.*, 1996; Rehmann *et al.*, 1998; Vila *et al.*, 2001; Khan *et al.*, 2002; Miller *et al.*, 2004; Kim *et al.*, 2005; Seo *et al.*, 2007; Hennessee *et al.*, 2009).

Cultural studies show that *M. vanbaalenii* PYR-1, when incubated with pyrene, produces a dihydroxylated metabolite, pyrene *cis*-4,5-dihydrodiol, by dioxygenation (Heitkamp *et al.*, 1988a; Heitkamp *et al.*, 1988b). Pyrene *trans*-4,5-dihydrodiol is also produced, although it is not the predominant metabolite produced from pyrene. The production of pyrene *trans*-4,5-dihydrodiol, the first *trans* metabolite identified from bacterial PAH metabolism, was later shown to be catalysed by a cytochrome P450 (CYP) monooxygenase (Brezna *et al.*, 2006). Another metabolite is produced from initial ring-hydroxylation at the C-1,2 positions of pyrene, forming pyrene *cis*-1,2-dihydrodiol. This metabolite was later shown to be transformed into dead-end methylated derivatives in what is proposed as an alternative detoxification process (Kim *et al.*, 2004b). Altogether, these results suggested the presence of multiple pathways for the initial oxidative attack of pyrene by the bacterium, which helped in establishing the future direction of the research to elucidate the mechanisms of pyrene degradation. These studies also identified pyrene ring cleavage metabolites, including phenanthrene-4-carboxylate, phthalate, and cinnamate, which were consistent with an earlier proposal for a pyrene degradation pathway (Cerniglia, 1992) via pyrene *cis*-4,5-dihydrodiol.

Since these initial studies with *M. vanbaalenii* PYR-1, the initiation of pyrene degradation, with either mono- or dioxygenation at the C-4,5 positions, was repeatedly confirmed in several other mycobacterial isolates (Dean-Ross and Cerniglia, 1996; Schneider *et al.*, 1996; Rehmann *et al.*, 1998; Vila *et al.*, 2001; Liang *et al.*, 2006). Metabolites produced from pyrene that were confirmed or newly identified included the first ring fission product of pyrene, phenanthrene 4,5-dicarboxylate (Dean-Ross and Cerniglia, 1996; Schneider *et al.*, 1996; Rehmann *et al.*, 1998), *cis*-3,4-dihydroxyphenanthrene-4-carboxylate (Rehmann *et al.*, 1998), 1-hydroxy-2-naphthoate (Rehmann *et al.*, 1998), and protocatechuate (Fritzsche, 1994; Rehmann *et al.*, 1998). Overall, the PAH metabolites identified from studies of *M. vanbaalenii* PYR-1 and other pyrene-degrading mycobacteria indicated that degradation of pyrene mostly occurs through formation of phthalate and protocatechuate.

Metabolites resulting from fluoranthene degradation have also been identified with species of *Mycobacterium*. 9-Fluorenone-1-carboxylate accumulates during growth of *M. vanbaalenii* PYR-1 with fluoranthene (Kelley *et al.*, 1991); this metabolite is a product of the oxidation and subsequent ring fission of fluoranthene. Based on this metabolite and others, such as 1-acenaphthenone, identified in a following study (Kelley *et al.*, 1993), a fluoranthene pathway initiated by dioxygenation of fluoranthene at the C-1,2 and C-7,8 positions was proposed (Cerniglia, 1992). A third route for the degradation of fluoranthene through dioxygenation of the C-2,3 positions was proposed from a culture study of *Mycobacterium* sp. KR20 (Rehmann *et al.*, 2001), with the identification of fluoranthene *cis*-2,3-dihydrodiol. This route was later reported as the predominant pathway for fluoranthene degradation (Kweon *et al.*, 2007). With the identification of 9-carboxymethylene-9*H*-fluorene-1-carboxylate, fluoranthene *cis*-2,3-dihydrodiol was proposed to be transformed into phthalate via 9-fluorenone-1-carboxylate and 1,2,3-benzenetricarboxylate (Rehmann *et al.*, 2001). The presence of the three ring-dihydroxylation pathways at C-1,2, C-2,3, and C-7,8 was confirmed in *M. holderi* EMI2 (Kleespies *et al.*, 1996), *M. flavescens* PYR-GCK (Dean-Ross *et al.*, 2002), and *Mycobacterium* spp. strains CP1, CP2, and AP1 (López *et al.*, 2005; Lopez *et al.*, 2006). A fourth pathway through initial ring-hydroxylation at C-7,8 was proposed in *Mycobacterium* sp. strain JS14 (Lee *et al.*, 2007). Later, a nonspecific monooxygenation followed by *O*-methylation of dihydroxyfluoranthene was also shown in the detoxification of fluoranthene by *M. vanbaalenii* PYR-1 (Kweon *et al.*, 2007). The identified metabolites indicated that the degradation of fluoranthene is initiated with ring-dioxygenation of multiple positions, which are channelled through phthalate and the β-ketoadipate pathway (Cerniglia, 1992; Kweon *et al.*, 2007).

9.2.3 Molecular genetics of PAH degradation

Compared to the wealth of information on the degradation of monocyclic aromatic hydrocarbons and LMW PAHs, relatively little is known about the genetics of bacterial HMW PAH degradation. One of the reasons for the lack of information is that relatively few bacteria have been isolated that degrade HMW PAHs. Another reason is that most of the bacteria known to play an important role in the degradation of HMW PAHs are in the Actinomycetales, and not many molecular tools and techniques are available for genetic manipulation in this group. They are refractory to genetic manipulation due to the presence of a thick and waxy, mycolic acid-rich, cell wall, a well-known trait of these bacteria. In addition, as information has increased, genes and genetic systems involved in the degradation of HMW PAHs have been revealed to have low homology, such as in gene sequence and order, with the widely characterized naphthalene-degrading *nah*-like genes from Gram-negative *Pseudomonas* species (Simon *et al.*, 1993). Attempts to identify HMW PAH-degrading genes with *nah*-like genes have been largely unsuccessful.

Studies of the genetics of biochemical pathways for the degradation of HMW PAHs started when the important ring-hydroxylating oxygenase (RHO) genes, *nidAB*, were identified from *M. vanbaalenii* PYR-1 (Khan *et al.*, 2001). The *nidAB* genes, encoding RHO α and β subunits, respectively, are induced by pyrene and are responsible for the initial ring-hydroxylation of pyrene, producing pyrene *cis*-4,5-dihydrodiol. Since then, the *nidAB* RHO genes have been identified in many other strains of PAH-degrading *Mycobacterium*, often after they are induced by pyrene or phenanthrene (Krivobok *et al.*, 2003; Miller *et al.*, 2004; Sho *et al.*, 2004; Stingley *et al.*, 2004b; Liang *et al.*, 2006). The *nidAB* genes have often been identified in culture-independent molecular studies for the detection of RHO genes in the environment (Cheung and Kinkle, 2001; Hall *et al.*, 2005; Marcos *et al.*, 2009; Ding *et al.*, 2010; DeBruyn *et al.*, 2011; Muangchinda *et al.*, 2015), which indicates that they play a critical role in the degradation of pyrene. Another gene, *nidD*, clustered with *nidAB*, was also identified and proposed to be involved in the conversion of 1-hydroxy-2-naphthaldehyde to 1-hydroxy-2-naphthoate in the pathway of pyrene degradation by *M. vanbaalenii* PYR-1 (Kim *et al.*, 2007). The genes *nidAB* are arranged in the *nidB-nidA* order, which is different from the order found in other bacterial RHOs. This atypical *nidBA* gene order was later proposed to be a unique feature of PAH-degrading mycobacteria (Miller *et al.*, 2004; Stingley *et al.*, 2004b).

Further genetic investigations with *M. vanbaalenii* PYR-1 revealed an operon involved in the degradation of phthalate (Stingley *et al.*, 2004a,b). They confirmed the previous metabolic results showing that phenanthrene, anthracene, pyrene, and fluoranthene are degraded through phthalate (Heitkamp *et al.*, 1988b; Kelley *et al.*, 1993; Moody *et al.*, 2001). The phthalate operon, positioned at 12 kb upstream of *nidAB*, consists of a putative regulatory protein (*phtR*), phthalate RHO α and β subunits (*phtAaAb*), a dihydrodiol dehydrogenase (*phtB*), an RHO ferredoxin (*phtAc*), and an RHO ferredoxin reductase (*phtAd*). It was shown to convert phthalate to 3,4-dihydroxyphthalate when expressed in *E. coli* (Stingley *et al.*, 2004a). This observation revealed, for the first time in a species of *Mycobacterium*, a genetic basis for the pathways of PAH degradation via phthalate and protocatechuate (Guerin and Jones, 1988; Moody *et al.*, 2001). The phthalate operon is also conserved in several other PAH-degrading *Mycobacterium* spp. (Stingley *et al.*, 2004a).

The metabolic studies showed multiple pathways for the degradation of HMW PAHs by *M. vanbaalenii* PYR-1 and the capability of the bacterium to degrade a broad range of aromatic substrates. This suggested that, although the aromatic substrates are usually converted into a limited number of central metabolic intermediates, this species might possess either several different RHO enzymes or an RHO enzyme with relaxed substrate specificity for initial hydroxylation of PAHs. Indeed, additional RHO α and β subunit genes, such as *orf25/26*, *pdoA2B2*, and *nidA3B3*, located in a separate region from *nidAB*, were identified in *Mycobacterium* spp. strains PYR-1, 6PY1, SNP11, and S65 (Krivobok *et al.*, 2003; Sho *et al.*, 2004; Stingley *et al.*, 2004b; Kim *et al.*, 2006; Pagnout *et al.*, 2007). Dozens more RHO genes were later identified, increasing, for example, the total number of RHO genes to 21 and 28 in *Mycobacterium* spp. strains PYR-1 and KMS, respectively, from the analysis of their whole genome sequences (Kim *et al.*, 2008; Zhang and Anderson, 2012).

These RHO enzymes have low PAH substrate specificity (Krivobok *et al.*, 2003; Kweon *et al.*, 2010b). They oxidize a wide range of related PAH substrates, producing diverse isomeric dihydrodiol metabolites at different conversion rates. However, RHO enzymes have an apparent PAH substrate preference (Krivobok *et al.*, 2003; Kweon *et al.*, 2010b), and the corresponding genes are often highly up-regulated by the presence of particular PAHs. For example, the NidAB and NidA3B3 RHO enzymes are involved in the ring-hydroxylation of toluene, *m*-xylene, naphthalene, phenanthrene, anthracene, fluoranthene, pyrene, benz[*a*]anthracene, and benzo[*a*]pyrene, but they show the highest conversion rates for pyrene and fluoranthene, respectively (Kweon *et al.*, 2010b). The PdoA2B2 enzyme is also involved in the oxidation of naphthalene, biphenyl, fluorene, anthracene, phenanthrene, pyrene, fluoranthene, and benzo[*a*]pyrene (Kweon *et al.*, 2010b). This enzyme, however, has a higher transformation efficiency towards fluorene, anthracene, and phenanthrene, and the gene *pdoA2B2* is particularly induced by these PAH substrates. The RHO catalytic activity and substrate specificity are correlated with the binding pocket size and shape in the active site of the enzyme (Kweon *et al.*, 2010b).

One of the notable findings in the genetic studies of HMW PAH degradation is the identification of the genes encoding CYP monooxygenases involved in the initial ring-hydroxylation of PAH substrates. The involvement of CYP enzymes in PAH degradation, together with epoxide hydrolase, which produces *trans*-dihydrodiol PAH metabolites, was expected. This was confirmed when three CYP genes were identified in *M. vanbaalenii* PYR-1 and their functions in the oxidation of PAHs were characterized (Brezna *et al.*, 2006). Later, 50 copies of CYP genes were identified, along with genes encoding epoxide hydrolases, in the genome of *M. vanbaalenii* PYR-1 (Kim *et al.*, 2008). This high number of CYP-encoding genes is considered to contribute to the catabolic versatility of the bacterium.

9.3 Functional genomics of an HMW PAH degrader, *M. vanbaalenii* PYR-1

The goal of genomics is to generate a nucleotide blueprint of an organism, which addresses all genes and their inter-relationships to figure out their combined influence on biological activities, whereas that of functional genomics is to extract the dynamic nature of gene function based on high-throughput approaches, including transcriptomics, proteomics, and metabolomics, from the static genomic sequence information.

Since the isolation of *M. vanbaalenii* PYR-1, the biochemistry and genetics approaches, which have identified PAH metabolites and characterized individual PAH-degrading genes, have produced important findings on the metabolic principles of bacterial PAH biodegradation (Cerniglia, 1992, 2003; Kanaly and Harayama, 2011). Despite the considerable success of these traditional approaches, however, many systems-level metabolic questions remained unanswered until the late 2000s. For example, little was known about how each of the RHO enzymes is involved in the degradation of various PAH substrates or what impact the degradation of PAHs has on the cellular physiology of the bacterium. When genomics and functional genomics of *M. vanbaalenii* PYR-1 became possible, approaches based on these omics methods opened the way for a systematic, genome-level understanding of the molecular basis of the structure, behaviour, and evolution in terms of PAH metabolism. Currently, complete genome sequences of about 10 mycobacterial strains capable of degrading PAHs, as well as alkanes, alkenes, and vinyl chloride, are available (https://img.jgi.doe.gov). The introduction of functional pan-genomic approaches, in which genomics and functional genomics information is integrated with a comparative pan-genomic concept of the genus, further enhanced the analytical resolution and solidity of the omics data.

9.3.1 Genomic analysis of PAH degradation

Considering the years of research on HMW PAH degradability, the genome project of *M. vanbaalenii* PYR-1 and four other PAH-degrading mycobacteria, including strains PYR-GCK, KMS, MCS, and JLS, was an important achievement towards genomic insight into the metabolic logic and strategies of the bacterium, such as any metabolic benefit that can be derived from biodegradation of HMW PAHs.

The genome of *M. vanbaalenii* PYR-1 was completely sequenced with 12× coverage, using whole-genome shotgun sequencing, by the US Department of Energy/Joint Genome Institute (JGI). An initial rough draft of the genome, consisting of a set of contigs separated by gaps, was released in 2005 and a gap-free genome was completed in 2006. The full genomic information, including the genome sequence of *M. vanbaalenii* PYR-1, is available at the Joint Genome Institute (JGI) (http://img.jgi.doe.gov) and NCBI with accession number CP000511. In 2008, the genes and open reading frames automatically generated by JGI were manually reanalysed with respect to PAH metabolism (Kim *et al.*, 2008). Previously functionally characterized proteins from *M. vanbaalenii* PYR-1 and orthologous genes and proteins, which have supporting functional evidence available from other strains, were taken into account for reassignment of functions. Paralogues in the genome have been identified, using the JGI system based on BLASTP hits, with E-value cut-off values of $<10^{-5}$. Missing genes have been checked with the JGI initial annotation and the start/stop positions of some genes were revised using the Artemis comparison tool (Carver *et al.*, 2005).

9.3.1.1 General genomic features

The genome of *M. vanbaalenii* PYR-1 consists of a single 6.49 Mbp circular chromosome with GC content of 68%. The complete genome has about 5979 predicted protein-coding sequences (CDSs) and 68 RNA genes, with a coding density of 91.6%. Among the 5979 CDSs, about 4256 (70.38%) were annotated with predicted functions, and about 3776 CDSs (62%) were functionally classified with clusters of orthologous groups (COGs). Of the 68 RNA genes, two sets of rRNA genes (5S rRNA, 16S rRNA, and 23S rRNA) are organized in two operons with the general structure 5S–23S–16S, and 49 tRNA genes contain 44 specific anticodon triplet sequences that are able to base-pair to the 20 amino acids. The genome of *M. vanbaalenii* PYR-1 contains about 202 genes likely to be associated with metabolism of aromatic hydrocarbons, including HMW PAHs (Kim *et al.*, 2008).

Comparative analysis of mycobacterial genomes reveals that the general genomic features of mycobacteria, such as size, GC content, and gene content, are strongly associated with their lifestyles (Kweon *et al.*, 2015). According to a mycobacterial phenotype network (MPN) proposed to represent the genotype–phenotype relationship within the genus *Mycobacterium*, *M. vanbaalenii* PYR-1 belongs to a distinct group whose members are fast-growing, non-pathogenic, free-living, and PAH-degrading (Kweon *et al.*, 2015). Compared with the host-dependent, slow-growing, pathogenic mycobacteria, such as *M. tuberculosis* and *M. leprae*, the free-living, fast-growing, and non-pathogenic mycobacteria generally exhibit bigger genome size (>~5.7 Mb and > 5400 genes), more rRNA genes (more than two sets of rRNA genes), and higher GC content (>67%). The numerical genomic indexes of free-living *M. vanbaalenii* PYR-1 well account for the observed phenotypic characteristics. Genome sizes of the PAH-degrading mycobacteria, including *M. vanbaalenii* PYR-1, are between 5.7 and 6.5 Mb, with an average of 6 Mb, which is larger than that of strains of *M. tuberculosis*, whose average genome size is ~4.4 Mb.

9.3.1.2 PAH-degrading genes in the genome

The genome of *M. vanbaalenii* PYR-1 addresses molecular genetic questions on bacterial PAH degradation, such as how many PAH-degrading genes does *M. vanbaalenii* PYR-1 have, or even evolutionary questions on how, when, and where the genes have been obtained. *M. vanbaalenii* PYR-1 harbours ~200 genes possibly involved in PAH biodegradation (Fig. 9.3). The most predictable genes involved in PAH degradation are localized in two genomic regions, A and B, at positions 494–643 kb and 4711–4741 kb, respectively, with others distributed all over the chromosome (Kim *et al.*, 2008). As identified in the analysis of horizontally transferred genomic islands, *M. vanbaalenii* PYR-1 seems to have acquired more than 400 genes, including PAH-degrading genes, by horizontal gene transfer (HGT) events (Kweon *et al.*, 2015). Out of the ~200 PAH-degrading genes, 29 PAH-degrading genes have been identified on genomic islands (Kweon *et al.*, 2015). The *M. vanbaalenii* PYR-1 genome shows a relatively low $\sigma_{\%G+C}$ score, which is a useful indication of the degree to which recent HGT may have shaped a genome, indicating relatively little recent genetic flux (Kweon *et al.*, 2011, 2015). Despite the high numbers of PAH-degrading genes in the genome, the facts that only 29 PAH-degrading genes (~15%) are on genomic islands and that the $\sigma_{\%G+C}$ score is relatively low suggest

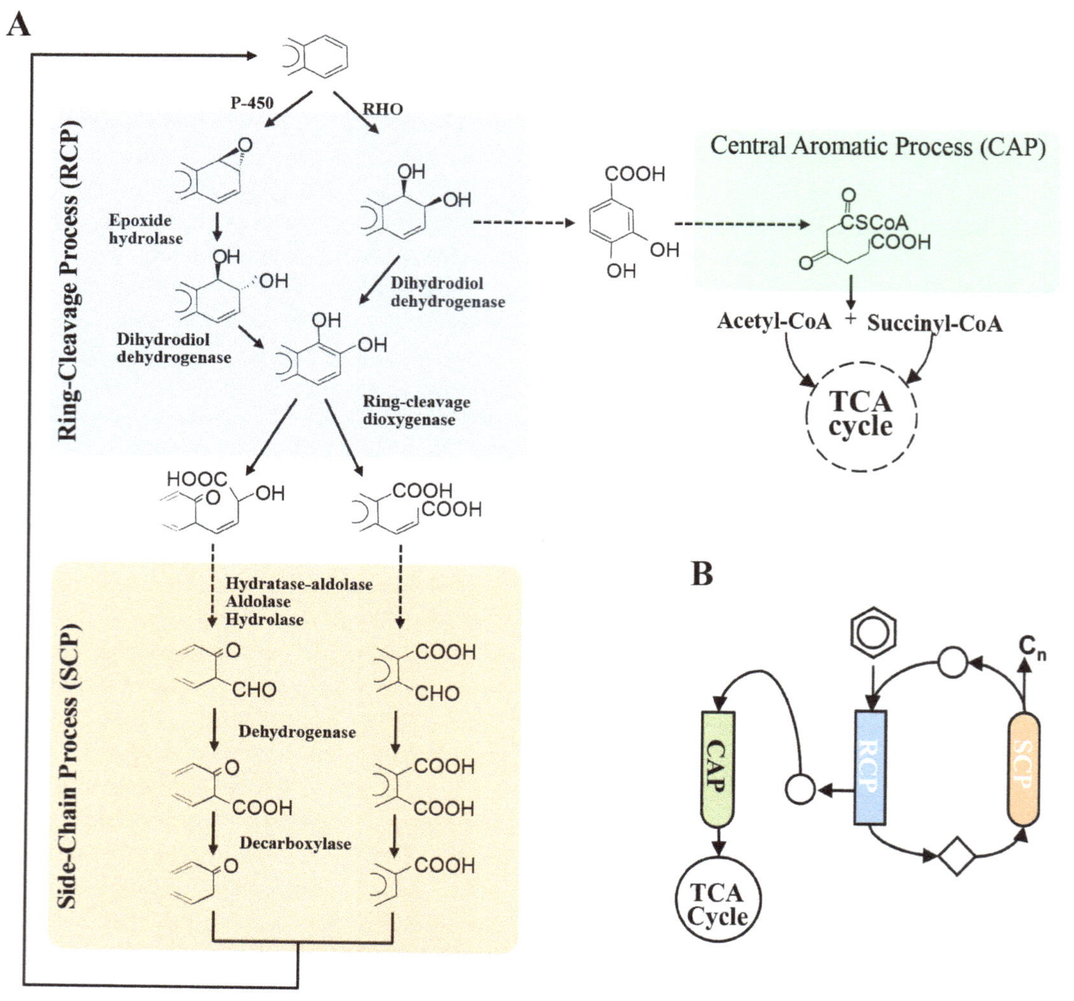

Figure 9.3 General scheme of PAH metabolism in terms of the level of functional module, with (A) and without (B) simplified chemical structures and enzymes involved. Solid and dashed arrows indicate one and two (or more) reaction steps, respectively.

that many genes responsible for PAH degradation were not acquired through recent HGT events. Sequence comparison, using a non-redundant protein database, shows that many PAH-degrading genes have high sequence similarity to the proteins associated with aromatic hydrocarbon degradation in other strains of *Mycobacterium* and other Actinomycetales, such as *Terrabacter, Arthrobacter, Nocardioides, Streptomyces, and Rhodococcus* (Kim *et al.*, 2008). In contrast, the highest BLAST hit to other well-known Gram-negative aromatic hydrocarbon degraders, such as strains of *Sphingomonas, Burkholderia, Comamonas,* and *Pseudomonas,* gave only 32 annotations. Collectively, *M. vanbaalenii* PYR-1 has ~200 genes responsible for PAH biodegradation, clustered into two regions, a ~150 kb catabolic gene cluster (region A) and a ~30 kb cluster (region B), respectively, which have been acquired by relatively few recent HGT events from neighbours with close physiological, biochemical, and phylogenetic relationships (Kweon *et al.*, 2011, 2015).

It appears that the 150 kb catabolic gene cluster in region A is specialized for PAH degradation, since it has all of the genes necessary for complete PAH degradation. It includes multiple copies of the genes encoding initial RHO enzymes and the genes encoding the β-ketoadipate pathway, which connects the catabolic pathways to the TCA cycle. As observed in the gene sequences, there is quite a difference from well-characterized Gram-negative aromatic hydrocarbon degraders, such as *Pseudomonas* and *Burkholderia,* in the gene organization. Whereas the catabolic genes in pseudomonads and closely related genera are usually well organized in a cluster (Assinder and Williams, 1990; van der Meer *et al.*, 1992), the catabolic genes in mycobacteria are organized as an atypical mosaic structure made of several complex gene clusters, in which genes involved in a particular degradative pathway are not positioned on the same operon but are dispersed throughout several gene clusters. For example, 27 genes predicted to be involved in the complete degradation of pyrene (Kim *et al.*, 2007), which will be discussed in detail in the following section, are arranged across 4 gene clusters scattered over at least 7 operons (Kim *et al.*, 2008).

Catabolic genes identified in the genome of *M. vanbaalenii* PYR-1 are often enriched with multiple paralogues with a range of diversity and involve all the functional categories required for aromatic hydrocarbon degradation, such as oxidation-reduction, hydrolysis, and CoA transfer. As shown in a generalized scheme for the degradation of PAHs (Fig. 9.3), these enzymatic reactions can be formulated into three functional modules or processes, ring cleavage processes (RCP), side chain processes (SCP), and central aromatic processes (CAP), respectively, based on the concept of a PAH-MN (Kweon *et al.*, 2011). These three functional processes consist of a series of enzymatic reactions, each of which is annotated with either a singular or a plural number of those catabolic genes identified in the genome of the bacterium. Briefly, degradation of aromatic hydrocarbons is initiated with aromatic ring oxygenation, followed by ring-cleavage in the RCP, to which 21 and 50 copies of genes, encoding RHO enzymes and CYPs, respectively, are annotated, together with the annotation of four, five and 10 copies of genes encoding epoxide hydrolases, dihydrodiol dehydrogenases, and ring-cleavage dioxygenases, respectively. Enzymatic reactions in the SCP that have genes involved in the production of intermediates acceptable to RCP, as well as other biological precursors including pyruvate from the ring-opened metabolites of RCP, are annotated. They include three copies of hydratase-aldolases, four copies of hydratases, two copies of aldolases, seven copies of alcohol dehydrogenases, and two copies of decarboxylases. A series of enzymatic reactions in CAP then converts protocatechuate to small aliphatic compounds, connecting the degradation pathway to the TCA cycle. In the genome of *M. vanbaalenii* PYR-1, a gene cluster encoding the β-ketoadipate pathway enzymes has been identified for the CAP. This gene cluster encodes all the enzymes for conversion of protocatechuate to succinyl-CoA and acetyl-CoA (Kim *et al.*, 2008, 2011).

9.3.2 Proteomic analysis of PAH degradation

Information on the *M. vanbaalenii* PYR-1 genome yields a molecular basis but has limitations to provide a dynamic and mechanistic understanding of the degradation of HMW PAHs. Compared with monocyclic and LMW PAHs, degradation of HMW PAHs requires more diverse pathways and complex biochemical reactions that involve a number of different types of genes and enzymes.

As genomic analysis reveals, a multiplicity of genes from the disorganized arrangement of the catabolic clusters is typically involved in HMW PAH degradation, which suggests that more complex molecular mechanisms or previously unknown pathways might be involved. Here, high-throughput proteomics in conjunction with bioinformatics offers powerful and effective tools, not only to define the current metabolic pathways, but also to elucidate, by detecting differential expression of enzymes, additional pathways in the complex network of HMW PAH degradation. It further provides a deeper understanding of the overall cellular responses involved in the degradation of HMW PAHs. For the functional genomic study of *M. vanbaalenii* PYR-1, a two-dimensional gel electrophoresis (2-DE) approach was initially used (Khan *et al.*, 2001; Wang *et al.*, 2000; Kim *et al.*, 2004a). This was followed by a one-dimensional SDS gel electrophoresis (1-DE) in combination with nanospray liquid chromatography-tandem mass spectrometry (GeLC-MS/MS), with the release of the whole genome sequence of the bacterium in 2007 (Kim *et al.*, 2007, 2012; Kweon *et al.*, 2007, 2011, 2014).

9.3.2.1 Proteomic analysis using 2-DE

Early proteomic study of the degradation of HMW PAHs relied mainly on 2-DE, in which proteins were separated based on isoelectric point and molecular weight. Subsequent identification of selected protein spots was by N-terminal sequencing (Wang *et al.*, 2000; Khan *et al.*, 2001) or by the later-developed protein mass spectrometry (Kim *et al.*, 2004a). In spite of the inherent analytical and technical shortcomings of this proteomics method, such as in detection of less abundant proteins and membrane proteins, this approach made major breakthroughs in the molecular study of bacterial HMW PAH biodegradation. The lysates of *M. vanbaalenii* PYR-1 cells incubated with different HMW PAHs were resolved on polyacrylamide gels and their profiles were compared to reveal differences in the level of protein expression. The experiments identified up-regulation of RHO α- and β-subunits (NidA and NidB3, respectively), a catalase peroxidase (KatG), a putative monooxygenase, and an aldehyde dehydrogenase (Rafii *et al.*, 1999; Wang *et al.*, 2000; Khan *et al.*, 2001; Kim *et al.*, 2004a). Identification of the two RHO enzyme subunits was directly linked to the elucidation of the entire NidAB and NidA3B3 genetic systems, which significantly contributed to the understanding of the molecular mechanisms of HMW PAH degradation. These studies also showed changes of profiles in the overall composition and abundance of expressed proteins from the PAH-exposed *M. vanbaalenii* PYR-1.

9.3.2.2 Proteomic analysis using GeLC-MS/MS

With the availability of the *M. vanbaalenii* PYR-1 whole genome sequence and the introduction of advanced high-resolution analytical chemistry, proteomics approaches started providing much more significant information on bacterial HMW PAH degradation. The GeLC-MS/MS, combined with previous metabolic, genetic, and enzyme biochemistry information, for the first time elucidated complete omics data-integrated pathways for the degradation of pyrene and fluoranthene in *M. vanbaalenii* PYR-1 (Kim *et al.*, 2007; Kweon *et al.*, 2007). Thousands of proteins were identified by mass spectrometry, from which 27 and 54 enzymes were proposed to be necessary for the degradation of pyrene and fluoranthene, respectively, through the phthalate and β-ketoadipate pathways.

These 1-DE-based proteomics, applied to *M. vanbaalenii* PYR-1, provided an innovative frame for the understanding of bacterial HMW PAH degradation. Most of all, they provided decisive evidence on the functions of enzymes in the pathways for HMW PAH degradation with respect to gene redundancy, multiple pathways, and relaxed enzyme-substrate specificity. For example, among 21 genes encoding RHO enzymes, the NidA and NidA3 enzymes were assigned to reaction steps for pyrene and fluoranthene, respectively. NidA was induced by pyrene but not by fluoranthene; however, NidA3 was induced by fluoranthene but not by pyrene (Kim *et al.*, 2007; Kweon *et al.*, 2007). These results indicate that the bacterium uses a different initial RHO enzyme in response to the different HMW PAHs, which corroborated earlier enzyme biochemical studies, in which both enzymes were proposed to be involved in the initial oxidation of both pyrene and fluoranthene (Kim *et al.*, 2006).

In addition to the molecular mechanisms directly involved in the degradation of HMW

PAHs, the GeLC-MS/MS results also revealed other biological reactions involved in ecophysiological responses during incubation of the bacterium with PAHs. For example, *M. vanbaalenii* PYR-1, incubated with seven different aromatic hydrocarbons, commonly up-regulates 189 proteins, which show a high degree of functional diversity across most of the COGs (clusters of orthologous groups) categories. They include proteins related to signal transduction/transcriptional regulation, membrane transporter proteins, cell wall/membrane biosynthesis, and proteins related to defence/detoxification processes. The treatment of aromatic compounds affects not only their direct metabolism but also the bacterial genome-wide complex global response involved in diverse cellular functions. This is essential information for exploring practical biotechnological applications of PAH biodegradation.

9.4 Systematic understanding of the PAH-MN in *M. vanbaalenii* PYR-1

M. vanbaalenii PYR-1 has been systematically studied with respect to the bacterial metabolism of HMW PAHs. The genome sequencing, along with three decades worth of detailed biochemical and enzymatic data, resulted in the proposal of a PAH-MN at the genome scale (Kweon *et al.*, 2011). In the study of bacterial PAH metabolism, reconstruction of the PAH-MN is an essential effort to organize the dispersed data into an integrated resource in a structured fashion for the multi-scale mining of topological, functional, and evolutionary information. Functional genomics data, obtained from proteome studies of *M. vanbaalenii* PYR-1 exposed to diverse PAH substrates, play a critical role in the analysis of the network. They not only bridge the gap between genome (genotype) and metabolites (phenotype), but also are essential for analysing functional dynamics of the PAH-MN with a holistic perspective. Proteomic data bring a qualitative change in the understanding of gene redundancy at the genomic level, which is transformed into functional redundancy, in terms of pleiotropic activity (functional contribution of an enzyme for two or more different substrates) and epistatic interaction (functional combination of two or more paralogous enzymes for an enzyme reaction step) of PAH-degrading enzymes (Kweon *et al.*, 2014, 2015). This pleiotropic and epistatic understanding is crucial for a rational decomposition of genes and enzymes into the functional modules and for enzyme-centric interpretation of the PAH-MN.

The PAH-MN is a structured format of such a *de facto* knowledge base. Systematic analysis of the PAH-MN provides answers to questions in the following categories: (i) structure, (ii) behaviour, and (iii) evolution of mycobacterial PAH metabolism (Fig. 9.4). Considering that biological networks behave in predictable ways (Cohen, 2002; Albert, 2005), the PAH-MN might additionally offer prediction tools for computational and quantitative queries to guide metabolic engineering of the organism.

9.4.1 Structure of the PAH-MN

A comprehensive and hierarchical PAH-MN from *M. vanbaalenii* PYR-1 has been reconstructed by a systematic integration of proteomic data, obtained from cellular responses to a range of different aromatic substrates, with genomic and metabolic data (Kweon *et al.*, 2011). The PAH-MN includes a set of 183 PAH metabolites identified from previous *M. vanbaalenii* PYR-1 metabolic studies of 10 aromatic hydrocarbons – biphenyl, naphthalene, acenaphthylene, acenaphthene, anthracene, phenanthrene, pyrene, fluoranthene, benz[*a*]anthracene, and benzo[*a*]pyrene – into a single graph, together with 183 chemical compounds connected by 224 chemical reactions (Fig. 9.5).

Topological properties of an MN are central to an understanding of behaviour and evolution of the network. To numerically describe the topological property, the graph theoretic measuring of the networks is initially required. In the study of a network, the degree of a node is the number of connections it has to other nodes, and the degree of distribution is the probability distribution of links per node in a network (Albert, 2005). In MNs, the degree distribution is non-random and can be fitted by a power law, reflecting a scale-free topology (Barabasi and Albert, 1999). Topological analysis of the PAH-MN shows it has typical scale-free properties. The log-log plots of probability distribution $P(k)$ versus the interactions k show a linear correlation, indicating that the degree of distribution follows a power law, $P(k) \approx k^{-\gamma}$, with $\gamma_{\text{in or out}} = 2.8$. The 'degree exponent ($\gamma$)', ranging between 2.0 and 2.8, indicates that

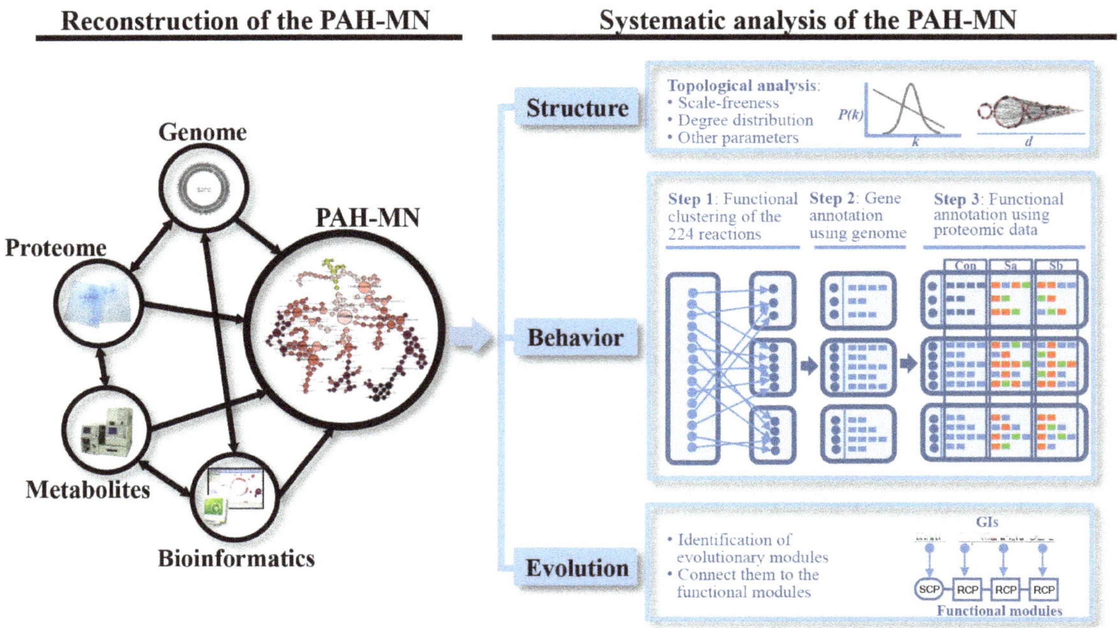

Figure 9.4 Systematic reconstruction and network analysis of the PAH-MN.

the PAH-MN is highly non-random, and a few nodes (hubs) dominate the overall connectivity in the network. Other structural parameters of the PAH-MN include 24 of a network diameter (the largest distance between two substrates), 7.2 of an average shortest path length, 2.265 of an average number of neighbours, and 0.007 of a clustering coefficient. These parameters of the PAH-MN are similar to those of other scale-free metabolic networks (Jeong *et al.*, 2000; Pazos *et al.*, 2003).

Some topological parameters are known to be specific to biodegradation networks. In the networks of biodegradation, distance of a given compound to the central metabolism represents properties of the chemical compounds (nodes) and the enzyme activities (edges) (Pazos *et al.*, 2003). Large and insoluble compounds tend to be far away from the central metabolism, while most 'ancient enzymes' are close to the central metabolism. Consistent with this common theme, molecular weight and water solubility of the aromatic compounds in the PAH-MN have a clear correlation with their distance from protocatechuate, the first product of the CAP, which generates intermediates connected to the TCA cycle. HMW PAHs are located in the periphery of the PAH-MN, while aromatic compounds with low molecular weights are crowded near the main hub metabolite, phthalate, which has the connection degree of 10 and the distance value of 3, respectively, to protocatechuate. As degradation proceeds, metabolites repeatedly fluctuate with molecular weight and hydrophobicity in only one way towards the central aromatic pathway, which is the β-ketoadipate pathway in *M. vanbaalenii* PYR-1. This directed repeating pattern of the chemical properties over the degradation process is one of the crucial structural observations to understand the functional modularity in the PAH-MN (Kweon *et al.*, 2011). In directed MNs, another interesting topological property is the number of incoming (c_i) and outgoing (c_o) connections for each compound. Like other biodegradation networks (Pazos *et al.*, 2003), the PAH-MN also has a 'concentrating' structure, in which several key compounds have larger in-degree (c_i) than out-degree (c_o). For example, the main hub metabolite, phthalate, has a connection degree of 10, with an in-degree of 7 and an out-degree of 3, respectively. In the PAH-MN, all the three outgoing routes of phthalate are channelled into protocatechuate, which has only one out-degree to β-carboxy-*cis,cis*-muconate, to produce acetyl-CoA, a common intermediate of the central metabolism.

A top-down structural view shows that the PAH-MN has a typical conical shape, in which several peripheral pathways converge into the

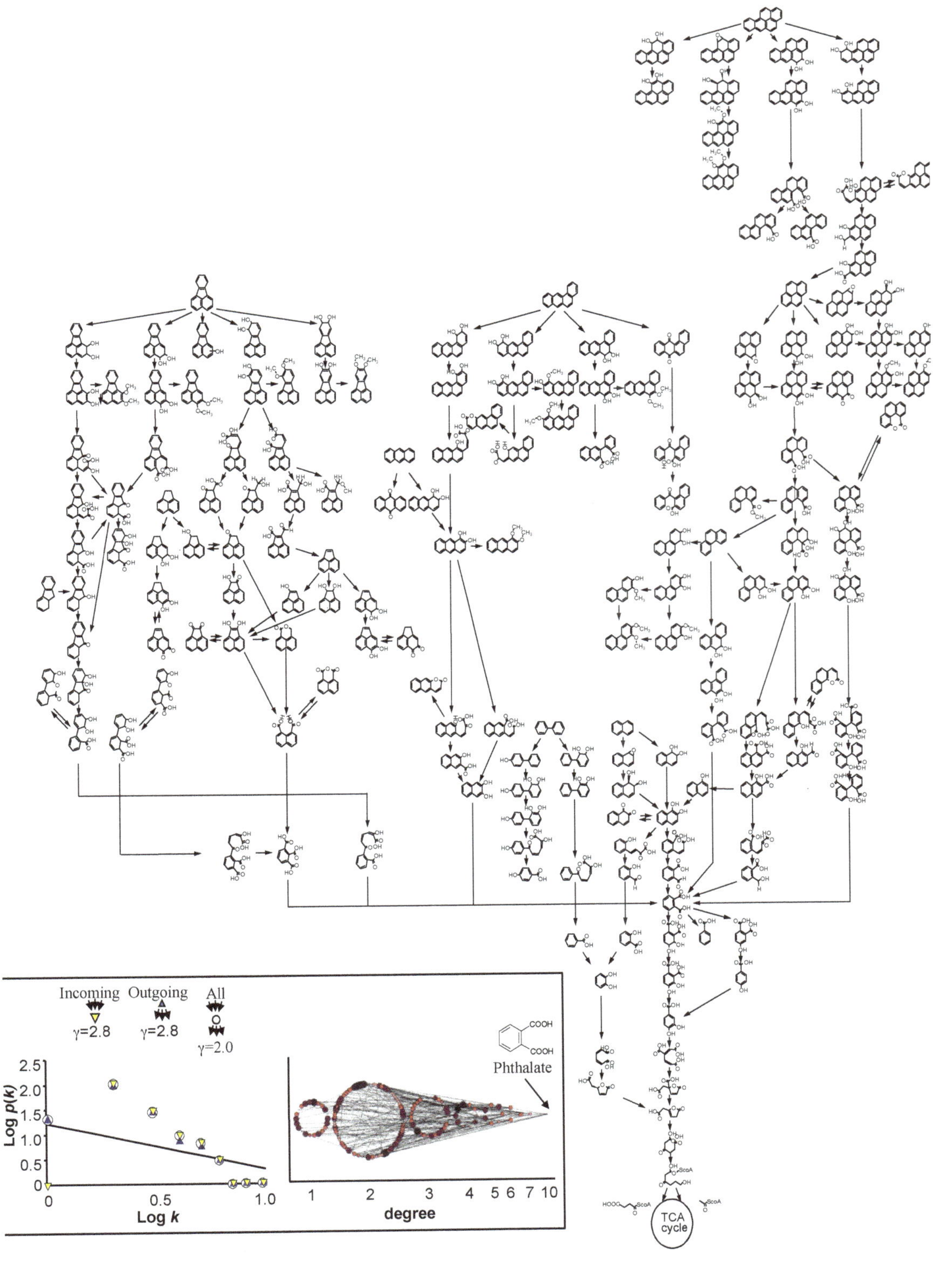

Figure 9.5 The PAH-MN in *M. vanbaalenii* PYR-1 showing 183 nodes and 224 edges. The log-log plot in the box shows the number of compounds versus connectivity. Connections involved either as a substrate or as a product are counted. The exponents of the power law distribution (γ) are shown. *k*, connectivity; *P*(*k*), number of compounds and degree distribution of the PAH-MN. The degree distribution of the PAH-MN is also shown in the box. The names of PAHs and their metabolites have been published previously (Kweon *et al.*, 2011).

central β-ketoadipate pathway via phthalate, the hub metabolite. The PAH-MN consists of two main subnetworks, the fluoranthene and pyrene subnetworks, to which the anthracene and benz[*a*]anthracene pathways and the benzo[*a*] pyrene pathway, respectively, are connected. In the fluoranthene subnetwork, pathways for four PAH substrates, fluoranthene, fluorene, acenaphthylene, and acenaphthene, are interconnected. *M. vanbaalenii* PYR-1 is able to dioxygenate the four-ringed PAH, fluoranthene, at the C-1,2, C-2,3, and C-7,8 positions. In the PAH-MN, the C-1,2 dioxygenation route is merged into the C-2,3 dioxygenation route via 9-fluorenone-1-carboxylic acid, which is further degraded via the fluorene pathway, whereas the C-7,8 (or C-9,10) dioxygenation route is joined to the pathway of the three-ringed PAHs, acenaphthene and acenaphthylene. In the fluoranthene subnetwork, all of the degradation routes of fluoranthene, fluorene, acenaphthylene, and acenaphthene eventually converge on 1,2,3-benzenetricarboxylic acid, which is decarboxylated to the hub metabolite, phthalate. Interestingly, the output metabolite of the fluoranthene subnetwork, 1,2,3-benzenetricarboxylic acid, only has a connection degree of 3 (an in-degree of 2 and an out-degree of 1), clearly indicating the funnel structure of the PAH-MN. On the other hand, the pyrene subnetwork is an interconnected pathway with other three PAH substrates, benzo[*a*]pyrene, phenanthrene, and naphthalene, which has a structurally conserved region. For the degradation of pyrene by *M. vanbaalenii* PYR-1, the pathway through C-4,5 dioxygenation is the productive route because it is channelled into the TCA cycle, whereas the other route, via C-1,2 dioxygenation, forms *O*-methylated derivatives as dead-end products (Kim *et al.*, 2004b). In the pyrene subnetwork, the C-4,5-dioxygenation route is connected with the C-3,4 route of the phenanthrene pathway, which is linked to phthalate directly or via naphthalene pathway. Therefore, phthalate is the key hub metabolite, connecting the two main subnetworks into the central metabolic pathway via CAP.

Overall, the topology parameters of the directed PAH-MN clearly indicate its scale-freeness with the appearance of super-hubs. They numerically depict the benefits of metabolic endeavours, which funnel the 'flux' of compounds to the central metabolism.

9.4.2 Behaviour of the PAH-MN

Although the structure of the PAH-MN proves its scale-freeness and provides a brief structural insight into the module-centric behaviour of the PAH-MN as well, graph-based analyses do not completely depict the dynamic behaviour of the PAH-MN. Here, identification of the functional modularity of the PAH-MN and consequent understanding of its metabolic benefits with evolutionary reason are also essential. In addition to the static structure, understanding the module-based metabolic behaviour of *M. vanbaalenii* PYR-1 further requires multi-layer information, including enzyme biochemistry data (Kweon *et al.*, 2010b), genomic and whole-cell proteomic data (Kim *et al.*, 2007, 2008; Kweon *et al.*, 2007, 2011), and bioinformatic data, to be systematically integrated into the structural information of the PAH-MN.

Fig. 9.4 shows a summary of the basic tasks and data used in the systematic process for the identification of functional modules, which is based on a rational decomposition of the chemical properties and reorganization of the enzymatic information of the PAH-MN (Kweon *et al.*, 2011). Initially, the process converts metabolic information into the corresponding enzyme reactions. According to the input/output compounds and the mode of reaction in the order of occurrence, the 224 reactions of the PAH-MN are clustered into three functional categories: (i) oxidoreduction of aromatic substrates to ring cleavage products via catechols; (ii) hydrolysis of the α-keto side chains of the ring-cleaved compounds to a biological precursor, pyruvate, and other metabolites with an aldehyde group, which is removed through oxidation/decarboxylation in the following reaction; and (iii) CoA transfer and subsequent degradative thiolase reactions to form acetyl-CoA and succinyl-CoA. These categories are defined as three functional processes, the RCP, SCP, and CAP, respectively (Fig. 9.3B). In the following step, enzyme reactions of the functional modules are annotated with the respective PAH-degrading genes, providing a more systematic genomic view, based on the functional modules in terms of PAH biodegradation. In the final step, proteomic data and traditional biochemistry data are combined. This converts gene redundancy, which is provided in a list of paralogous genes for each enzyme reaction, to functional redundancy, which is provided in a list of enzymes responsible for each reaction, with a concept of pleiotropy and epistasis (Kweon *et al.*, 2015).

According to the common underlying metabolic logic in the degradation of PAHs, which is the activation of benzene rings followed by ring opening, PAH substrates are transformed in the RCP into dihydrodiols either by monooxygenation with the aid of epoxide hydrolase or by dioxygenation, and the dihydrodiols are ring cleaved via the corresponding catechols. The outputs of the RCP, the ring cleavage compounds, then go through the SCP, which converts them to enter another round of RCP or CAP and produce active biological precursors. The SCP has a number of different enzymatic steps in the following functional order: (i) hydroxylation of side chains of the ring cleavage compounds, (ii) oxidation of the hydroxylation-generated aldehyde group to a carboxylic acid, and (iii) decarboxylation of the carboxyl group from the aromatic nucleus. The functional linearity for the rearrangement of metabolites to enter another round of RCP or CAP results in relatively low in- and out-degrees of the metabolic compounds. If more benzene rings still remain, the RCP and SCP are repeated until the pathways produce an intermediate that can enter the CAP. Thus, the number of RCP-SCP repetitions depends on the number of benzene rings in HMW PAHs. Diverse PAHs are transformed into the common metabolic intermediate, protocatechuate, which is then funnelled into the CAP. The CAP consists of a series of reactions for conversion of protocatechuate to the aliphatic compounds, acetyl-CoA and succinyl-CoA, which can directly enter central metabolism. In *M. vanbaalenii* PYR-1, the β-ketoadipate pathway functions to complete the pathway to the TCA cycle intermediates.

Table 9.1 shows characteristics of the functional modules (or processes) for the PAH-MN. Structures and functions of the RCP and SCP modules apparently are funnel-shaped, with a wide conical mouth and a narrow stem, whereas the CAP shows almost the same input and output diameters in terms of its function (Fig. 9.5). As clearly revealed in the proteomic data, the function of the RCP enzymes is substrate-dependent, and their expression is tightly regulated, whereas the CAP enzymes are relatively loosely regulated and functionally shared in common. In contrast, the SCP enzymes are dispersed in modularity and redundant in function. Therefore, the functional modularity of the PAH-MN is in a funnel-shaped structure with a hierarchical functional interaction, which supports the metabolite-centric, scale-free characteristics of the network. In addition, functional module-specific regulation allows more metabolic benefits from the funnel effect. Practical metabolic benefits of the funnel effect are as follows: (i) it enhances input diversity with the controlled production

Table 9.1 Characteristics of functional processes for the degradation of HMW PAHs

	Characteristics of		
Parameter	RCP	SCP	CAP
Input	(HMW) PAHs	Ring cleavage compounds	Protocatechuate
Output	Ring cleavage compounds	Pyruvate Metabolites acceptable to RCP	Acetyl-CoA/ succinyl-CoA
Metabolite centric	Relatively highly branched pathways High-degree metabolites Gradual increase in MW	Linear or less branched pathways Low-degree metabolites Gradual decrease in MW	Partial RCP and SCP Linear pathway
Enzyme centric	O_2 dependent Non-productive metabolites Tightly regulated Diverse substrate specificity Pathway determining Preparative for SCP Mostly belong to Enzyme Commission 1	Productive Constitutively/loosely regulated Generally broad substrate specificities Preparative for RCP or CAP Belong to Enzyme Commission 1–5	Productive Constitutive expression Entering TCA cycle Belong to Enzyme Commission 1–5
Enzymes	RHO, dihydrodiol dehydrogenase, ring-cleavage dioxygenase	Hydratase-aldolase Aldolase Hydrolase Decarboxylase	Enzymes for β-ketoadipate pathway

of limited outputs, which concentrates the flux of intermediates to central metabolism; (ii) it increases connectivity between functional modules, which decreases the epimetabolome, or freely diffusible metabolic pool (de Lorenzo, 2008); (iii) it allows more coordinated regulation; and (iv) it enhances linearity of the metabolic pathways, which reduces dissipation of metabolites, ensuring a more efficient metabolic flow (Kweon *et al.*, 2011). Conclusively, the metabolic behaviour of *M. vanbaalenii* PYR-1 coordinates all the metabolic activities, so that it can keep the funnel effects in the biodegradation of HMW PAHs, which are metabolically ambiguous and may serve either as nutrients or as toxicants. Such a channel management, with an apparent preferred route(s) or pathway(s) for each substrate in the PAH-MN, evidently allows more metabolic and toxicological benefits from HMW PAHs that have metabolic ambiguity (Kweon *et al.*, 2015).

Recently, using a gene perturbation model of *M. vanbaalenii* PYR-1, an attractive approach has been used to reconstruct an RHO-centric functional map, which provides numerical scores of the relative contribution of each RHO enzyme in the PAH-MN (Kweon *et al.*, 2015; Kweon *et al.*, 2014). According to the model, the total numerical score of each enzyme at the level of the network represents the pleiotropic and epistatic numerical scores. Fig. 9.6 shows the RHO-centric functional maps for pyrene and phenanthrene pathways in the PAH-MN from wild-type *M. vanbaalenii* PYR-1 and a *pdoA2* mutant designated strain 6G11. The functional maps numerically describe the pleiotropic and epistatic contribution of the initial RHO enzyme systems to the degradation of PAH substrates. *M. vanbaalenii* PYR-1 oxidizes pyrene and phenanthrene through different degradation routes, the productive C-4,5 dioxygenation route and the non-productive C-1,2 dioxygenation route for pyrene; and the productive C-3,4 and C-9,10 dioxygenation routes and the non-productive C-1,2 dioxygenation route for phenanthrene (Kweon *et al.*, 2010b). According to the proposed functional map, the three type V RHO enzymes, NidAB, NidA3B3, and PdoA2B2, work together in terms of epistatic functional interaction. In the regiospecific dioxygenation of pyrene at C-4,5, the major contribution comes from NidAB, while the NidA3B3 RHO system only contributes to the C-1,2 hydroxylation, which leads to the non-productive route of pyrene degradation.

In the phenanthrene pathway, two RHO systems, NidAB and PdoA2B2, work together (that is, in the pleiotropic and epistatic functions) for both productive dioxygenation routes, the main C-3,4 route and the minor C-9,10 route, whereas NidA3B3 only catalyses the non-productive phenanthrene C-1,2 hydroxylation route. Interestingly, pyrene and phenanthrene induce NidAB and PdoA2B2 but not NidA3B3. Considering the pleiotropic and epistatic contributions to the biodegradation of pyrene and phenanthrene, regulation of the RHO systems in such a pattern is reasonable in terms of channel management. Avoiding up-regulation of NidA3B3 in the presence of pyrene and phenanthrene may reduce the chance of non-productive C-1,2 dioxygenation. The NidA3B3 system regiospecifically guides the degradation of fluoranthene exclusively into the productive C-2,3 dioxygenation route among the five possible degradation routes (Kim *et al.*, 2012; Kweon *et al.*, 2014). Genetic perturbation in the PdoA2B2 system also has apparent metabolic perturbation impacts, by losing the enzyme's epistatic function, resulting in a lower rate of pyrene degradation than in the wild-type (Kweon *et al.*, 2014). As a whole, metabolic feedback in the PAH-MN exists through successful functional interactions, in epistatic and pleiotropic combinations, among catabolic enzymes at the levels of functional modules and the network (Kim *et al.*, 2012; Kweon *et al.*, 2014). The pleiotropic and epistatic control of the key enzymes in the PAH-MN might be a key of the module-centric metabolic behaviour to enhance the funnel effects, that is, channel management. From the perspective of the degradation of a wide range of PAHs, pleiotropic and epistatic functional effects of catabolic enzymes are the only way to control the metabolic quality and quantity of the PAH-MN.

9.4.3 Evolution of the PAH-MN

Since phylogenetic network modules are evolutionarily conserved as functional units in the metabolic network (Yamada *et al.*, 2006), a probable evolutionary pathway of the PAH-MN was traced by an investigation of the structures of evolutionary modules and their relationships to the functional modules (Kweon *et al.*, 2011). Analysis of the genomes of 23 bacteria capable of degrading aromatic hydrocarbons, including *M. vanbaalenii* PYR-1 and four other PAH-degrading

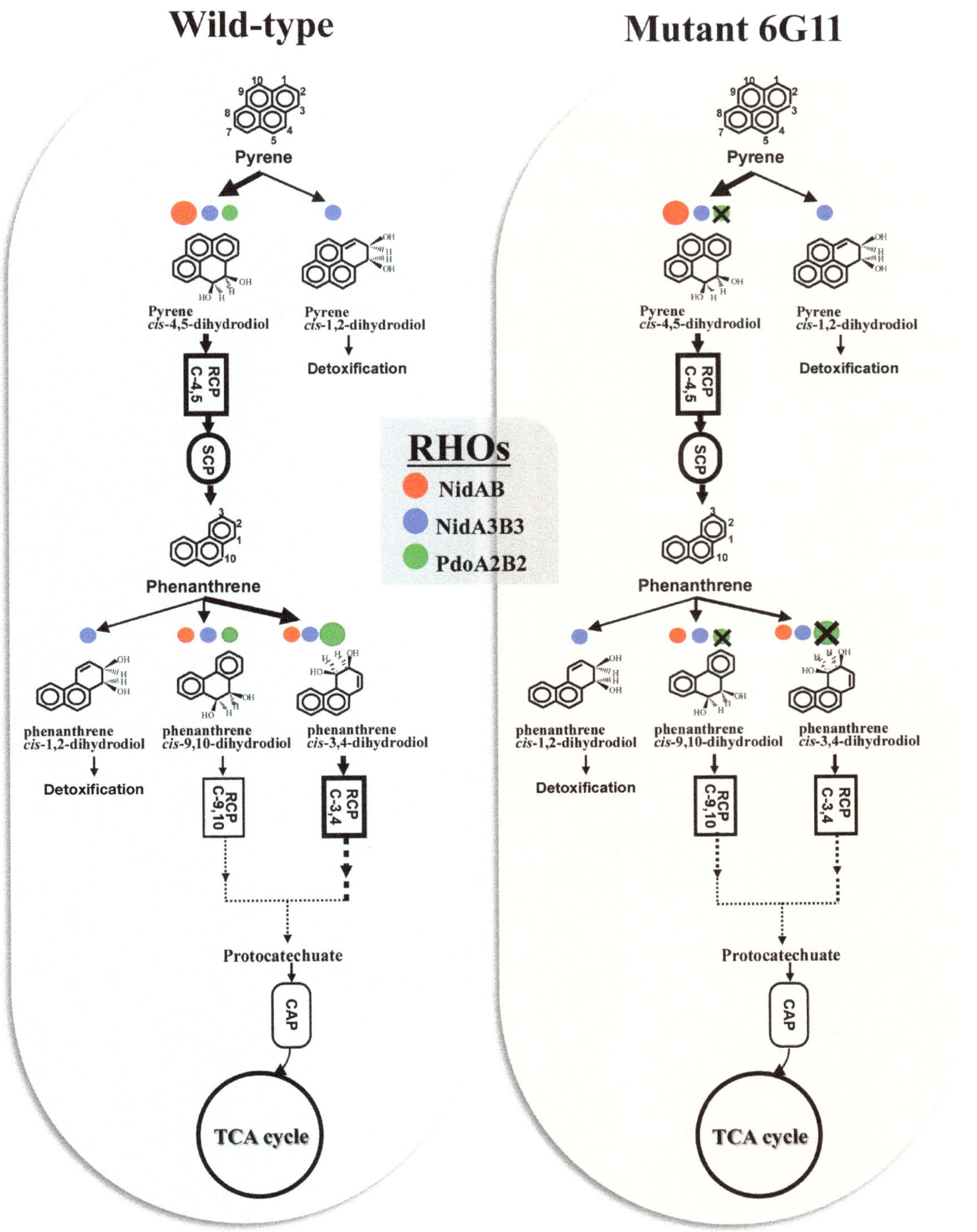

Figure 9.6 RHO-centric functional maps of wild-type *M. vanbaalenii* PYR-1 and a *pdoA2* mutant, 6G11, in the PAH-MN. They show the relative contributions of the three RHO enzyme systems to the pathways of phenanthrene and pyrene degradation. Coloured circles indicate RHO enzymes, with the sizes being proportional to the degrees of functional contribution. The arrow indicates direction of degradation and the arrow thickness represents the difference in the efficiency of transformation.

mycobacteria, with respect to HGT and the standard deviations ($\sigma_{\%G+C}$) (McLeod *et al.*, 2006), shows the contribution of the genomic islands to the evolution of PAH metabolism and the degree to which recent HGT shapes the genomes. All five PAH-degrading mycobacteria, isolated from geographically diverse locations, share the conserved ~150-kb catabolic gene region A, which is mainly responsible for PAH degradation in *M. vanbaalenii* PYR-1, with slightly different genetic configurations and values of $\sigma_{\%G+C}$ ranging from 2.99 to 3.17. This suggests that mycobacterial PAH-degraders may have acquired the gene cluster from a common ancestor by occasional ancient HGT events rather than by vertical descent, promoting rapid pathway evolution that likely conferred immediate selective advantage to the recipients. Analysis of the conserved ~150 kb region A also suggests that the PAH-degrading genes have been recruited from separate HGT events from neighbours with close physiological, biochemical, and phylogenetic relationships. Three genomic islands, GI-I, II, and III, have been identified in region A, occupying 57% (85 kb) of the 150 kb, and show a high sequence similarity to the catabolic genes of other Actinomycetales, such as *Terrabacter* and *Nocardioides*.

The GI-I contains the genes responsible for the RCP and SCP (Fig. 9.7). It includes enzymes responsible for the degradation of the hub metabolite phthalate to protocatechuate, such as the type V RHO system, *phtAaAb*, encoding a phthalate dioxygenase, and a type V electron transfer chain (ETC) system, *phtAcAd*, encoding a [3Fe–4S]-type ferredoxin and a GR-type reductase, respectively. The GI-II contains an RCP gene, *nidAB*, which is mainly involved in the productive regiospecific C-4,5 dioxygenation of the pyrene pathway (Kweon *et al.*, 2011). The GI-III contains four oxygenases involved in RCP and other enzymes involved in SCP. Two of these oxygenase genes encode the type V RHO systems, NidA3B3 and PdoA2B2, which are responsible mainly for the productive dioxygenation of fluoranthene and the three-ringed PAHs, phenanthrene, fluorene, and anthracene, respectively (Kweon *et al.*, 2011). The other two oxygenase genes belong to type X (Kweon *et al.*, 2008). The gene cluster for CAP (the *pca* operon for the β-ketoadipate pathway) is not in the catabolic regions of the genomic islands. This strongly suggests that the *pca* operon was not acquired through recent HGT at about the same time as the three genomic islands. Its acquisition probably precedes

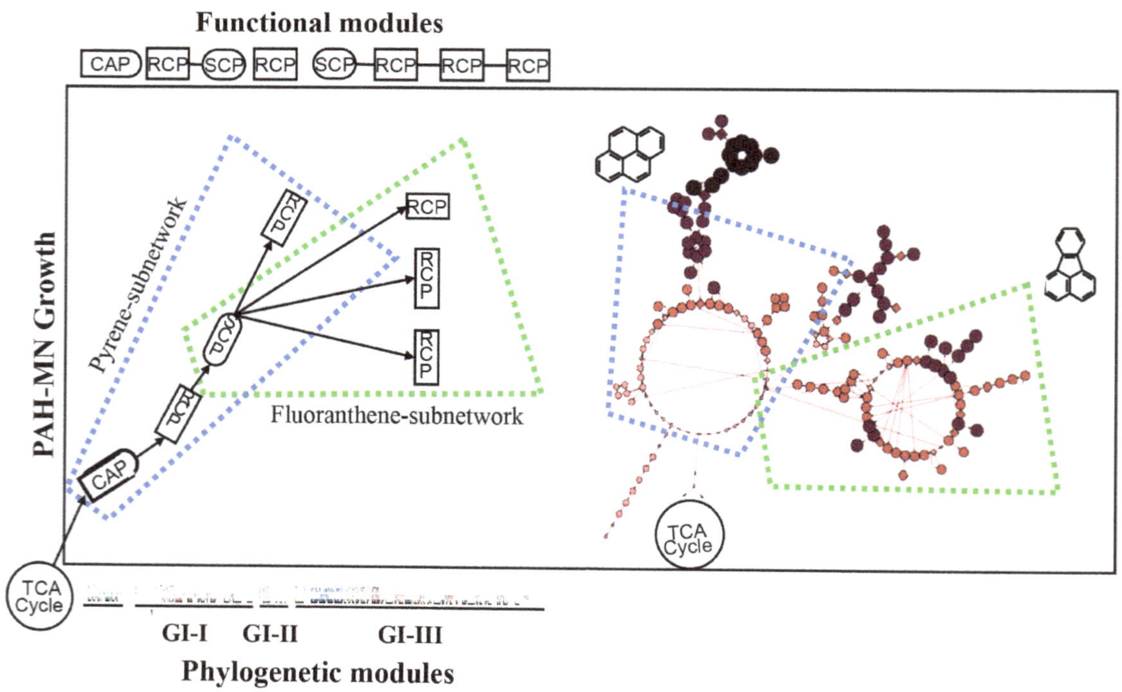

Figure 9.7 The relationships between phylogenetic and functional modules and the growth of the PAH-MN.

the three genomic islands, whose functions are mainly in the RCP and SCP. As discussed below, structural stability of the *pca* operon is strongly correlated with the PAH-degrading ability of the mycobacterial genome. All species of PAH-degrading mycobacteria have the *pca* gene cluster, which is resistant to operon breakage, whereas all non-PAH-degrading mycobacteria have the *pca* operon in a drastically reduced structure, which has lost several genes or has even been completely dispersed on the chromosome (Kweon *et al.*, 2015). Some PAH catabolic genes identified on genomic islands have been located on the plasmids of PAH-degrading mycobacteria, such as *Mycobacterium* spp. MCS and KMS, which further supports the hypothesis of separate HGT events (Kweon *et al.*, 2015).

Although GI-II and GI-III function mainly in the pyrene and fluoranthene subnetworks, respectively, each phylogenetic module cannot provide all the enzymes required for the pyrene and/or fluoranthene subnetwork (Fig. 9.7). For example, phylogenetic modules GI-I and GI-II cannot support all enzymes for the degradation steps from pyrene to protocatechuate, due to the absence of two enzymes, decarboxylase (Mvan_0543) and dihydrodiol dehydrogenase (Mvan_0544). These genes, located on GI-III, show low genetic and functional redundancy, which indicates that all three genomic islands work together in the complete pyrene subnetwork. The GI-III RHO enzymes NidA3B3, Mvan_0539/0540, and Mvan_0546/0547, which function in the lateral or angular dioxygenation of fluoranthene, 9-fluorenone, acenaphthylene, and naphthalene, are also essential for the fluoranthene subnetwork. As mentioned earlier, two oxygenases, NidA3B3 and Mvan_0539/0540, are up-regulated by fluoranthene but not by pyrene, and the enzyme NidA3B3 dioxygenates fluoranthene regiospecifically to fluoranthene *cis*-2,3-dihydrodiol with the highest conversion rate (Kweon *et al.*, 2007). Collectively, this indicates that functional interaction and complementation among the three genomic islands are necessary for complete pyrene and fluoranthene subnetworks. The recruitment of metabolic capabilities towards pyrene and fluoranthene probably occurred at about the same time, and it indicates that the two HMW PAHs are the preferential choices of *M. vanbaalenii* PYR-1, which is consistent with the metabolic behaviour of the PAH-MN.

Functional association of the phylogenetic modules analysed in the PAH-MN also explains a fundamental building principle of the PAH-MN, which is a patchwork assembly of the phylogenetic modules, with a backward evolutionary direction from CAP via SCP to RCP (Kweon *et al.*, 2015). Considering the ambiguous metabolic effects of HMW PAHs, the build-up of the PAH-MN is to be in the best interests of the mycobacterium from both architectural and functional aspects. It increases architectural and functional connectivity among phylogenetic modules and decreases the epimetabolome, including toxic intermediates, such as diol epoxides. Initially, a common ancestor progressively recruited the phylogenetic modules in a way that functions co-ordinately, which promoted rapid pathway evolution, and which eventually enabled mycobacteria to overcome evolutionary pressure, enhancing functional interactions among the functional modules. In this respect, mycobacterial PAH metabolism is an epistasis-centric phenotype, resulting from functional combination of two or more genetic modules. Evidently, it requires continuous active investment and sophisticated management efforts to obtain and keep the catabolic capability.

9.5 Functional pan-genomics of the genus *Mycobacterium*

The PAH metabolism phenotype can be an attractive backup option for an alternative carbon-energy source for free-living mycobacteria, although it seems to have been dispensable for the host-dependent, parasitic species. However, due to the metabolic ambiguity of PAH substrates as nutrients or toxicants, evolution of the phenotype might have been neither simple nor completely random. Recently, a full evolutionary story of the PAH-degrading phenotype in the genus *Mycobacterium* has been proposed to address this question (Kweon *et al.*, 2015).

As shown in Fig. 9.8, this story is based on a new approach, network-based functional pan-genomics (NBFPG), which employs three key concepts: networks, pan-genomics, and functional genomics. Initially, genomic and phenotypic information of 27 completely genome-sequenced mycobacteria were collected and then systematically integrated to generate a practical mycobacterial compendium,

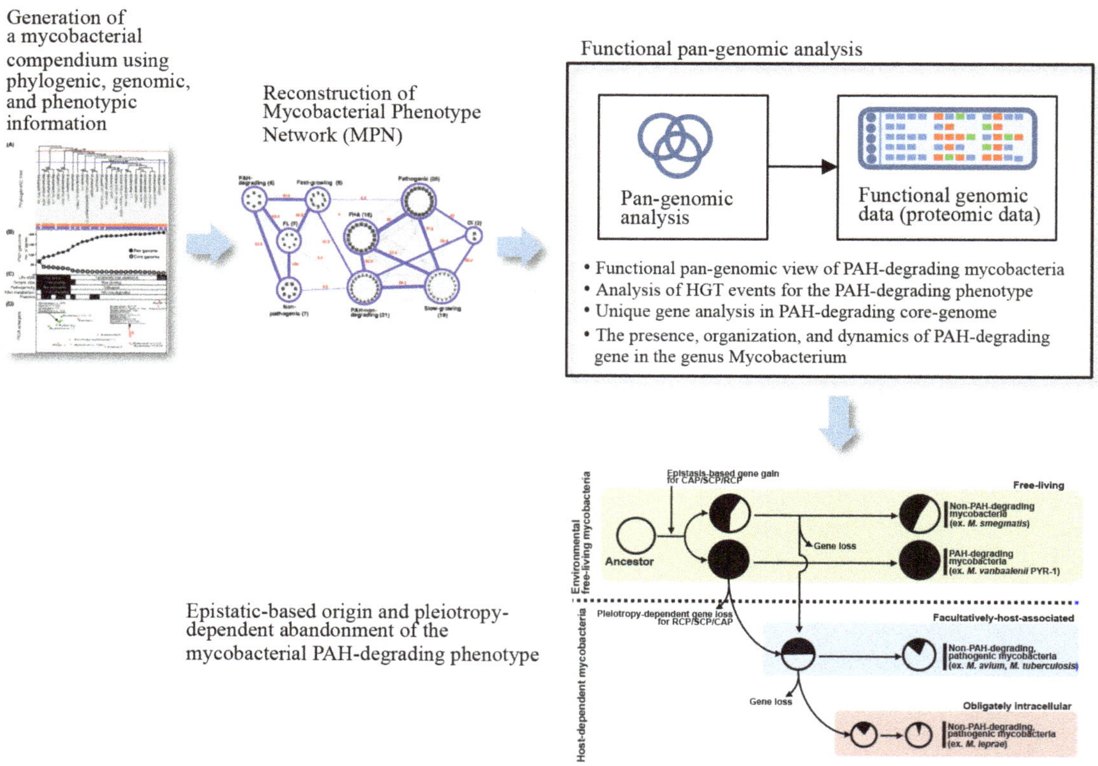

Figure 9.8 Functional pan-genomic approach employing three key concepts, networks, pan-genomics, and functional genomics, for the full evolutionary story of the PAH-degrading phenotype in the genus *Mycobacterium*.

which provides well-organized information of phylogeny, pan-genome, phenotypic and functional clustering. In a second step, based on the mycobacterial compendium, a mycobacterial phenotype network (MPN) was reconstructed, which represents phenotypic relationships in the genus *Mycobacterium*. The MPN is scale-free, showing apparent connection preference. Interestingly, the 'PAH-degrading' phenotype has the lowest connection degree of 3 and shows strong connection to the free-living phenotype nodes, the 'fast-growing' and the 'non-pathogenic' node. However, no connection of the 'PAH-degrading' phenotype to the pathogenic phenotype nodes was observed. The following MPN-based functional pan-genomic analyses provided conserved and unique genomic evidence for strong epistatic and pleiotropic pressures on evolution of PAH metabolism in the genus *Mycobacterium*. This multidisciplinary approach provides functional pan-genomic evidence at the level of the network for the nature, the underlying mechanisms, and the phenotypic feedbacks of epistatic and pleiotropic evolutionary effects of PAH metabolism. Under strong selection pressure, the bottom-up view of PAH-degrading genes from HGT/deletion events hypothesized a plausible evolutionary trajectory, an epistasis-based gene gain and pleiotropy-dependent loss for the PAH-degrading phenotype in the genus *Mycobacterium* (Fig. 9.9) (Kweon *et al.*, 2015).

9.6 Conclusions

During the last three decades, studies of *M. vanbaalenii* PYR-1 and related species have increased our knowledge of the mechanism of bacterial degradation of HMW PAHs. Molecular genetics and biochemical studies have characterized PAH metabolic intermediates, genes and enzymes involved in the pathways of HMW PAH degradation. However, new systems biology approaches with bioinformatics are particularly important for gaining a deeper understanding of the mechanisms of HMW PAH degradation. As shown in this review, the functions of genes, enzymes, metabolites, and all other cellular factors are understood in a cellular dynamic

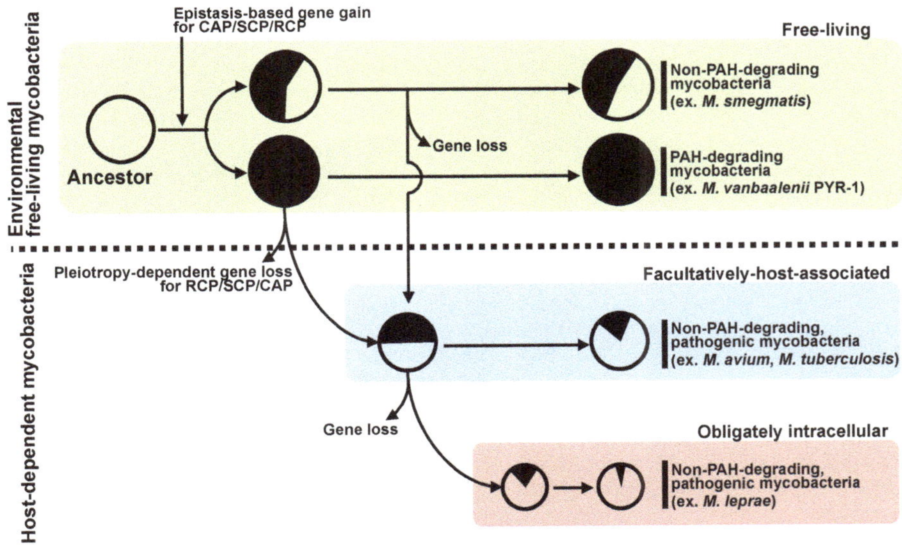

Figure 9.9 A plausible evolutionary pathway of the PAH-degrading phenotype in the genus *Mycobacterium*.

context, providing information on how each of these factors involved in the various metabolic pathways is controlled at the system-wide level and how they evolve for different PAH substrates and environments.

Disclaimer

The views presented in this article do not necessarily reflect those of the Food and Drug Administration.

References

Albert, R. (2005). Scale-free networks in cell biology. J. Cell Sci. *118*, 4947–4957.

Assinder, S.J., and Williams, P.A. (1990). The TOL plasmids: determinants of the catabolism of toluene and the xylenes. Adv. Microb. Physiol. *31*, 1–69.

Atlas, R.M. (1988). Biodegradation of hydrocarbons in the environment. Basic Life Sci. *45*, 211–222.

Atlas, R.M., Stoeckel, D.M., Faith, S.A., Minard-Smith, A., Thorn, J.R., and Benotti, M.J. (2015). Oil biodegradation and oil-degrading microbial populations in marsh sediments impacted by oil from the Deepwater Horizon well blowout. Environ. Sci. Technol. *49*, 8356–8366.

ATSDR (2011). Priority list of hazardous substances (Agency for Toxic Substances and Disease Registry).

Barabasi, A.L., and Albert, R. (1999). Emergence of scaling in random networks. Science *286*, 509–512.

Blumer, M. (1976). Polycyclic aromatic compounds in nature. Sci. Am. *234*, 35–45.

Boldrin, B., Tiehm, A., and Fritzsche, C. (1993). Degradation of phenanthrene, fluorene, fluoranthene, and pyrene by a *Mycobacterium* sp. Appl. Environ. Microbiol. *59*, 1927–1930.

Brezna, B., Kweon, O., Stingley, R.L., Freeman, J.P., Khan, A.A., Polek, B., Jones, R.C., and Cerniglia, C.E. (2006). Molecular characterization of cytochrome P450 genes in the polycyclic aromatic hydrocarbon degrading *Mycobacterium vanbaalenii* PYR-1. Appl. Microbiol. Biotechnol. *71*, 522–532.

Carver, T.J., Rutherford, K.M., Berriman, M., Rajandream, M.A., Barrell, B.G., and Parkhill, J. (2005). ACT: the Artemis Comparison Tool. Bioinformatics *21*, 3422–3423.

Cerniglia, C.E. (1992). Biodegradation of polycyclic aromatic hydrocarbons. Biodegradation 3, 351–368.

Cerniglia, C.E. (2003). Recent advances in the biodegradation of polycyclic aromatic hydrocarbons by *Mycobacterium* species. In The utilization of bioremediation to reduce soil contamination: problems and solutions, V. Šašek, ed. (Netherlands: Kluwer Academic Publishers), pp. 51–73.

Cheung, P.Y., and Kinkle, B.K. (2001). *Mycobacterium* diversity and pyrene mineralization in petroleum-contaminated soils. Appl. Environ. Microbiol. *67*, 2222–2229.

Cheung, P.Y., and Kinkle, B.K. (2005). Changes in *Mycobacterium* spp. population structure and pyrene mineralization in polycyclic aromatic hydrocarbon-amended soils. Soil Biol. Biochem. *37*, 1929–1937.

Cohen, P. (2002). Protein kinases – the major drug targets of the twenty-first century? Nat. Rev. Drug Discov. *1*, 309–315.

de Lorenzo, V. (2008). Systems biology approaches to bioremediation. Curr. Opin. Biotechnol. *19*, 579–589.

Dean-Ross, D., and Cerniglia, C.E. (1996). Degradation of pyrene by *Mycobacterium flavescens*. Appl. Microbiol. Biotechnol. *46*, 307–312.

Dean-Ross, D., Moody, J., and Cerniglia, C.E. (2002). Utilization of mixtures of polycyclic aromatic hydrocarbons by bacteria isolated from contaminated sediment. FEMS Microbiol. Ecol. *41*, 1–7.

Debruyn, J.M., Chewning, C.S., and Sayler, G.S. (2007). Comparative quantitative prevalence of Mycobacteria and functionally abundant *nidA*, *nahAc*, and *nagAc* dioxygenase genes in coal tar contaminated sediments. Environ. Sci. Technol. *41*, 5426–5432.

DeBruyn, J.M., Mead, T.J., and Sayler, G.S. (2011). Horizontal transfer of PAH catabolism genes in *Mycobacterium*: evidence from comparative genomics and isolated pyrene-degrading bacteria. Environ. Sci. Technol. *46*, 99–106.

Ding, G.C., Heuer, H., Zuhlke, S., Spiteller, M., Pronk, G.J., Heister, K., Kogel-Knabner, I., and Smalla, K. (2010). Soil type-dependent responses to phenanthrene as revealed by determining the diversity and abundance of polycyclic aromatic hydrocarbon ring-hydroxylating dioxygenase genes by using a novel PCR detection system. Appl. Environ. Microbiol. *76*, 4765–4771.

Fritzsche, C. (1994). Degradation of pyrene at low defined oxygen concentrations by a *Mycobacterium* sp. Appl. Environ. Microbiol. *60*, 1687–1689.

Fuentes, S., Mendez, V., Aguila, P., and Seeger, M. (2014). Bioremediation of petroleum hydrocarbons: catabolic genes, microbial communities, and applications. Appl. Microbiol. Biotechnol. *98*, 4781–4794.

Grosser, R.J., Warshawsky, D., and Vestal, J.R. (1991). Indigenous and enhanced mineralization of pyrene, benzo[*a*]pyrene, and carbazole in soils. Appl. Environ. Microbiol. *57*, 3462–3469.

Guerin, W.F., and Jones, G.E. (1988). Mineralization of phenanthrene by a *Mycobacterium* sp. Appl. Environ. Microbiol. *54*, 937–944.

Hall, K., Miller, C.D., Sorensen, D.L., Anderson, A.J., and Sims, R.C. (2005). Development of a catabolically significant genetic probe for polycyclic aromatic hydrocarbon-degrading mycobacteria in soil. Biodegradation *16*, 475–484.

Hartmans, S., De Bont, J.A., and Stackebrandt, E. (2006). The Genus *Mycobacterium*-Nonmedical, 3rd edn (New York, NY: Springer Verlag).

Heitkamp, M.A., and Cerniglia, C.E. (1988). Mineralization of polycyclic aromatic hydrocarbons by a bacterium isolated from sediment below an oil field. Appl. Environ. Microbiol. *54*, 1612–1614.

Heitkamp, M.A., Franklin, W., and Cerniglia, C.E. (1988a). Microbial metabolism of polycyclic aromatic hydrocarbons: isolation and characterization of a pyrene-degrading bacterium. Appl. Environ. Microbiol. *54*, 2549–2555.

Heitkamp, M.A., Freeman, J.P., Miller, D.W., and Cerniglia, C.E. (1988b). Pyrene degradation by a *Mycobacterium* sp.: identification of ring oxidation and ring fission products. Appl. Environ. Microbiol. *54*, 2556–2565.

Hennessee, C.T., Seo, J.S., Alvarez, A.M., and Li, Q.X. (2009). Polycyclic aromatic hydrocarbon-degrading species isolated from Hawaiian soils: *Mycobacterium crocinum* sp. nov., *Mycobacterium pallens* sp. nov., *Mycobacterium rutilum* sp. nov., *Mycobacterium rufum* sp. nov. and *Mycobacterium aromaticivorans* sp. nov. Int. J. Syst. Evol. Microbiol. *59*, 378–387.

Hites, R.A., Laflamme, R.E., and Farrington, J.W. (1977). Sedimentary polycyclic aromatic hydrocarbons: the historical record. Science *198*, 829–831.

IARC (1983). Monograph on the evaluation of carcinogenic risk of chemicals to man: Polynuclear aromatic compounds. Part 1: chemical, environmental and experimental data. 32 (World Health Organization. Lyon, France).

Jeong, H., Tombor, B., Albert, R., Oltvai, Z.N., and Barabasi, A.L. (2000). The large-scale organization of metabolic networks. Nature *407*, 651–654.

Kanaly, R.A., and Harayama, S. (2011). Advances in the field of high-molecular-weight polycyclic aromatic hydrocarbon biodegradation by bacteria. Microb. Biotechnol. *3*, 136–164.

Kelley, I., Freeman, J.P., Evans, F.E., and Cerniglia, C.E. (1991). Identification of a carboxylic acid metabolite from the catabolism of fluoranthene by a *Mycobacterium* sp. Appl. Environ. Microbiol. *57*, 636–641.

Kelley, I., Freeman, J.P., Evans, F.E., and Cerniglia, C.E. (1993). Identification of metabolites from the degradation of fluoranthene by *Mycobacterium* sp. strain PYR-1. Appl. Environ. Microbiol. *59*, 800–806.

Khan, A.A., Kim, S.J., Paine, D.D., and Cerniglia, C.E. (2002). Classification of a polycyclic aromatic hydrocarbon-metabolizing bacterium, *Mycobacterium* sp. strain PYR-1, as *Mycobacterium vanbaalenii* sp. nov. Int. J. Syst. Evol. Microbiol. *52*, 1997–2002.

Khan, A.A., Wang, R.F., Cao, W.W., Doerge, D.R., Wennerstrom, D., and Cerniglia, C.E. (2001). Molecular cloning, nucleotide sequence, and expression of genes encoding a polycyclic aromatic ring dioxygenase from *Mycobacterium* sp. strain PYR-1. Appl. Environ. Microbiol. *67*, 3577–3585.

Kim, S.J., Jones, R.C., Cha, C.J., Kweon, O., Edmondson, R.D., and Cerniglia, C.E. (2004a). Identification of proteins induced by polycyclic aromatic hydrocarbon in *Mycobacterium vanbaalenii* PYR-1 using two-dimensional polyacrylamide gel electrophoresis and *de novo* sequencing methods. Proteomics *4*, 3899–3908.

Kim, S.J., Kweon, O., and Cerniglia, C.E. (2009). Proteomic applications to elucidate bacterial aromatic hydrocarbon metabolic pathways. Curr. Opin. Microbiol. *12*, 301–309.

Kim, S.J., Kweon, O., and Cerniglia, C.E. (2010). Degradation of polycyclic aromatic hydrocarbons by *Mycobacterium* strains. In Handbook of Hydrocarbon and Lipid Microbiology, K.N. Timmis, ed. (Braunschweig, Germany: Springer), pp. 1865–1880.

Kim, S.J., Kweon, O., Freeman, J.P., Jones, R.C., Adjei, M.D., Jhoo, J.W., Edmondson, R.D., and Cerniglia, C.E. (2006). Molecular cloning and expression of genes encoding a novel dioxygenase involved in low- and high-molecular-weight polycyclic aromatic hydrocarbon degradation in *Mycobacterium vanbaalenii* PYR-1. Appl. Environ. Microbiol. *72*, 1045–1054.

Kim, S.J., Kweon, O., Jones, R.C., Edmondson, R.D., and Cerniglia, C.E. (2008). Genomic analysis of polycyclic aromatic hydrocarbon degradation in *Mycobacterium vanbaalenii* PYR-1. Biodegradation *19*, 859–881.

Kim, S.J., Kweon, O., Jones, R.C., Freeman, J.P., Edmondson, R.D., and Cerniglia, C.E. (2007). Complete and integrated pyrene degradation pathway in *Mycobacterium vanbaalenii* PYR-1 based on systems biology. J. Bacteriol. *189*, 464–472.

Kim, S.J., Song, J., Kweon, O., Holland, R.D., Kim, D.W., Kim, J.N., Yu, L.R., and Cerniglia, C.E. (2012). Functional robustness of a polycyclic aromatic hydrocarbon metabolic network examined in a *nidA* aromatic ring hydroxylating oxygenase mutant of *Mycobacterium vanbaalenii* PYR-1. Appl. Environ. Microbiol. *78*, 3715–3723.

Kim, Y.H., Engesser, K.H., and Cerniglia, C.E. (2005). Numerical and genetic analysis of polycyclic aromatic hydrocarbon-degrading mycobacteria. Microbial Ecol. *50*, 110–119.

Kim, Y.H., Moody, J.D., Freeman, J.P., Engesser, K.H., and Cerniglia, C.E. (2004b). Evidence for the existence of PAH-quinone reductase and catechol-*O*-methyltransferase in *Mycobacterium vanbaalenii* PYR-1. J. Ind. Microbiol. Biotechnol. *31*, 507–516.

Kleespies, M., Kroppenstedt, R.M., Rainey, F.A., Webb, L.E., and Stackebrandt, E. (1996). *Mycobacterium hodleri* sp. nov., a new member of the fast-growing mycobacteria capable of degrading polycyclic aromatic hydrocarbons. Int. J. Syst. Bacterio.l *46*, 683–687.

Krivobok, S., Kuony, S., Meyer, C., Louwagie, M., Willison, J.C., and Jouanneau, Y. (2003). Identification of pyrene-induced proteins in *Mycobacterium* sp. strain 6PY1: evidence for two ring-hydroxylating dioxygenases. J. Bacteriol. *185*, 3828–3841.

Kweon, O., Kim, S.J., Baek, S., Chae, J.C., Adjei, M.D., Baek, D.H., Kim, Y.C., and Cerniglia, C.E. (2008). A new classification system for bacterial Rieske non-heme iron aromatic ring-hydroxylating oxygenases. BMC Biochem. *9*, 11.

Kweon, O., Kim, S.J., Blom, J., Kim, S.K., Kim, B.S., Baek, D.H., Park, S.I., Sutherland, J.B., and Cerniglia, C.E. (2015). Comparative functional pan-genome analyses to build connections between genomic dynamics and phenotypic evolution in polycyclic aromatic hydrocarbon metabolism in the genus *Mycobacterium*. BMC Evol. Biol. *15*, 21.

Kweon, O., Kim, S.J., and Cerniglia, C.E. (2010a). Genomic view of mycobacterial high molecular weight polycyclic aromatic hydrocarbon degradation. In Handbook of Hydrocarbon and Lipid Microbiology, K.N. Timmis, ed. (Braunschweig, Germany: Springer), pp. 1165–1178.

Kweon, O., Kim, S.J., Freeman, J.P., Song, J., Baek, S., and Cerniglia, C.E. (2010b). Substrate specificity and structural characteristics of the novel Rieske nonheme iron aromatic ring-hydroxylating oxygenases NidAB and NidA3B3 from *Mycobacterium vanbaalenii* PYR-1. MBio *1*.

Kweon, O., Kim, S.J., Holland, R.D., Chen, H., Kim, D.W., Gao, Y., Yu, L.R., Baek, S., Baek, D.H., Ahn, H., *et al.* (2011). Polycyclic aromatic hydrocarbon–metabolic network in *Mycobacterium vanbaalenii* PYR-1. J. Bacteriol. *193*, 4326–4337.

Kweon, O., Kim, S.J., Jones, R.C., Freeman, J.P., Adjei, M.D., Edmondson, R.D., and Cerniglia, C.E. (2007). A polyomic approach to elucidate the fluoranthene degradative pathway in *Mycobacterium vanbaalenii* PYR-1. J. Bacteriol. *189*, 4635–4647.

Kweon, O., Kim, S.J., Kim, D.W., Kim, J.M., Kim, H.L., Ahn, Y., Sutherland, J.B., and Cerniglia, C.E. (2014). Pleiotropic and epistatic behavior of a ring-hydroxylating oxygenase system in the polycyclic aromatic hydrocarbon metabolic network from *Mycobacterium vanbaalenii* PYR-1. J. Bacteriol. *196*, 3503–3515.

Leahy, J.G., and Colwell, R.R. (1990). Microbial degradation of hydrocarbons in the environment. Microbiol. Rev. *54*, 305–315.

Lee, S.E., Seo, J.S., Keum, Y.S., Lee, K.J., and Li, Q.X. (2007). Fluoranthene metabolism and associated proteins in *Mycobacterium* sp. JS14. Proteomics *7*, 2059–2069.

Liang, Y., Gardner, D.R., Miller, C.D., Chen, D., Anderson, A.J., Weimer, B.C., and Sims, R.C. (2006). Study of biochemical pathways and enzymes involved in pyrene degradation by *Mycobacterium* sp. strain KMS. Appl. Environ. Microbiol. *72*, 7821–7828.

López, Z., Vila, J., and Grifoll, M. (2005). Metabolism of fluoranthene by mycobacterial strains isolated by their ability to grow in fluoranthene or pyrene. J. Ind. Microbiol. Biotechnol. *32*, 455–464.

Lopez, Z., Vila, J., Minguillon, C., and Grifoll, M. (2006). Metabolism of fluoranthene by *Mycobacterium* sp. strain AP1. Appl. Microbiol. Biotechnol. *70*, 747–756.

Marcos, M.S., Lozada, M., and Dionisi, H.M. (2009). Aromatic hydrocarbon degradation genes from chronically polluted subantarctic marine sediments. Lett. Appl. Microbiol. *49*, 602–608.

McLeod, M.P., Warren, R.L., Hsiao, W.W., Araki, N., Myhre, M., Fernandes, C., Miyazawa, D., Wong, W., Lillquist, A.L., Wang, D., *et al.* (2006). The complete genome of *Rhodococcus* sp. RHA1 provides insights into a catabolic powerhouse. Proc. Natl. Acad. Sci. U.S.A. *103*, 15582–15587.

Miller, C.D., Hall, K., Liang, Y.N., Nieman, K., Sorensen, D., Issa, B., Anderson, A.J., and Sims, R.C. (2004). Isolation and characterization of polycyclic aromatic hydrocarbon-degrading *Mycobacterium* isolates from soil. Microb. Ecol. *48*, 230–238.

Moody, J.D., Freeman, J.P., Doerge, D.R., and Cerniglia, C.E. (2001). Degradation of phenanthrene and anthracene by cell suspensions of *Mycobacterium* sp. strain PYR-1. Appl. Environ. Microbiol. *67*, 1476–1483.

Muangchinda, C., Chavanich, S., Viyakarn, V., Watanabe, K., Imura, S., Vangnai, A.S., and Pinyakong, O. (2015). Abundance and diversity of functional genes involved in the degradation of aromatic hydrocarbons in antarctic soils and sediments around Syowa Station. Environ. Sci. Pollut. Res. Int. *22*, 4725–4735.

Pagnout, C., Frache, G., Poupin, P., Maunit, B., Muller, J.F., and Ferard, J.F. (2007). Isolation and characterization of a gene cluster involved in PAH degradation in *Mycobacterium* sp. strain SNP11: expression in *Mycobacterium smegmatis* mc^{2}155. Res. Microbiol. *158*, 175–186.

Pazos, F., Valencia, A., and De Lorenzo, V. (2003). The organization of the microbial biodegradation network from a systems-biology perspective. EMBO Rep. *4*, 994–999.

Rafii, F., Lunsford, P., Hehman, G., and Cerniglia, C.E. (1999). Detection and purification of a catalase-peroxidase from *Mycobacterium* sp. Pyr-1. FEMS Microbiol. Lett. *173*, 285–290.

Rehmann, K., Hertkorn, N., and Kettrup, A.A. (2001). Fluoranthene metabolism in *Mycobacterium* sp. strain KR20: identity of pathway intermediates during degradation and growth. Microbiology *147*, 2783–2794.

Rehmann, K., Noll, H.P., Steinberg, C.E., and Kettrup, A.A. (1998). Pyrene degradation by *Mycobacterium* sp. strain KR2. Chemosphere *36*, 2977–2992.

Saviola, B., and Bishai, W. (2006). The genus *Mycobacterium*-medical. In The Prokaryotes, S. Falkow, E. Rosenberg, K.-H. Schleifer, and E. Stackebrandt, eds. (New York, NY: Springer Verlag), pp. 919–933.

Schneider, J., Grosser, R., Jayasimhulu, K., Xue, W., and Warshawsky, D. (1996). Degradation of pyrene, benz[*a*]anthracene, and benzo[*a*]pyrene by *Mycobacterium* sp. strain RJGII-135, isolated from a former coal gasification site. Appl. Environ. Microbiol. *62*, 13–19.

Seo, J.S., Keum, Y.S., Harada, R.M., and Li, Q.X. (2007). Isolation and characterization of bacteria capable of degrading polycyclic aromatic hydrocarbons (PAHs) and organophosphorus pesticides from PAH-contaminated soil in Hilo, Hawaii. J Agric Food Chem *55*, 5383–5389.

Sho, M., Hamel, C., and Greer, C.W. (2004). Two distinct gene clusters encode pyrene degradation in *Mycobacterium* sp. strain S65. FEMS Microbiol Ecol *48*, 209–220.

Simon, M.J., Osslund, T.D., Saunders, R., Ensley, B.D., Suggs, S., Harcourt, A., Suen, W.C., Cruden, D.L., Gibson, D.T., and Zylstra, G.J. (1993). Sequences of genes encoding naphthalene dioxygenase in *Pseudomonas putida* strains G7 and NCIB 9816-4. Gene *127*, 31–37.

Stackebrandt, E., Rainey, F., and Ward-Rainey, N.L. (1997). Proposal for a new hierarchic classification system, *Actinobacteria* classis nov. Int. J. Syst. Bacteriol. *47*, 479–491.

Stahl, D.A., and Urbance, J.W. (1990). The division between fast- and slow-growing species corresponds to natural relationships among the mycobacteria. J. Bacteriol. *172*, 116–124.

Stingley, R.L., Brezna, B., Khan, A.A., and Cerniglia, C.E. (2004a). Novel organization of genes in a phthalate degradation operon of *Mycobacterium vanbaalenii* PYR-1. Microbiology *150*, 3749–3761.

Stingley, R.L., Khan, A.A., and Cerniglia, C.E. (2004b). Molecular characterization of a phenanthrene degradation pathway in *Mycobacterium vanbaalenii* PYR-1. Biochem. Biophys. Res. Commun. *322*, 133–146.

van der Meer, J.R., de Vos, W.M., Harayama, S., and Zehnder, A.J. (1992). Molecular mechanisms of genetic adaptation to xenobiotic compounds. Microbiol. Rev. *56*, 677–694.

Van Hamme, J.D., Singh, A., and Ward, O.P. (2003). Recent advances in petroleum microbiology. Microbiol Mol Biol Rev *67*, 503–549.

Vila, J., López, Z., Sabaté, J., Minguillón, C., Solanas, A.M., and Grifoll, M. (2001). Identification of a novel metabolite in the degradation of pyrene by *Mycobacterium* sp. strain AP1: actions of the isolate on two- and three-ring polycyclic aromatic hydrocarbons. Appl. Environ. Microbiol. *67*, 5497–5505.

Wang, R.F., Wennerstrom, D., Cao, W.W., Khan, A.A., and Cerniglia, C.E. (2000). Cloning, expression, and characterization of the *katG* gene, encoding catalase-peroxidase, from the polycyclic aromatic hydrocarbon-degrading bacterium *Mycobacterium* sp. strain PYR-1. Appl. Environ. Microbiol. *66*, 4300–4304.

Yamada, T., Kanehisa, M., and Goto, S. (2006). Extraction of phylogenetic network modules from the metabolic network. BMC Bioinformatics *7*, 130.

Zhang, C., and Anderson, A.J. (2012). Multiplicity of genes for aromatic ring-hydroxylating dioxygenases in *Mycobacterium* isolate KMS and their regulation. Biodegradation *23*, 585–596.

Biosurfactant Enhancement Factors in Microbial Degradation Processes

10

Katarzyna Paraszkiewicz

Contents

Abstract

The efficiency of remediation processes is often limited by a strong hydrophobicity of contaminant compounds. This characteristic facilitates the pollutant sorption to solid particles causing an additional decrease in the contaminant concentration in the aqueous phase. Consequently, hydrophobic compounds are characterized by poor bioavailability for organisms with potent biodegradation activity. Owing to their amphiphilic structure, molecules of surface active agents (surfactants) accumulate at interfaces between immiscible phases and cause a reduction in the interfacial tension. Surfactants also exhibit emulsifying, foaming, and dispersing properties; act as detergents facilitating desorption processes; and increase the apparent water solubility and mobility of hydrophobic compounds. Moreover, surfactants can modify microbial cell properties/activity and have the same influence on contaminant bioavailability and the kinetics of microbial biodegradation processes. Biosurfactants (surfactants of biological origin, mainly produced by microorganisms) exhibit not only multidirectional activity but are also eco-friendly. The full potential of biosurfactants in remediation technologies is not utilized mostly due to the limited understanding of interactions between these compounds and the environment (organisms, contaminants, and environmental abiotic elements), as well as their high production cost.

10.1 A summary of biosurfactant characteristics

10.1.1 Surface properties

Biosurfactants are surface active agents (surfactants) of a biological origin that are synthesized by various organisms, especially microorganisms. A characteristic feature of biological and synthetic surfactants is their amphiphilic (amphipathic) structure. This means that the hydrophilic (polar) and hydrophobic (nonpolar) moieties (described as a head and a tail, respectively) are present simultaneously in a surfactant molecule (Fig.10.1). Surfactants accumulate at the interface between two immiscible phases (e.g. fluid/fluid, air/fluid, solid/fluid) and reduce the interfacial tension, which is described as surface tension in the case of the boundary between air and liquid. It causes a decrease in forces between two dissimilar phases and allows these phases to mix and interact more easily (Desai and Banat, 1997). Depending on the chemical structure, the interface type, and environmental conditions, biosurfactants are able to generate several, sometimes opposing, processes as follows: (1) the reduction of the surface tension, foaming, or the destabilization of foams at the air/liquid interface; (2) the reduction of interfacial tension, stabilization, or destabilization of emulsions at the liquid/liquid interface; (3) wetting, coagulation, or dispersion at the solid/liquid interface; and 4) wetting or anti-static action at the solid–gas interface. As a result, biosurfactants are used in a wide variety of applications, including bioremediation, agriculture, and food-processing, as well as in the petroleum industry and various fields of biomedicine and cosmetics (Cameotra and Makkar, 2004; Banat *et al.*, 2010; Soberón-Chávez and Maier, 2011; Marchant and Banat, 2012).

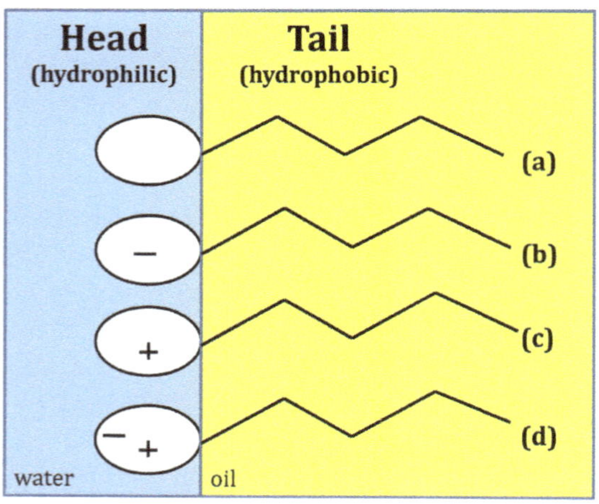

Figure 10.1 Surfactant amphiphilic structure and surfactant classification according to the head part polarity: (a) non-ionic (no charged groups); (b) anionic (negatively charged); (c) cationic (positively charged); (d) amphoteric (zwitterionic – containing two oppositely charged groups).

In addition, these compounds interact with components present on cell envelopes. Consequently, surfactants may have a multidirectional impact on contaminant availability as well as the metabolic activity of microorganisms (Pacwa-Płociniczak *et al.*, 2011; Silva *et al.*, 2014).

10.1.2 Parameters of surface activity

Parameters used most often to characterize and compare the activity of surfactants include (1) the critical micelle concentration (CMC), (2) surface/interfacial tension, (3) hydrophilic–lipophilic balance (HLB) and (4) emulsification activity (Satpute *et al.*, 2010).

Before reaching the CMC value, surfactant molecules are present as monomers and when their concentration reaches CMC, the lowest value of the surface tension is also achieved (Fig. 10.2). After reaching CMC, surfactant molecules tend towards self-assembly and associate forming micelles, bilayers, or vesicles (Hamley, 2015; Liu *et al.*, 2015). In an aqueous medium, hydrophilic parts of surfactant molecules are oriented to the exterior. An inverse (reverse) type of micelle is formed in a non-polar solvent system. Under such conditions, the hydrophilic groups are sequestered in the micelle core and the hydrophobic groups protrude away from the centre (Satpute *et al.*, 2010). Surfactants have their own individual CMC values commonly used to compare surface efficiency. Compounds with a high surface activity have a low CMC value, which means that fewer molecules are required to decrease the surface tension. The CMC values of low-weight biosurfactants are usually lower than that of their synthetic counterparts (Table 10.1). Thus, biosurfactants can be used in much lower concentrations. The most active surfactants (e.g. surfactin) reduce the surface tension to 27–30 mN/m, as compared with 72 mN/m between water and air. Surfactant concentrations in aqueous solutions above the CMC cause an increase in nonpolar compound dispersion (pseudo-solubilization). This phenomenon results in the incorporation of contaminant molecules inside the hydrophobic core of micelles.

The interfacial tension is determined most often between the aqueous solution of the surfactant at CMC and n-hexadecane. The hydrophilic–lipophilic balance (HLB) describes the hydrophobicity

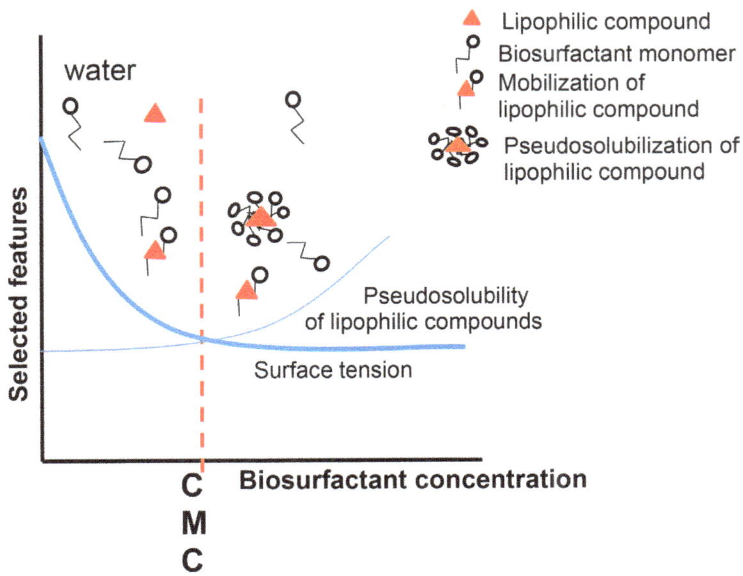

Figure 10.2 The relationship between low-weight biosurfactant concentration, surface tension, and micelles formation with mechanisms of lipophilic compounds removal (adapted from Rosenberg and Ron, 1999; Urum and Pekdemir, 2004; Pacwa-Płociniczak *et al*., 2011).

Table 10.1 Microbial and synthetic surfactants selected parameters values (adapted from Desai and Banat, 1997; Fuguet *et al*., 2005; van Bogaert *et al*., 2007; Arab and Mulligan, 2014)

Surfactant	CMC (mg/l)	Surface tension[a] (mN/m)	Interfacial tension[b] (mN/m)
Surfactin (*B. subtilis*)	13–24	27–32	1
Rhamnolipids (*P. aeruginosa*)	25–30	25–30	0.1–1
Sophorolipids (*C. bombicola*)	11–250	30–40	1.8
Sodium dodecyl sulfate (SDS)	2331	37	0.02

[a]Surface tension at water/air interface (at 25°C) 72 mN/m.
[b]Interfacial tension measured between water/n-hexadecane.

of an amphiphilic compound; this parameter is usually used with synthetic surfactants. Oleic acid and sodium oleate have been chosen as models with the lowest and highest values of HLB (1 and 20, respectively). Surfactants with an HLB value lower than 7 form 'water-in-oil' (reversed) emulsions whereby oil forms the continuous phase in which water droplets are suspended. In the case of compounds with high HLB values, denoting better water solubility, 'oil-in-water' (direct) emulsion is apparent (Satpute *et al.*, 2010).

Emulsifying agents stabilize emulsions (two-phase dispersion systems formed after mixing immiscible liquids). Many surfactants exhibit emulsifying activity leading to the formation of microemulsions (colloidal dispersion of one liquid droplet in another liquid, where the size of droplets is about 10–100 nm in diameter) (Satpute *et al.*, 2010).

10.1.3 Biosurfactant classification

Biosurfactants can be classified according to biological origin (the producing organism), chemical composition, molecular weight, the head group polarity (dissociation pattern in water), physico-chemical properties, and mode of action.

Various organisms (including animals and plants) are able to produce surface active agents, but the highest diversity and widest commercial potential is recognized for microbial biosurfactants, secreted mainly by bacteria and yeast. This recognition is not surprising because most microorganisms tend to attach to surfaces and concentrate at interfaces. Secreted surfactants, due to accumulation at interfaces, alter the conditions prevailing at the boundaries. Generally, bacteria with large surface-to-volume ratios are the most common producers of a wide variety of biosurfactants.

The hydrophilic (head) part of microbial surfactants may contain a phosphate, carboxylic acid, a carbohydrate, amino acid, a peptide, or an alcohol. The hydrophobic domain usually comprises a fatty acid residue, and, in some cases, hydrophobic amino acids of the polypeptide or protein. Based on the chemical structure of the hydrophilic portion, microbial surfactants are arranged into the following classes: (1) glycolipids with many subclasses, e.g. rhamnolipids, sophorolipids, trehalolipids, cellobiose lipids, and mannosylerytritol lipids (MELs); (2) lipopeptides and lipoproteins; (3) fatty acids, phospholipids, and neutral lipids; (4) polymeric surfactants, e.g. peptidoglycolipids, glycoproteins; and (5) polymeric biosurfactants (Table 10.2). Bodour *et al.* described in 2004 a new class of biosurfactants, flavolipids (produced by soil isolate *Flavobacterium* sp.), which have a polar moiety containing citric acid and two cadaverine molecules.

Microbial surfactants are divided into two groups according to molecular weight. The first group comprises low-weight compounds with a molecular weight less than 1600 Da (e.g. lipopeptides, glycolipids, MELs, phospholipids, and fatty acids) that strongly reduce surface and interfacial tension, but exhibit rather poor emulsifying properties. The second group includes high-molecular-weight polymers (e.g. proteins, and lipoproteins), collectively referred to as bioemulsans or bioemulsifiers due to their strong ability to stabilize emulsions (Rosenberg and Ron, 1999).

Based on the head group polarity, non-ionic, anionic, cationic, and zwitterionic surfactants can be distinguished (Fig. 10.1). Microbial surfactants are neutral or anionic and, at the same time, less toxic to organisms than their synthetic (in most cases cationic) counterparts (Paria, 2008).

Table 10.2 Biosurfactants produced by microorganisms (adapted from Bodour, 2004; Mulligan, 2005; van Hamme *et al.*, 2006; Soberón-Chávez and Maier, 2011; Morita *et al.*, 2015)

Biosurfactant class	Type of biosurfactant/biosurfactant example	Example of microbial producer
Glycolipids	Rhamnolipids Sophorolipids Trehalolipids Cellobiose lipids Mannosylerythritol lipids (MELs)	*Pseudomonas aeruginosa*, *Serratia* sp. *Starmerella bombicola (former Candida bombicola)* *Rhodococcus erythropolis*, *Mycobacterium* sp., *Ustilago maydis* *Pseudozyma* sp. *Ustilago* sp.
Flavolipids		*Flavobacterium* sp.
Lipopeptides Lipoproteins	Surfactin, iturin, fengycin Kurstakin Polymyxin Gramicidin Viscosin Putisolvin Syringomycin Serrawetin Lipoprotein	*Bacillus subtilis*, *B. thuringiensis* *B. polymyxa* *B. brevis* *Pseudomonas aeruginosa* *P. putida* *P. syringae* *Serratia marcescens* *Bacillus* sp.
Fatty acids, phospholipids and neutral lipids		*Corynobacterium lepus*, *Penicillium spiculisporum* *Acidithiobacillus thiooxidans* *Nocardia erythropolis*
Polymeric surfactants	Emulsan Biodispersan Liposan Alasan Mannoprotein	*Acinetobacter calcoaceticus* *A. calcoaceticus* *Candida lipolytica* *A. radioresistens* *Saccharomyces cerevisiae*
Particulate surfactants	Extracellular vesicles	*A. calcoaceticus*

Recently, many published reviews have been devoted to the detailed characterization and application potential of microbial surfactants from particular classes or subclasses, e.g. yeast glycolipids (comprising sophorolipides, cellobiose lipids, and MELs) (van Bogaert *et al.*, 2007; Morita *et al.*, 2015; Yu *et al.*, 2015), rhamnolipids (Abdel-Mawgoud *et al.*, 2010; Pornsunthorntawee *et al.*, 2010; Müller *et al.*, 2012), lipopeptides (Raaijmakers *et al.*, 2010; Ines and Dhouha, 2015; Meena and Kanwar, 2015; Mnif and Ghribi, 2015; Walia and Cameotra, 2015), exopolysaccharide-type biosurfactants, and bioemulsifiers produced by macro- and microalgae (Paniagua-Michel *et al.*, 2014).

10.1.4 Natural roles of biosurfactants

Biosurfactants play several physiological and ecological roles in producing microbial cells/cell populations due to their impacts on producer cells and other microorganisms/microbial communities. The most important processes recognized to be involved with biosurfactants include (1) swarming motility, signalling of differentiation, and biofilm formation; (2) the increase of surfaces and nutrient hydrophobic substrate availability; and (3) the modulation of the toxic effect of compounds and heavy metal ions. Many biosurfactants can serve as cytotoxic, antimicrobial (antiviral, antibacterial, antifungal), antitumour, and/or immunosuppressant agents (Raaijmakers *et al.*, 2010).

Van Hamme *et al.* (2006) proposed the classification of biosurfactant roles as follows: (1) intracellular (e.g. nutrient uptake and storage, genes, and enzymes activity regulation); (2) extracellular (e.g. cellular differentiation, cell hydrophobicity changes, and motility); and (3) intracellular (e.g. antagonism, pathogenesis, biofilm formation, and stability). For example, *Bacillus* lipopeptides belong to the surfactin, iturin, and fengycin families exhibit antagonistic activity for a wide range of bacteria, fungi, and oomycetes, and play a role in interactions of *Bacillus* species with plants by stimulating host defence mechanisms against different plant pathogens (Ongena and Jacques, 2008; Moryl *et al.*, 2015).

The natural functions (roles in interactions with coexisting organisms including microorganisms, protozoa, nematodes, and plants) of lipopeptide biosurfactants from *Pseudomonas* and *Bacillus* were recently reviewed by Raaijmakers *et al.* (2010). The intensification of research in areas such as genomics, proteomics, and metabolomics will provide deeper insight into the structures and functions of biosurfactants and will help better understand the physiological importance of biosurfactants to microbial life (van Hamme *et al.*, 2006; Płaza *et al.*, 2015a).

10.1.5 Biosurfactant commercial potential

Synthetic surfactants are mainly petroleum-derived compounds and, due to the low cost and high surface activity, constitute an important group of chemicals widely used in several industrial sectors. As compared with synthetic counterparts, biosurfactants exhibit higher surface activity and better emulsifying and foaming properties, and some of these compounds are able to retain their activity at extreme temperatures, pH, and salinity (Sakthipriya *et al.*, 2015a; Bezza and Chirwa, 2015; Ayed *et al.*, 2015). Moreover, biosurfactants are much more eco-friendly due to the lower toxicity and faster biodegradation. Widespread applicability of microbial surfactants comprise their use as detergents and compounds applied in cosmetics and pharmaceuticals (as antimicrobial or immunomodulatory agents); in the food, petroleum, agrochemical mining, and metallurgical industries; in some processes of dyes, textiles, paper, and ceramics manufacturing; and in environmental management and enhanced oil recovery. Currently, the most advanced level of biosurfactant commercialization has been achieved for the following: (1) rhamnolipids from *P. aeruginosa* used as a fungicide for agricultural purposes and as an additive to enhance bioremediation processes; (2) sophorolipids from *Starmerella bombicola* (former *Candida bombicola*) and *S. apicola* utilized on a wide scale mostly by the cosmetic industry; (3) bacilli lipopeptides, a surfactin, iturin, and fengycin with a broad spectrum of applications due to very high surface and antimicrobial activities; and (4) RAG-1 emulsan from *Acinetobacter calcoaceticus* RAG-1 (Singh *et al.*, 2007; Suthar *et al.*, 2008; Banat *et al.*, 2010; Müller *et al.*, 2012; Van Bogaert *et al.*, 2013; Roelants *et al.*, 2014; Lovaglio *et al.*, 2015; Walia and Cameotra, 2015; Dobler *et al.*, 2016).

Major challenges in the commercialization of microbial surfactants involve the high production cost. According to Müller *et al.* (2012), a sustainable bioeconomic approach combining a holistic X-omics strategy with metabolic engineering can lead to the next step in biosurfactant production mainly based on non-food renewable resources. For example, a variety of cheap growth media based on low-cost agro-industrial raw materials and industrial wastes or by-products have been described (Das and Mukherjee, 2007a; Khondee *et al.*, 2015; Marti *et al.*, 2015; Płaza *et al.*, 2015b).

10.2 Biosurfactants in remediation processes

10.2.1 Introduction

Currently, one of the greatest global problems is an increase in environmental pollution, which severely damages the earth by disturbing the ecological balance and posing serious threats to the health of all living organisms. Common contaminants of water, soil, and sediments are hydrocarbons (including polycyclic hydrocarbons (PAHs)), heavy metals, and a variety of industrially manufactured xenobiotics. Most of these contaminants exhibit toxic, mutagenic, and/or endocrine disrupting activity, exist in the environment as mixtures, and, due to their highly hydrophobic nature, tend to permanent accumulation by sorption to solid particles of soil and sediments. Consequently, there are significant limitations in contaminant recovery through washing techniques, the bioavailability for degrading organisms, and the susceptibility to oxidative/reductive chemicals (Bachmann *et al.*, 2014; Galabova *et al.*, 2014). Required technologies are expected to be not only effective but also environmentally compatible. Therefore, the use of biosurfactants (alone or in combination with other technologies) is suggested as an innovative, economical, and eco-friendly solution (Agarwal and Liu, 2015). Many recently published review papers summarizing current and recently developed remediation techniques (including the use of biosurfactants) have focused on particular polluted ecosystems (e.g. soil, sediments, and seawater) or on methods suitable for the removal of selected pollutant groups (Calvo *et al.*, 2009; Bustamante *et al.*, 2012; Chakraborty and Das, 2014; Silva *et al.*, 2014; Souza *et al.*, 2014; Agarwal and Liu, 2015; Jamaly *et al.*, 2015; Mao *et al.*, 2015). Reviews devoted especially to the environmental application potential of biosurfactants are also available (Mulligan, 2005; Franzetti *et al.*, 2010a; Pacwa-Płociniczak *et al.*, 2011).

10.2.2 The influence of biosurfactants on contaminant (bio)availability

The acceleration of organic pollutants removal/biodegradation mediated by biosurfactants include enhanced desorption, solubilization, emulsification, and pseudo-solubilization. Below the biosurfactant CMC, the mobilization mechanism occurs due to the reduction of interfacial tension followed by the facilitation of desorption processes (Fig. 10.2). Surfactant monomers accumulate at the soil-contaminant and soil–water interfaces and alter the wettability of the system by increasing the contact angle between the soil and hydrophobic contaminants (Urum and Pekdemir, 2004; Mao *et al.*, 2015).

Biosurfactants interact with both abiotic particles and bacterial cells and influence contaminant distribution by altering the conditions at interfaces. Soil interfaces are especially abundant and comprise borders between several components: microbial cells, water, air, particles of soil, and, in the case of contaminants, solid particles or oil droplets. Thus, hydrophobic compound interactions in the soil ecosystem can include (1) solubilization in the water phase; (2) absorption to soil particles; (3) adsorption to cell surfaces; and (4) the formation of a distinct insoluble phase. The soil type can strongly influence remediation efficiency. The characterization of hydrophobic pollutant adsorption/desorption processes is critical for the selection of an effective remediation strategy. Previously reported results indicate that the efficiency of desorption is dependent on the hydrocarbon structure, type of solid fraction, as well as the kind of biosurfactant used in the washing solution (Portet-Koltalo *et al.*, 2013). According to Kavitha *et al.* (2014), the solubilization of crude oil (mediated by the addition of various surfactants) was strongly influenced by soil type, with clay soil exhibiting a higher oil sorption capacity.

In contrast to synthetic surfactants, surface active agents of soil microbes are produced *in situ* and often have a positive effect with less follow-up management, and are therefore effective with regard to technique and cost. Additionally, biosurfactants can stimulate various soil organisms (e.g. microbes and plants) to absorb and decompose contaminants (Mao *et al.*, 2015).

Many reports reveal that the utilization of crude oil (a very complex mixture of alkanes, aromatics, naphthenes, and resins) by a single bacterial strain as well as the microbial consortia is accompanied by an enhanced production of surfactants with high emulsifying activity causing a reduction in both oil viscosity and surface tension and thus improving the microbial adherence to oil (Sakthipriya at al., 2015b). Emulsification resulting in the formation of an emulsion increases the surface area of oils, thus extending the surface area available to biodegraders. Indeed, many studies confirm that the microbial capability to degrade crude oil is accompanied by the production of surfactants exhibiting emulsifying properties (Ibrahim *et al.*, 2013; Ismail *et al.*, 2013). For example, a new, efficient biosurfactant-producing bacterium (identified using partial sequenced 16S rDNA analysis as *Bacillus methylotrophicus*) was isolated from a petroleum reservoir. This bacterium was able to produce a biosurfactant with a high emulsification activity and high foam-forming properties (Chandankere *et al.*, 2013). Pacwa-Płociniczak *et al.* (2014) reported isolation of a hydrocarbon-degrading strain of *Pseudomonas* sp. from soil with heavy petroleum hydrocarbon contamination and proved in that bacterium genome the presence of gene-encoding enzymes responsible for the degradation of alkanes and naphthalene as well as the gene rhl, which is involved in the biosynthesis of rhamnolipid. Results obtained from ^{1}H and ^{13}C nuclear magnetic resonance (NMR), Fourier transform-infrared spectroscopy (FTIR), and mass spectrum analyses indicated that the extracted biosurfactant was in fact affiliated with rhamnolipid. Another microbial isolate reported to produce biosurfactants with strong application potential in environmental remediation is the strain *Bacillus amyloliquefaciens* An6, which secretes a surface active agent with a high emulsifying activity, solubilization efficiency towards diesel oil, and good stability over a wide range of pH, temperatures, and salinity (Ayed *et al.*, 2015). Isolated from creosote-contaminated soil, the strain CN5, identified as *Paenibacillus dendritiformis*, has been recognized as a producer of a novel lipopeptide biosurfactant that is an effective emulsifier with a high thermal, pH, and saline stability and, moreover, has the ability to desorb PAHs and motor oil from spiked soils (Bezza and Chirwa, 2015). Luna and co-workers (2015) proposed the biosurfactant *C. sphaerica* UCP 0995 as a complement to remediation processes involving contaminated water. The yeast biosurfactant proved to be an excellent solubilizer of motor oil in seawater, while nontoxic to indigenous marine microbiota. A thermophilic isolate of *Bacillus subtilis* able to produce biosurfactants and utilize various hydrocarbon substrates was proposed by Sakthipriya *et al.* (2015a) as a potent microorganism for tackling oil spills, wax degradation, flow assurance, and enhanced oil recovery (EOR).

In summary, biosurfactant impacts on remediation processes of organic compounds result from the increase in contaminant (bio)availability and mobility, which are important in bioremediation processes and washing/flushing techniques, respectively (Mulligan, 2014). The application of biosurfactants in the removal of heavy metal ions is targeted at chelating and removing such ions during a washing/flushing step (by the application of a surfactant water solution and foam generated and stabilized by the surfactant present) (Banat *et al.*, 2010).

10.2.3 Biosurfactant application in the petroleum industry

Petroleum is a primary energy source. Yet, at the same time, compounds and by-products of petroleum are major environmental pollutants due to uncontrolled release (losses) during petroleum extraction, transport, and processing. Sources of petroleum contamination might include accidents during transportation by ships and trucks, corrosion leakages of underground storage tanks, and inadequate release of industrial petroleum by-products (Silva *et al.*, 2014). It is also worth mentioning that global petroleum production (around three billion tons a year) relies on transportation by ships and hydrocarbon contamination of the oceans results from routine tanker washing, seabed petroleum leakages, and accidents during petroleum exploration and transportation (Silva *et al.*, 2014; Souza *et al.*, 2014). In the petroleum industry, biosurfactants have been applied for the following: (1) extraction of crude oil from reservoirs (MEOR technique); (2) facilitation of oil mobility during pipeline transport (by the formation of heavy oil–water emulsions); and (3) cleaning of oil storage tanks (Silva *et al.*, 2014). Within these applications, various properties of biosurfactants are utilized (e.g. wetting of solid surfaces, reduction of oil viscosity, reduction of interfacial tension, emulsification, de-emulsification, and oil solubilization) (Pacwa-Płociniczak *et al.*, 2011; Reis *et al.*, 2013).

A significant quantity of crude oil is trapped in reservoirs and conventional enhanced oil recovery (EOR) methods often prove insufficient. Legal mandates increasingly place economic and technical pressure on refineries to improve crude oil recovery using environmentally friendly methods. Therefore, increased attention has been paid to biotechnological solutions such as microbial enhanced oil recovery (MEOR). This technique, in which biosurfactant-producing bacteria or sole biosurfactant solutions are applied, has recently been reviewed in detail (Bachmann *et al.*, 2014; Patel *et al.*, 2015). MEOR technology uses microbes and their biosurfactants to mobilize residual oil in oil reservoirs that is trapped in porous rocks by capillary forces. Biosurfactants reduce capillary forces by reducing the oil–water interfacial tension, altering wettability, and forming micelles (Liu *et al.*, 2015).

10.2.4 Remediation studies

Microbial surfactants display a wide range of chemical structures and properties. This diversity makes a general description of the possible impacts of biosurfactants on remediation processes difficult. Several natural roles of biosurfactants in the growth of producers as well as interactions with other organisms are hypothetically possible (van Hamme *et al.*, 2006). Mechanisms of biosurfactant interactions with various pollutants (e.g. petroleum products, PAHs, and heavy metal ions) as well as the biosurfactant effects on cells and enzyme activities were recently reviewed by Galabova *et al.* (2014).

In bioremediation, a special role is played by the increased bioavailabity of pollutants, based on previously mentioned processes, but the process becomes much more complicated by an additional biosurfactant influence on degrader organism structures/activity as well as by a modulation of contaminant toxicity towards organisms. Thus, the presence of biosurfactants may in some cases limit or even inhibit degradation processes. The addition of surfactin, contrary to the performance of synthetic surfactant Triton X-100, restored (in the case of Cd^{2+} and Pb^{2+} ions) and improved (in the case of Co^{2+}, Cu^{2+}, Ni^{2+} and Mn^{2+}) the swarming motility of *Bacillus cereus* CM100B. Thus, it was suggested that surfactin could improve motility not only by reducing the surface tension of swarming medium but also by binding to metal ions and, consequently, reducing metal ion stress (Singh *et al.*, 2014). Therefore, the effect of biosurfactants, especially on microbial soil bioremediation, is difficult to predict and must be evaluated experimentally on a case-by-case basis (Peng *et al.*, 2007; Franzetti *et al.*, 2008a,b). Biosurfactants even at concentrations below CMC may interact with components of microbial cell envelopes and, in consequence, modulate cell surface hydrophobicity increasing the transport of hydrophobic compounds into microbial cells. For example, a strain of *P. aeruginosa* during growth on hexadecane produced rhamnolipids, causing a lipopolysaccharide (LPS) release from degrader cell outer membranes that resulted in an increase in the cell surface hydrophobicity (Al-Tahhan *et al.*, 2000). On the other hand, Sotirova *et al.* (2009) demonstrated that rhamnolipid instead of LPS changed the composition of outer membrane proteins in cells of biosurfactant-producing *P. aeruginosa*, but that modification also resulted in an increase in the cell surface hydrophobicity. It is suggested that high cell hydrophobicity facilitates direct contact between microorganisms and oil drops or solid hydrocarbons, while low cell hydrophobicity facilitates cell adhesion to biosurfactant micelles (Franzetti *et al.*, 2010a,b). Cameotra and Singh (2009) characterized the mode of hexadecane uptake by a *P. aeruginosa* strain and demonstrated that rhamnolipid produced by cells caused the dispersion of hexadecane into microdroplets, increasing the availability of the hydrocarbon. Electron microscopic studies revealed the uptake of biosurfactant-coated hydrocarbon droplets through a mechanism similar to active pinocytosis.

Microbial biodegradation of hydrophobic contaminants accompanied by biosurfactant production, as reported by many scientists, was excellently reviewed by Pacwa-Płociniczak *et al.* (2011). For example, Obayori *et al.* (2009) described the very efficient removal of crude and diesel oil by a biosurfactant-producing strain of *Pseudomonas* sp. Reddy *et al.* (2010) reported that the biosurfactant-producing *Brevibacterium* sp. PDM-3 strain degraded PAHs such as phenanthrene, anthracene, and fluorene. Kang *et al.* (2010) proved that sophorolipid added to soil contaminated by aliphatic and aromatic hydrocarbons accelerated the rate of contaminant biodegradation. Bezza and Chirwa (2016) studied PAH biodegradation of creosote-contaminated soil using *ex situ* bio-slurry reactors. The addition of a biosurfactant enhanced the degradation of PAHs compared with controls with no biosurfactant and nutrient amendments. A slight decrease in the degradation rate was observed in the simultaneous treatments with a biosurfactant and with nutrient microcosm supplementation, most likely due to preferential microbial consumption of the supplemented biosurfactant. The overall removal of PAHs was determined to be mass transport-limited since the dissolution rate caused by the biosurfactant enhanced the bioavailability of the PAHs to the microorganisms.

Despite many reports describing the isolation of bacterial surfactant producers able to utilize crude oil or aromatic hydrocarbons, only a few reports concern strains for which pesticide degradation is accompanied by surfactant synthesis. One example is a bacterial isolate designated as *Pseudomonas* sp. strain GBS.5 that exhibits the ability to degrade carbazole and at the same time produce a surface active agent with emulsifying activity. Moreover, analysis of the carbazole-degrading genes sequence revealed changes in six different amino acids as compared to other well-established strains (Singh *et al.*, 2013). Also, *Methylobacterium* sp. GPE1, isolated from a former gasworks site, was suggested as a bacterium able to produce surface-active compounds during the utilization of carbazole as a sole source of carbon and energy. According to the measured surface tension, the biodegradation process was initiated by the production of the biosurfactant (Pasternak and Kołwzan, 2013)

Hydrocarbon-polluted sites are often remediated by augmentation with bacterial strains possessing an ability to produce biosurfactants, but without detailed characterization of secreted microbial metabolites. In addition, culture supernatants containing surface active compounds can be applied at the contaminated sites. In environmental practices, surfactant solutions and surfactant microfoams can be applied for contaminant removal (Fig. 10.3). Microfoams offer the advantage of improving contact with the contaminated environment due to their surface properties. Da Rosa and colleagues (2015) studied petroleum and diesel oil removal from soil using microfoams of biological (rhamnolipid) and chemical [sodium dodecyl sulfate (SDS) and cetyltrimethylammonium bromide (CTAB)] surfactants. For both contaminants, the CTAB solution and microfoam had the lowest removal efficiency. Solubilization for petroleum and mobilization for diesel oil were proposed as mechanisms involved in the remediation process.

Continued industrialization has led to widespread contamination of the environment by heavy metals. Among the most toxic heavy metals (as included in the list of priority pollutants published by the US Environmental Protection Agency (US EPA)) are cadmium, copper, lead, mercury, and chromium. The non-degradable nature and toxicity of heavy metals make their remediation crucial for the health and survival of organisms. The mobilization and removal of metal ions facilitated by biosurfactant–metal complexes has recently been reviewed (Chakraborty and Das, 2014; Akcil *et al.*, 2015). The removal of heavy metals and radionuclides from soil involves the mechanisms of dissolution, surfactant-associated complexation, and ionic exchange (Singh and Cameotra, 2004; Mulligan, 2009; Mao *et al.*, 2015). Because most microbial surfactants are anionic, they form complexes with metal ions and the created bonds are stronger than the bonds between metals and soil particles. Therefore, in the case of lowered interfacial tension, metal–biosurfactant complexes are easily desorbed from the soil solid particles to biosurfactant solutions and removed from soil by washing treatments. Metal ions can also be removed from soil surfaces by biosurfactant micelles.

10.2.5 Mixed techniques

Washing solutions containing mixtures of surfactant types (e.g. ionic and non-ionic) are commonly used in soil remediation due to a synergetic effect of surfactants, providing a more efficient desorption and solubilization of pollutants (Mao *et al.*, 2015). A combination of surface active agents and other additives (e.g. organic solvents, chelating agents, and ligand ions) can also enhance the removal of contaminants from soil (Mao *et al.*, 2015).

For example, marine sediment contaminated with heavy metals and PAHs was subjected to electrokinetic treatments and a mixture containing

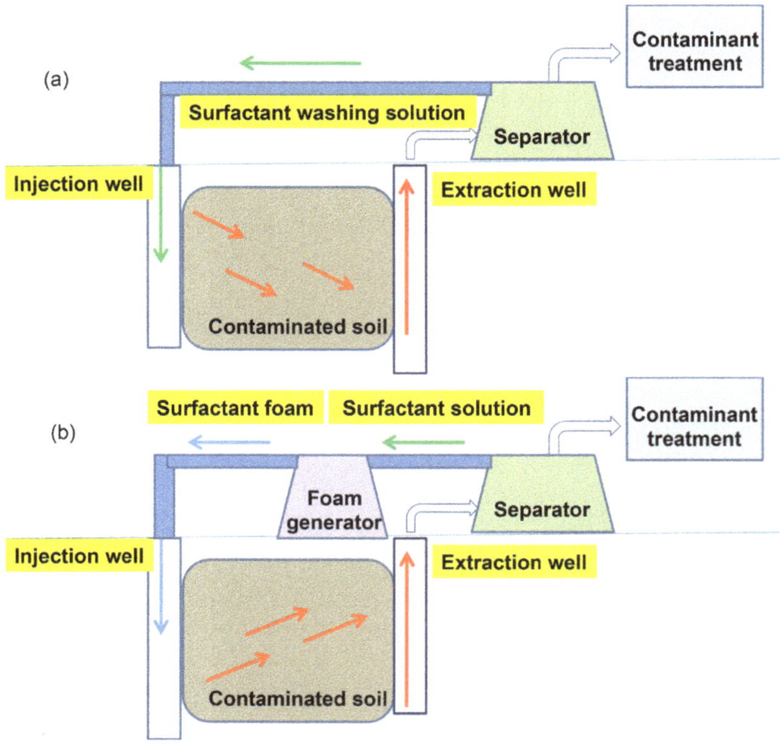

Figure 10.3 Contaminated soil remediation through (a) biosurfactant solution and (b) biosurfactant foam (adapted from US EPA, 1996; Wang and Mulligan, 2004).

citric acid (as a chelating agent) and synthetic or microbial surfactants in the processing fluids. Promising results were obtained with solutions containing rhamnolipids (0.028%) and a viscosin-like biosurfactant produced by *Pseudomonas fluorescens* Pfa7B (0.025%), associated with a periodic voltage gradient (Ammami *et al.*, 2015). A combined application of iodide (I(−)) ligand and surfactants produced by different bacterial species was investigated by Lima *et al.* (2011) for the simultaneous removal of heavy metals and WWA from polluted soil. Microbial surfactants used included flavolipids from *Flavobacterium* and lipopeptides produced by three microbial strains: *Bacillus subtilis, Arthrobacter oxydans,* and *Bacillus* sp. The results indicated efficient removal of organic contaminants and heavy metals by the combined surfactant-ligand treatment. A mixture of surfactin and fengycin (obtained from *Bacillus subtilis* A21) was used by Singh and Cameotra (2013) to remediate soil samples collected from an industrial dumping site. The soil washing technique with a solution of biosurfactants (at a concentration above CMC) removed a significant amount of petroleum hydrocarbon (64.5%) and ions of cadmium (44.2%), cobalt (35.4%), lead (40.3%), nickel (32.2%), copper (26.2%), and zinc (32.1%). Although about half of the biosurfactant amount was sorbed to the soil particles, decreasing the effective concentration during the washing process, the lipopeptide-washed soil exhibited a 100% mustard seed germination. Based on the results, soil washing with a mixture of lipopeptide biosurfactants at concentrations above its CMC was proposed as an efficient and environmentally friendly approach for removing pollutants (petroleum hydrocarbon and heavy metals) from contaminated soil. Chromium ions were removed from an aqueous solution by a two-stage process including the use of ferrous sulfate ($FeSO_4$) for the precipitation of Cr(VI) to Cr(III) followed by heavy metal ion flotation by *P. aeruginosa* rhamnolipid (Abyaneh and Fazaelipoor, 2016). Viisimaa at al. (2013) developed an integrated remediation technique to apply a biosurfactant combined with a chemical–biological treatment for soil contaminated by polychlorinated biphenyl (PCB). Application of the *Pseudomonas* sp. biosurfactant combined with a natural consortium of microorganisms and oxidizing chemicals increased soil respiration and dehydrogenase activity, indicating stimulation of microflora through the integrated process.

Bioleaching is an environmentally friendly and economical technology for removing heavy metals from contaminated soils. Yang and colleagues (2016) addressed bioleaching remediation of heavy metal-contaminated soil using the *Burkholderia* sp. Z-90 strain isolated from cafeteria sewer sludge that exhibited a capacity of both glycolipid biosurfactant and alkaline production. The removal efficiency was determined as 44.0% for Zn^{2+}, 32.5% for Pb^{2+}, 52.2% for Mn^{2+}, 37.7% for Cd^{2+}, 24.1% for Cu^{2+}, and 31.6% for As^{2+}. The heavy metal removal from soil resulted from the adhesion of heavy metal-contaminated soil minerals to strain Z-90 cells and the formation of metal complexes with biosurfactant molecules/micelles.

Phytoremediation is believed to be one of the safest, most innovative, and effective tools for the removal of heavy metals from soil. This technique is enhanced by the assistance of plant growth-promoting (PGP) bacteria that transforms metals into bioavailable and soluble forms through their capability of producing siderophores, organic acids, and biosurfactants (Ullah *et al.*, 2015). Phytobial remediation, integrating the phytoremediation and bioremediation potential of plants and microbes, is considered an eco-friendly and efficient solution, especially useful for the removal of heavy metals from soil and sediment ecosystems (Roy *et al.*, 2015).

Plants also secrete biosurfactants into soil, for example as intermediates during the production of mucilage at the root tip. Sun and colleagues (2015) established that a cowpea (*Vigna unguiculata*) line producing a high amount of root mucilage selectively enriched the phenantrene-degrading bacteria population in the rhizosphere, suggesting that differences in mucilage production may be an important criterion for the selection of the optimal plant species in the remediation of PAH-contaminated soils in the combined phytobial technique.

10.3 Conclusions and future prospects

Experiments conducted to date indicate that microbial biosurfactants can greatly impact the removal of both organic and inorganic contaminants by an involvement in both physicochemical and biological processes. Therefore, biosurfactant activity could be exploited in various applications designed to clean the environment (bioremediation, phytoremediation, and contaminants removal by washing/flushing techniques). Although biosurfactants are environmentally friendly, their commercial potential is still limited by relatively high costs, not fully understood impacts on environmental components (especially organisms existing in soil), and a limited number of field and full-scale remediation studies (Mao *et al.*, 2015). Cheap renewable substrates used as culture media components (e.g. agro-industrial wastes) and novel, efficient methods for biosurfactant isolation are more economically feasible. Another requirement of applying microbial surfactants in real environmental technologies is developing the knowledge of their *in situ* production. One promising clean-up technique seems to be the use of microbial strains exhibiting both biosurfactant production and plant growth promotion properties (Blyth *et al.*, 2015). More information is needed on the diversity and properties of biosurfactants produced by contaminant-degrading organisms present in polluted sites. Conventional screening procedures allow characterizing many new microbial producers of biosurfactants but an advanced methodology such as functional metagenomics could be useful in exploring novel biosurfactants produced from uncultured microbes (Sachdev and Cameotra, 2013).

References

Abdel-Mawgoud, A.M., Lépine, F., and Déziel, E. (2010). Rhamnolipids: diversity of structures, microbial origins and roles. Appl. Microbiol. Biotechnol. *86*, 1323–1336.

Abyaneh, A.S., and Fazaelipoor, H.M. (2016). Evaluation of rhamnolipid (RL) as a biosurfactant for the removal of chromium from aqueous solutions by precipitate flotation. J. Environ. Manage. *165*, 184–187.

Akcil, A., Erust, C., Ozdemiroglu, S., Fonti, V., and Beolchini, F.,(2015). A review of approaches and techniques used in aquatic contaminated sediments: metal removal and stabilization by chemical and biotechnological processes. J. Cleaner Production. *86*, 24–36.

Agarwal, A., and Liu, Y. (2015). Remediation technologies for oil-contaminated sediments. Mar. Pollut. Bull. *101*, 483–90.

Al-Tahhan, R., Sandrin, T.R., Bodour, A.A., and Maier, R.M. (2000). Rhamnolipid-iduced removal of lipopolysaccharide from *Pseudomonas aeruginosa*: effect on cell surface properties and interaction with hydrophobic substrates. Appl. Environ. Microbiol. *66*, 3262–3268.

Ammami, M.T., Portet-Koltalo, F., Benamar, A., Duclairoir-Poc, C., Wang, H., and Le Derf, F. (2015). Application of biosurfactants and periodic voltage gradient for enhanced electrokinetic remediation of metals and PAHs in dredged marine sediments. Chemosphere. *125*, 1–8.

Arab, F., and Mulligan, C.N. (2014). Rhamnolipids: Characteristics, production applications and analysis. In Biosurfactants: Research Trends and Applications, C.N. Mulligan, S.K. Sharma and A. Mudhoo, eds. (CRC Press, New York), pp. 49–104.

Ayed, H.B., Jemil, N., Maalej, H., Bayoudh, A., Hmidet, N., and Nasri, M. (2015). Enhancement of solubilization and biodegradation of diesel oil by biosurfactant from *Bacillus amyloliquefaciens* An6. Int. Biodeterior. Biodegradation. *99*, 8–14.

Bachmann, R.T., Johnson, A.C., and Edyvean, R.G.J. (2014). Biotechnology in the petroleum industry: An overview. Int. Biodeterior. Biodegradation. *86*, 225–237.

Banat, I.M., Franzetti, A., Gandolfi, I., Bestetti, G., Martinotti, G., Fracchia, L., Smyth, T.J., and Marchant, R. (2010). Microbial biosurfactants production, applications and future potential. Appl. Microbiol. Biotechnol. *87*, 427–444.

Bezza, F.A., and Chirwa, E.M.N. (2015). Biosurfactant from *Paenibacillus dendritiformis* and its application in assisting polycyclic aromatic hydrocarbon (PAH) and motor oil sludge removal from contaminated soil and sand media. Process Saf. Environ. Prot. *98*, 354–364.

Bezza, F.A., and Chirwa, E.M.N. (2016). Biosurfactant-enhanced bioremediation of aged polycyclic aromatic hydrocarbons (PAHs) in creosote contaminated soil. Chemosphere. *144*, 635–644.

Blyth, W., Shahsavari, E., Morrison, P.D., and Ball, A.S. (2015). Biosurfactant from red ash trees enhances the bioremediation of PAH contaminated soil at a former gasworks site. J. Environ. Manage. *162*, 30–36.

Bodour, A.A., Guerrero-Barajas, C., Jiorle, B.V., Malcomson, M.E., Paull, A.K., Somogyi, A., Trinh, L.N., Bates, R.B., and Maier, R.M. (2004). Structure and characterization of flavolipids, a novel class of biosurfactants produced by *Flavobacterium* sp. strain MTN11. Appl. Environ. Microbiol. *70*, 114–20.

Bustamante, M., Durán, N., and Diez, M.C. (2012). Biosurfactants are useful tools for the bioremediation of contaminated soil: a review. J. Soil Sci. Plant Nutr. *12*, 667–687.

Calvo, C., Manzanera, M., Silva-Castro, G.A., Uad, I., and González-López, J. (2009). Application ofbioemulsifiers in soil oil bioremediation processes. Future prospects Sci. Total Environ. *407*, 3634–3640.

Cameotra, S.S., and Makkar, R.S. (2004). Recent applications of biosurfactants as biological and immunological molecules. Curr. Opin. Microbiol. *7*, 262–6.

Cameotra, S.S., and Singh, P. (2009). Synthesis of rhamnolipid biosurfactant and mode of hexadecane uptake by *Pseudomonas* species. Microbial. Cell Fact. *8*, 1–7.

Chakraborty, J., and Das, S. (2014). Biosurfactant-based bioremediation of toxic metals. In: Microbial Biodegradation and Bioremediation, S. Das, ed. (Elsevier), pp. 167–201.

Chandankere, R., Yao, J., Choi, M.M.F., Masakorala, K., and Chan, Y. (2013). An efficient biosurfactant-producing and crude-oil emulsifying bacterium *Bacillus methylotrophicus* USTBa isolated from petroleum reservoir. Biochem. Eng. J. *74*, 46–53.

da Rosa, C.F.C., Freire, D.M.G., and Ferraz, H.C. (2015). Biosurfactant microfoam: application in the removal of pollutants from soil. J. Environ. Chem. Eng. *3*, 89–94.

Das, K., and Mukherjee, A.K. (2007)a. Comparison of lipopeptide biosurfactants production by *Bacillus subtilis* strains in submerged and solid state fermentation systems using a cheap carbon source: some industrial applications of biosurfactants. Process Biochem. *42*, 1191–1199.

Desai, J.D., and Banat, I.M. (1997). Microbial production of surfactants and their commercial potential. Microbiol. Mol. Biol. *61*, 47–64.

Dobler, L., Vilela, L.F., Almeida, R.V., and Neves, B.C. (2016). Rhamnolipids in perspective: gene regulatory pathways, metabolic engineering, production and technological forecasting. N. Biotechnol. *25*, 123–35.

Franzetti, A., Bestetti, G., Caredda, P., La Colla, P., and Tamburini, E. (2008a). Surface-active compounds and their role in the access to hydrocarbons in *Gordonia* strains. FEMS Microbiol. Ecol. *63*, 238–248.

Franzetti, A., Di Gennaro, P., Bestetti, G., Lasagni, A., Pitea, D., and Collina, E. (2008b) Selection of surfactants for enhancing diesel hydrocarbons-contaminated media bioremediation. J. Hazard. Mater. *152*, 1309–1316.

Franzetti, A., Tamburini, E., and Banat, I.M. (2010a). Applications of biological surface active compounds in remediation technologies. Adv. Exp. Med. Biol. *672*, 121–34.

Franzetti, A., Gandolfi, I., Bestetti, G., Smyth, T.J., and Banat, I.M. (2010b). Production and applications of trehalose lipid biosurfactants. Eur. J. Lipid. Sci. Tech. *112*, 617–627.

Fuguet, E., Ràfals, C., Rosés, M., and Bosch, E. (2005). Critical micelle concentration of surfactants in aqueous buffered and unbuffered systems. Anal. Chim. Acta *548*, 95–100.

Galabova, D., Sotirova, A., Karpenko, E., and Karpenko, O. (2014). Role of microbial surface-ctive compounds in environmental protection. In: The role of colloidal systems in environmental protection, M. Fanun, ed. (Elsevier), pp. 41–84.

Hamley, I.W. (2015). Lipopeptides: from self-assembly to bioactivity. Chem. Commun. (Camb). *21*, 8574–83.

Ibrahim, M.L., Ijah, U.J.J., Manga, S.B., Bilbis, L.S., and Umar, S. (2013). Production and partial characterization of biosurfactant produced by crude oil degrading bacteria. Int. Biodeterior. Biodegradation. *81*, 28–34.

Inès, M., and Dhouha, G. (2015). Lipopeptide surfactants: Production, recovery and pore forming capacity. Peptides. *71*, 100–12.

Ismail, W., Al-Rowaihi, I.S., Al-Humam, A.A., Hamza, R.Y., El Nayal, A.M., and Bououdina, M. (2013).

Characterization of a lipopeptide biosurfactant produced by a crude-oil-emulsifying *Bacillus* sp. I-15. Int. Biodeterior. Biodegradation. *84*, 168–178.

Jamaly, S., Giwa, A., and Hasan, S.W. (2015). Recent improvements in oily wastewater treatment: Progress, challenges, and future opportunities. J. Environ. Sci. *37*, 15–30.

Kang, S.W., Kim, Y.B., Shin, J.D., and Kim, E.K. (2010). Enhanced biodegradation of hydrocarbons in soil by microbial biosurfactant, sophorolipid. Appl. Biochem. Biotechnol. *160*, 780–790.

Kavitha, V., Mandal, A.B., and Gnanamani, A. (2014). Microbial biosurfactant mediated removal and/or solubilization of crude oil contamination from soil and aqueous phase: An approach with *Bacillus licheniformis* MTCC 5514. Int. Biodeterior. Biodegradation. *94*, 24–30.

Khondee, N., Tathong, S., Pinyakong, O., Müller, R., Soonglerdsongpha, S., Ruangchainikom, C., Tongcumpou, C., and Luepromchai, E. (2015). Lipopeptide biosurfactant production by chitosan-immobilized *Bacillus* sp. GY19 and their recovery by foam fractionation. Biochem. Eng. J. *93*, 47–54.

Lima, T.M., Procópio, L.C., Brandão, F.D., Carvalho, A.M., Tótola, M.R., and Borges, A.C. (2011). Simultaneous phenanthrene and cadmium removal from contaminated soil by a ligand/biosurfactant solution. Biodegradation. *22*, 1007–15.

Liu, J.F., Mbadinga, S.M., Yang, S.Z., Gu, J.D., and Mu, B.Z. (2015). Chemical structure, property and potential applications of biosurfactants produced by *Bacillus subtilis* in petroleum recovery and spill mitigation. Int. J. Mol. Sci. *3*, 4814–37.

Lovaglio, R.B., Silva, V.L., Ferreira, H., Hausmann, R., and Contiero, J. (2015). Rhamnolipids know-how: Looking for strategies for its industrial dissemination. Biotechnol Adv. 15. pii: S0734-9750(15)30034-3. doi: 10.1016/j.biotechadv.2015.09.002. [Epub ahead of print]

Luna, J.M., Rufino, R.D., Jara, A.M.A.T., Brasileiro, P.P.F., and Sarubbo, L.A. (2015). Environmental applications of the biosurfactant produced by *Candida sphaerica* cultivated in low-cost substrates. Colloids Surf. A Physicochem. Eng. Asp. *480*, 413–418.

Mao, X., Jiang, R., Xiao, W., and Yu, J. (2015). Use of surfactants for the remediation of contaminated soils: a review. J. Hazard. Mater. *285*, 419–35.

Marchant, R., and Banat, I.M. (2012). Microbial biosurfactants: challenges and opportunities for future exploitation. Trends Biotechnol. *30*, 558–65.

Marti, M.E., Colonna, W.J., Reznik, G., Pynn, M., Jarrell, K., Lamsal, B., and Glatz, C.E. (2015). Production of fatty-acyl-glutamate biosurfactant by *Bacillus subtilis* on soybean co-products. Biochem. Eng. J. *95*, 48–55.

Meena, K.R., and Kanwar, S.S. (2015). Lipopeptides as the antifungal and antibacterial agents: applications in food safety and therapeutics. Biomed. Res. Int. 2015, ID 473050, 9 pages, doi: 10.1155/2015/473050.

Mnif, I., and Ghribi, D. (2015). Review lipopeptides biosurfactants: Mean classes and new insights for industrial, biomedical, and environmental applications. Biopolymers. *104*, 129–47.

Morita, T., Fukuoka, T., Imura, T., and Kitamoto, D. (2015). Mannosylerythritol lipids: production and applications. J. Oleo. Sci. *64*, 133–41.

Moryl, M., Spętana, M., Dziubek, K., Paraszkiewicz, K., Różalska, S., Płaza, G.A., and Różalski, A. (2015). Antimicrobial, antiadhesive and antibiofilm potential of lipopeptides synthesised by *Bacillus subtilis*, on uropathogenic bacteria. Acta Biochim. Pol. *62*, 725–32.

Müller, M.M., Kügler, J.H., Henkel, M., Gerlitzki, M., Hörmann, B., Pöhnlein, M., Syldatk, C., and Hausmann, R. (2012). Rhamnolipids-next generation surfactants? J. Biotechnol. *31*, 366–80.

Mulligan, C.N. (2005). Environmental applications for biosurfactants. Environ. Pollut. *133*, 183–98.

Mulligan, C.N. (2009). Recent advances in the environmental applications of biosurfactants. Curr. Opin. Colloid Interface Sci. *14*, 372–378.

Mulligan, C.N. (2014). Enhancement of remediation technologies with biosurfactants. In Biosurfactants: Research Trends and Applications, C.N. Mulligan, S.K. Sharma and A. Mudhoo, eds. (CRC Press Taylor and Francis Group), pp. 231–276.

Obayori, O.S., Ilori, M.O., Adebusoye, S.A., Oyetibo, G.O., Omotayo, A.E., and Amund, O.O. (2009). Degradation of hydrocarbons and biosurfactant production by *Pseudomonas* sp. strain LP1. World J. Microbiol. Biotechnol. *25*, 1615–1623.

Ongena, M., and Jacques, P. (2008). *Bacillus* lipopeptides: versatile weapons for plant disease biocontrol. Trends Microbiol. *16*, 15–25.

Pacwa-Płociniczak, M., Płaza, G.A., Piotrowska-Seget, Z., and Cameotra, S.S. (2011). Environmental applications of biosurfactants: Recent advances. Int. J. Mol. Sci. *12*, 633–654.

Pacwa-Płociniczak, M., Płaza, G.A., Poliwoda, A., and Piotrowska-Seget, Z. (2014). Characterization of hydrocarbon-degrading and biosurfactant-producing *Pseudomonas* sp. P-1 strain as a potential tool for bioremediation of petroleum-contaminated soil. Environ. Sci. Pollut. Res. Int. *21*, 9385–9395.

Paniagua-Michel, J.J., Olmos-Soto, J., and Morales-Guerrero, E.R. (2014). Algal and microbial exopolysaccharides: new insights as biosurfactants and bioemulsifiers. Adv. Food Nutr. Res. 73, 221–57.

Paria, S. (2008). Surfactant enhanced remediation of organic contaminated soil and water. Adv. Colloid Interface Sci. *138*, 24–58.

Pasternak, G., and Kołwzan, B. (2013). Surface tension and toxicity changes during biodegradation of carbazole by newly isolated methylotrophic strain *Methylobacterium* sp. GPE1. Int. Biodeterior. Biodegradation. *84*, 143–149.

Patel, J., Borgohain, S., Kumar, M., Rangarajan, V., Somasundaran, P., and Sen, R. (2015). Recent developments in microbial enhanced oil recovery. Renew. Sust. Energ. Rev. *52*, 1539–1558.

Płaza, G., Chojniak, J., Rudnicka, K., Paraszkiewicz, K., and Bernat, P. (2015a). Detection of biosurfactants in *Bacillus* species: genes and products identification. J. Appl. Microbiol. *119*, 1023–34.

Płaza, G.A., Chojniak, J., Mendrek, B., Trzebicka, B., Kvitek, L., Panacek, A., Prucek, R., Zboril, R., Paraszkiewicz, K., and Bernat, P. (2015b). Synthesis of silver nanoparticles by *Bacillus subtilis* T-1 growing on agro-industrial wastes and producing biosurfactant. IET Nanobiotechnology DOI: 10.1049/iet-nbt.2015.0016. Online ISSN 1751-875X. Available online: 27 October 2015.

Peng, F., Liu, Z., Wang, L., and Shao, Z. (2007). An oil-degrading bacterium: *Rhodococcus erythropolis* strain 3C-9 and its biosurfactants. J. Appl. Microbiol. *102*, 1603–1611.

Pornsunthorntawee, O., Wongpanit, P., and Rujiravanit, R. (2010). Rhamnolipid biosurfactants: production and their potential in environmental biotechnology. Adv. Exp. Med. Biol. *672*, 211–21.

Portet-Koltalo, F., Ammami, M.T., Benamar, A., Wang, H., Le Derf, F., and Duclairoir-Poc, C. (2013). Investigation of the release of PAHs from artificially contaminated sediments using cyclolipopeptidic biosurfactants. J. Hazard. Mater. *261*, 593–601.

Raaijmakers, J.M., De Bruijn, I., Nybroe, O., and Ongena, M. (2010). Natural functions of lipopeptides from *Bacillus* and *Pseudomonas*: more than surfactants and antibiotics. FEMS Microbiol. Rev. *34*, 1037–1062.

Reddy, M.S., Naresh, B., Leela, T., Prashanthi, M., Madhusudhan, N.C., Dhanasri, G., and Devi, P. (2010). Biodegradation of phenanthrene with biosurfactant production by a new strain of *Brevibacillus* sp. Bioresource Technol. *101*, 7980–7983.

Reis, R.S., Pacheco, G.J., Pereira, A.G., and Freire, D.M.G. (2013). Biosurfactants: production and applications. In: Biodegradation – life of science, R. Chamy, ed. ISBN: 978-953-51-1154-2, InTech, DOI: 10.5772/56144.

Roelants, S.L., De Maeseneire, S.L., Ciesielska, K., Van Bogaert, I.N., and Soetaert, W. (2014). Biosurfactant gene clusters in eukaryotes: regulation and biotechnological potential. Appl. Microbiol. Biotechnol. *98*, 3449–61.

Rosenberg, E., and Ron, E.Z. (1999). High- and low-molecular-mass microbial surfactants. Appl. Microbiol. Biotechnol. 52, 154–162.

Roy, M., Giri, A.K., Dutta, S., and Mukherjee. P. (2015). Integrated phytobial remediation for sustainable management of arsenic in soil and water. Environ. Int. 75, 180–198.

Sachdev, D.P., and Cameotra, S.S. (2013). Biosurfactants in agriculture. Appl. Microbiol. Biotechnol. *97*, 1005–16.

Sakthipriya, N., Doble, M., and Sangwai, J.S. (2015a). Action of biosurfactant producing thermophilic *Bacillus subtilis* on waxy crude oil and long chain paraffins. Int. Biodeterior. Biodegradation. *105*, 168–177.

Sakthipriya, N., Doble, M., and Sangwai, J.S. (2015b). Bioremediation of coastal and marine pollution due to crude oil using a microorganism *Bacillus subtilis*. Procedia Eng. *116*, 213–220.

Satpute, S.K., Banpurkar, A.G., Dhakephalkar, P.K., Banat, I.M., and Chopade, B.A. (2010). Methods for investigating biosurfactants and bioemulsifiers: a review. Crit. Rev. Biotechnol. *30*, 127–44.

Silva, R.C.F.S., Almeida, D.G., Rufino, R.D., Luna, J.M., Santos, V.A., and Sarubbo, L.A. (2014). Application of biosurfactants in the petroleum industry and the remediation of oil spills. Int. J. Mol. Sci. *15*, 12523–42.

Singh, P., and Cameotra, S.S. (2004). Enhancement of metal bioremediation by use of microbial surfactants. Biochem. Biophy. Res. Commun. *319*, 291–297.

Singh, A., van Hamme, J.D., and Ward, O.P. (2007). Surfactants in microbiology and biotechnology: Part 2. Applications aspects. Biotech. Adv. *25*, 99–121.

Singh, A.K., and Cameotra, S.S. (2013). Efficiency of lipopeptide biosurfactants in removal of petroleum hydrocarbons and heavy metals from contaminated soil. Environ. Sci. Pollut. Res. Int. *20*, 7367–76.

Singh, G.B., Gupta, S., and Gupta, N. (2013). Carbazole degradation and biosurfactant production by newly isolated *Pseudomonas* sp. strain GBS.5. Int. Biodeterior. Biodegradation. *84*, 35–43.

Singh, A.K., Dhanjal, S., and Cameotra, S.S. (2014). Surfactin restores and enhances swarming motility under heavy metal stress. Colloids Surf., B. Biointerfaces, *116*, 26–31.

Soberón-Chávez, G., and Maier, R.M. (2011). Biosurfactants: A General Overview. In: Biosurfactants: From Genes to Applications, Microbiology Monographs. G. Soberón-Chávez, ed. (Springer Science and Business Media, Springer Verlag Berlin Heidelberg), pp. 2–11.

Sotirova, A., Spasova, D., Vasileva-Tonkova, E., and Galabova, D. (2009). Effects of rhamnolipid-biosurfactant on cell surface of *Pseudomonas aeruginosa*. Microbiol. Res. *164*, 297–303.

Suthar, H., Hingurao, K., Desai, A., and Nerurkar, A. (2008). Evaluation of bioemulsifier mediated microbial oil recovery using sand pack column. J. Microbiol. Meth. *75*, 225–230.

Souza, E.C., Vessoni-Penna, T.C., and de Souza Oliveira, R.P. (2014). Biosurfactant-enhanced hydrocarbon bioremediation: An overview. Int. Biodeterior. Biodegrad. *89*, 88–94.

Sun, R., Belcher, R.W., Liang, J., Wang, L., Thater, B., Crowley, D.E., and Wei, G. (2015). Effects of cowpea (*Vigna unguiculata*) root mucilage on microbial community response and capacity for phenanthrene remediation. J. Environ. Sci. *33*, 45–59.

Ullah, A., Heng, S., Munis, M.F.H., Fahad, S., and Yang, X. (2015). Phytoremediation of heavy metals assisted by plant growth promoting (PGP) bacteria: A review. Environ. Exper. Bot. *117*, 28–40.

Urum, K., and Pekdemir, T. (2004). Evaluation of biosurfactants for crude oil contaminated soil washing. Chemosphere 57, 1139–1150.

US EPA (1996). A citizen's guide to *in situ* soil flushing. Office of Solid Waste and Emergency Response (5102G). EPA 542-F-96-006.

Van Bogaert, I.N., Saerens, K., De Muynck, C., Develter, D., Soetaert, W., and Vandamme, E.J. (2007). Microbial production and application of sophorolipids. Appl. Microbiol. Biotechnol. *76*, 23–34.

Van Bogaert, I.N., Holvoet, K., Roelants, S.L., Li, B., Lin, Y.C., Van de Peer, Y., and Soetaert, W. (2013). The biosynthetic gene cluster for sophorolipids: a biotechnological interesting biosurfactant produced by *Starmerella bombicola*. Mol. Microbiol. *88*, 501–509.

Van Hamme, J.D., Singh, A., and Ward, O.P. (2006). Physiological aspects: Part 1 in a series of papers devoted to surfactants in microbiology and biotechnology. Biotechnol. Adv. *24*, 604–620.

Viisimaa, M., Karpenko, O., Novikov, V., Trapido, M., and Goi, A. (2013). Influence of biosurfactant on combined

chemical–biological treatment of PCB-contaminated soil. Chem. Eng. J. *220*, 352–359.
Walia, N.K., and Cameotra, S.S. (2015). Lipopeptides: biosynthesis and applications. J. Microb. Biochem. Technol. 7, 103–107.
Wang, S., and Mulligan, C.N. (2004). An evaluation of surfactant foam technology in remediation of contaminated soil. Chemosphere 57, 1079–1089.
Yang, Z., Zhang, Z., Chai, L., Wang, Y., Liu, Y., and Xiao, R. (2016). Bioleaching remediation of heavy metal-contaminated soils using *Burkholderia* sp. Z-90. J. Hazard. Mat. *301*, 145–152.
Yu, M., Liu, Z., Zeng, G., Zhong, H., Liu, Y., Jiang, Y., Li, M., He, X., and He, Y. (2015). Characteristics of mannosylerythritol lipids and their environmental potential. Carbohydr. Res. *407*, 63–72.

11 Microorganisms Application for Volatile Compounds Degradation

Christian Kennes, Haris N. Abubackar, Jianmeng Chen and María C. Veiga

Contents

Abstract

The emission of volatile pollutants to the atmosphere, and air pollution have become a major concern. Volatile pollutants are emitted from either stationary sources or mobiles sources. Common air pollutants are particulate matter as well as volatile organic and inorganic compounds. Different technologies can be used for their removal, including bioprocesses in case of volatile compounds. Both bacteria and fungi are able to degrade a rather wide range of natural as well as anthropogenic substrates. Although less research has been done on fungi, the latter have recently proven to be efficient biocatalysts for the removal of some of those pollutants, tolerating acidic conditions and environments with limited moisture content. Rather than simply mineralizing pollutants into harmless compounds, recent approaches have also considered the possibility to convert them into end metabolites with added value and commercial interest, e.g. fuels or platform chemicals. The biodegradation or bioconversion of volatile pollutants is performed in bioreactors. The most common bioreactors used in full-scale installations include the biofilter, the biotrickling filter, the bioscrubber and systems based on gas diffusion through suspended-growth bioreactors. Some other bioreactor configurations have been tested at laboratory scale as well. In most cases, such bioprocesses allow for a complete and efficient removal of volatile organic and inorganic compounds.

11.1 Introduction

In recent years concern related to air pollution problems as well as the scarcity of fossil fuels has increased. The expansion of industrial activities, the combustion of fossil fuels, and road traffic are some of the major sources contributing to air pollution; damaging the environment, affecting human health and being a possible cause of premature death. Therefore, it has become necessary to take actions in order to mitigate such pollution problems and their adverse effects. Several industrialized countries, among others in the European Union and America, have taken efficient measures to prevent and control air pollution; but air quality is also still a large problem in new developing countries characterized by an exponential industrial growth over the past few years. Nowadays, most developed countries have effective regulations in place aimed at lowering atmospheric pollution either through reduction of emissions at the source or through their treatment once released to the atmosphere; but improvements are still necessary.

Different alternatives are available for the removal of volatile pollutants including physical and chemical methods as well as bioprocesses. Bioprocesses are based on the ability of microorganisms to biodegrade a range of volatile pollutants. Over the past few decades several bacterial and fungal species have been identified to be able to degrade volatile compounds, some of them still considered to be recalcitrant pollutants a few decades ago. Some microorganisms seem to have developed enzymes capable of degrading anthropogenic volatile compounds not originally found in the atmosphere. Thus, microbial biocatalysts can be used in bioreactors for the removal of such volatile pollutants. Different bioreactor configurations have been optimized over the past decades, based either on the activity of suspended-growth biomass or attached cells. Those systems can efficiently remove volatile compounds up to concentrations of about 4–5 g/m^3 (air). Another recent alternative has considered the possibility to convert those pollutants into useful metabolites rather than simply biodegrading them into innocuous end products.

11.2 Types of volatile pollutants

Different types of pollutants can be released to the atmosphere depending on the emission source. Those pollutants can be categorized into two major groups, namely particulate matter and volatile compounds. Particulate matter (PM) is a solid or liquid mass in suspension in the atmosphere, with typical sizes ranging from submicron up to several micrometres. Their concentration in the atmosphere is generally expressed as PM2.5 for particles that are smaller than 2.5 μm in diameter, or PM10 for particles that are larger than 2.5 μm but smaller than 10 μm in diameter. It includes dust, fume, smoke, soot, or liquid droplets, among others. The smallest particles pose the highest health risk. They can be inhaled and retained in the lungs. Different equipment is suitable for the removal of particulate matter, including cyclone collectors, scrubbers, filtration units and electrostatic precipitators, among the most common ones (Kennes and Veiga, 2001). Bioprocesses are so far not suitable for the removal of such pollutants.

On the other side, volatile pollutants can be grouped into volatile organic compounds (VOC) and volatile inorganic compounds (VIC). They usually have high vapour pressures and those who are in liquid form at room temperature do easily volatilize. Besides being volatile, as pollutants VOC are considered to be involved in photochemical reactions in the atmosphere. The presence of such VOC enhances undesirable ozone formation from NOx. Odour nuisance is the result of the presence of one, or more often several, volatile pollutants – either organic or inorganic – that produce a smell. Although some volatile pollutants are odourless, the level of toxicity is not related to such characteristic, and non odorous compounds may be highly toxic or detrimental to the environment.

Volatile organic and inorganic compounds are thus released in gaseous form to the atmosphere while particulate matter is either solid or liquid. Table 11.1 summarizes the ratios of different VIC, VOC and particulate matter according to the emission sources (Kennes and Veiga, 2013).

Different technologies can be used for the removal of volatile pollutants from waste gases. They include absorption, adsorption, thermal and catalytic oxidation, and condensation (Kennes and Veiga, 2001), but also bioprocesses as described in section 'Biodegradation of volatile compounds' (Kennes and Thalasso, 1998; Kennes *et al.*, 2009). Gas flow rates of industrial emissions of volatile compounds can typically range from a few hundreds of cubic metres per hour up to several hundreds of thousands cubic metres per hour. On the other side, pollutant concentrations can range

Table 11.1 Distribution (%) of different groups of pollutants according to their emission sources

%	PM2.5	CO	NH_3	VOC	NOx	SO_2
Stationary sources: combustion processes	50.6	43.7	0.5	12.1	42.8	90.2
Stationary sources: non-combustion processes	20.6	10	2.7	52.7	4.3	6.4
Mobile sources	18.7	42.6	1.4	29.9	51.5	3.25
Waste disposal	3.5	3.3	1.9	1.0	0.2	0.1
Miscellaneous (incl. agriculture)	6.6	0.4	93.5	4.3	1.1	0.05

from a few $\mu g/m^3$(air) up to several g/m^3. Even at quite low concentrations, some pollutants can already be harmful or generate odour nuisance. This is the case, for example, of hydrogen sulfide with an odour threshold limit of hardly 0.014 mg/m^3. The most suitable technology for the removal of volatile pollutants will depend, among others, on the gas flow rate and pollutant concentration, besides other additional characteristics. Bioprocesses are most efficient for concentrations of up to about 5 g/m^3 (air) and over a quite wide range of gas flow rates that can occasionally reach values as high as 500,000 m^3/h (Fig. 11.1). However, high pollutant concentrations can only be combined with low to moderate flow rates as the overall loading rate is the product of the pollutant concentration and the gas flow rate, as shown in equation (11.1), meaning that the loading rate would be excessively high if both concentration and flow rate are high.

$$LR = \frac{Q.C}{V} \quad (11.1)$$

where LR is the loading rate, Q is the gas flow rate, C is the inlet pollutant concentration and V is the reactor volume.

11.3 Sources of emission of volatile pollutants

Volatile pollutants are emitted to the atmosphere mainly from two different sources, namely mobile sources and stationary sources. Mobile sources include all vehicles and other major means of transport in which a combustion process takes place, as air pollution is mainly due to fuel combustion (Table 11.1). This includes sources such as cars, trucks, buses, trains, airplanes, boats, but does also include other ones not related to transportation such as fuel-powered agricultural and construction equipment or lawn tools, among others. On the other hand, stationary sources of air pollution are sources that do not move, such as industrial facilities, power plants, wastewater treatment plants, or composting facilities. Stationary sources may emit volatile pollutants through combustion processes, similarly as mobile sources, but also from industrial unit operations, reactors or other industrial activities. It is worth reminding that some natural sources do also emit volatile compounds to the atmosphere, such as natural fires or volcanic activities to cite only few examples. Different technologies can be used for waste gas treatment, depending on the emission

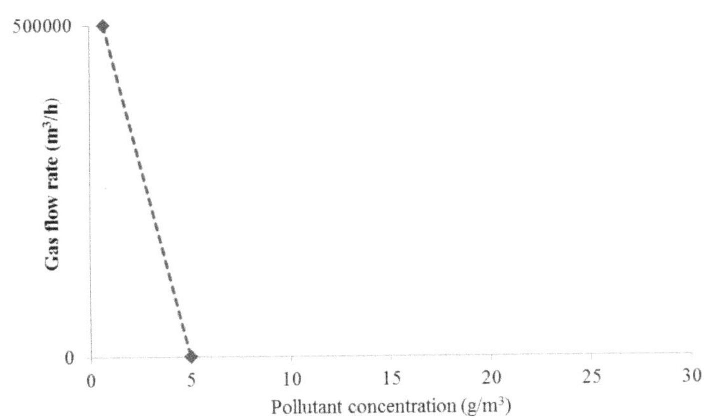

Figure 11.1 Suitable operation range for bioreactors used for the removal of volatile compounds.

source. Bioprocesses have so far only been applied to the removal of volatile pollutants from stationary sources. The treatment of gaseous effluents takes then place in bioreactors with different possible configurations. The most common ones are the biofilter, the biotrickling filter, bioscrubbers and suspended-growth bioreactors with gas diffusion, as described in section 'Bioreactors for the removal of volatile compounds'.

11.4 Biodegradation of volatile compounds

One main requirement for the microbial removal of volatile compounds is that they should be biodegradable. Although many industrial air pollutants are of anthropogenic origin and did thus not appear in nature before they were emitted to the atmosphere as a result of industrial activities, some microorganisms are often able to metabolize such anthrogenic compounds. Over the past few decades, several bacteria and fungi have been discovered and studied for their ability to biodegrade substrates generally considered to be toxic or recalcitrant.

11.4.1 Biodegradation processes

Both bacteria and fungi have been found to metabolize a wide range of volatile substrates. Here we will mainly focus on volatile pollutants and their biodegradation and complete mineralization to end products such as water or carbon dioxide, although some microorganisms can also bioconvert volatile compounds into valuable products (see section 11.5). Bioconversion will then also briefly be addressed in this chapter, mainly in the case of the bioconversion of pollutants, which is a rather new research field.

The main biodegradation reactions found in microorganisms, either bacteria or fungi, are summarized hereafter for both typical VIC and VOC (Kennes *et al.*, 2009).

In case of VOC containing only carbon, hydrogen and/or oxygen atoms ($C_xH_yO_z$), the biodegradation products will be water and carbon dioxide, as shown hereafter.

$$a\,C_xH_yO_z + b\,O_2 \rightarrow c\,CO_2 + d\,H_2O$$

where a, b, c and d are stoichiometric coefficients.

In fact, besides using the VOC as a potential carbon and/or energy source, microorganisms will also need some nutrients, such as a nitrogen source for maintenance and growth of the biomass. If ammonium chloride is added as a nitrogen source, VOC biodegradation and microbial growth can be summarized as in the stoichiometric equation shown hereafter.

$$aC_xH_yO_z + bO_2 + cNH_4Cl \rightarrow dCO_2 + eH_2O + fHCl + gC_kH_lN_mO_n$$

where $C_kH_lN_mO_n$ represents the biomass, and a, b, c, d, e, f and g are stoichiometric coefficients.

A typical formula for the biomass could be $C_5H_7NO_2$, although this can vary depending on the microorganism. Note that in the above equation hydrogen chloride does also appear as an end-product simply because ammonium chloride was chosen as the nitrogen source.

When additional atoms are found in the VOC, besides C, H or O, then additional end-products will generally appear. A common case is when dealing with halogenated pollutants ($C_xH_yO_zCl_t$) such as chloroethylenes or chlorobenzenes, as shown hereafter.

$$aC_xH_yO_zCl_t + bO_2 \rightarrow cCO_2 + dH_2O + eHCl$$

Here also biomass formation may need to be taken into account, as appears hereafter.

$$aC_xH_yO_zCl_t + bO_2 + cNH_4Cl \rightarrow dCO_2 + eH_2O + fHCl + gC_kH_lN_mO_n$$

In this case, hydrogen chloride is an end-product from the biodegradation of the chlorinated VOC besides its production from ammonium chloride.

For inorganic volatile compounds, it is more difficult to write a general stoichiometric equation as there are many different VIC. The most common one is hydrogen sulfide, which is degraded in two steps under aerobic conditions as shown hereafter (Jin *et al*, 2005a).

$$H_2S + 0.5O_2 \rightarrow S^0 + H_2O$$

$$S^0 + H_2O + 1.5O_2 \rightarrow H_2SO_4$$

And if biomass growth is taken into account (Jin *et al.*, 2005b):

$$0.444H_2S + 0.4HS^- + 1.2555O_2 + 0.0865H_2O + 0.346CO_2 + 0.0865HCO_3^- + 0.0865NH_4^+ \rightarrow 0.844SO_4^{2-} + 1.288H^+ + 0.0865C_5H_7NO_2$$

11.4.2 Bacterial degradation of volatile compounds

Bacteria can degrade both VOC and VIC. Some examples of biodegradable volatile organic and inorganic compounds and bacterial species involved in their biodegradation appear in Table 11.2. Listing all of them would be quite difficult, but still Table 11.2 shows that many volatile pollutants are biodegradable by a quite wide range of different bacterial species.

Most VOC degrading bacteria exhibit their optimal activity at near neutral or slightly acidic pH (e.g. pH 6–7) in case of compounds containing C, H, and/or O atoms. Bacteria degrading halogenated volatile compounds may have a somewhat different optimal pH. For example, dichloromethane can be degraded by *Hyphomicrobium* strains (Bailón *et al.*, 2009), which have been shown to grow best under neutral pH conditions or even in slightly alkaline media (e.g. pH 7–8) (Bailón *et al.*, 2007). Besides, the dichloromethane dehalogenase of *Hyphomicrobium* DM2 appears to have an optimal pH of 8.5 (Kohler-Staub and Leisinger, 1985). Interestingly, many bacteria degrading halogenated compounds or other VOC pollutants yielding acid end-products, do often not tolerate low-pH environments. Conversely, the biodegradation of sulfur containing compounds, mainly VIC, does also produce acids and many bacteria degrading such sulfur-containing VIC are tolerant to very low pH values. This is the case of many *Thiobacillus* strains and other similar bacteria growing in presence of hydrogen sulfide. The biodegradation of H_2S yields sulfuric acid as an end metabolite, resulting in a quick pH drop from near neutral values down to pH 2 or lower (Fig. 11.2) (Jin *et al.*, 2005b). In a gas-phase bioreactor, i.e. biotrickling filter, used for the removal of H_2S from waste gases, without any pH regulation, the pH was observed to drop from an original value of 6.5 down to pH 2 in just a few hours (Jin *et al.*, 2005b). However, this did not affect the microbial activity. Even at pH values below 1, such type of bacterial-reactor systems have been shown the continue degrading the volatile sulfur pollutants very efficiently, as observed for industrial-scale bioreactors treating mixtures of CS_2 and H_2S in viscose industries, among others (Willers *et al.*, 2013). All bacteria listed in Table 11.2 are mesophilic strains. Hardly any research has been done on thermophilic pure cultures degrading such compounds.

Table 11.2 Examples of aerobic bacteria degrading typical common volatile air pollutants

Volatile pollutant	Bacteria	References
Alkylbenzenes (toluene, ethylbenzene and/or xylenes)	*Pseudomonas putida*	Shim and Yang (1999)
	Pseudomonas fluorescens	Shim and Yang (1999)
	Acinetobacter sp.	Kim and Jeon (2009)
	Bacillus subtilis	Mukherjee and Bordoloi (2012)
	Rhodococcus rhodochrous	Rodrigues *et al.* (2015)
	Nocardia farcinica	Rodrigues *et al.* (2015)
Benzene	*Pseudomonas putida*	Shim and Yang, 1999
	Pseudomonas fluorescens	Shim and Yang (1999)
	Pseudomonas aeruginosa	Kim *et al.* (2003)
	Acinetobacter sp.	Kim and Jeon (2009)
Carbon disulfide (CS_2)	*Thiobacillus thioparus*	Smith and Kelly (1988)
	Thiomonas sp.	Pol *et al.*,(2007)
Dichloroethane	*Xanthobacter autotrophicus*	Beschkov *et al.* (2008)

Table 11.2 Continued

Volatile pollutant	Bacteria	References
Dichloromethane	*Ancilobacter dichloromethanicus*	Firsova *et al.* (2009)
	Bacillus circulans	Wu *et al.* (2007)
	Hyphomicrobium sp.	Bailón *et al.* (2009)
	Methylobacterium dichloromethanicum	Kayser *et al.* (2002)
	Methylobacterium rhodesianum	Chen *et al.* (2014)
	Pandoraea pnomenusa	Yu *et al.* (2014)
	Xanthobacter sp.	Emanuelsson *et al.* (2009)
Dimethyl (di)sulfide (DM(D)S)	*Hyphomicrobium* sp.	Smet *et al.* (1996)
	Thiobacillus thioparus	Tanagawa and Mikami (1989)
Formaldehyde	*Bacillus* sp.	Xiao *et al.* (2015)
	Paracoccus sp.	Zhao *et al.* (2013)
	Ralstonia eutropha	Habibi and Vahabzadeh (2013)
Hexane	*Rhodococcus* sp.	Lee and Cho (2008)
	Rhodococcus erythropolis	De Carvalho *et al.* (2005)
Hydrogen sulfide (H_2S)	*Thiobacillus thioparus*	Tanagawa and Mikami (1989)
	Thiobacillus denitrificans	Ma *et al.* (2006)
Methyl-*t*-butyl-ether (MTBE)	*Pseudomonas aeruginosa*	Garnier *et al.* (1999)
Styrene	*Corynebacterium* sp.	Itoh *et al.* (1996)
	Pseudomonas fluorescens	Baggi *et al.* (1983)
	Pseudomonas putida	O´Connor *et al.* (1995)
	Rhodococcus sp.	Toda *et al.* (2012)
Trichloroethylene (TCE) (usually degraded cometabolically)	*Methylosinus trichosporium*	Oldenhuis *et al.* (1991)
	Bacillus spp.	Kim *et al.* (2010)
	Pseudomonas putida	Kim *et al.* (2010)
	Pseudomonas cepacia	Folsom *et al.* (1990)

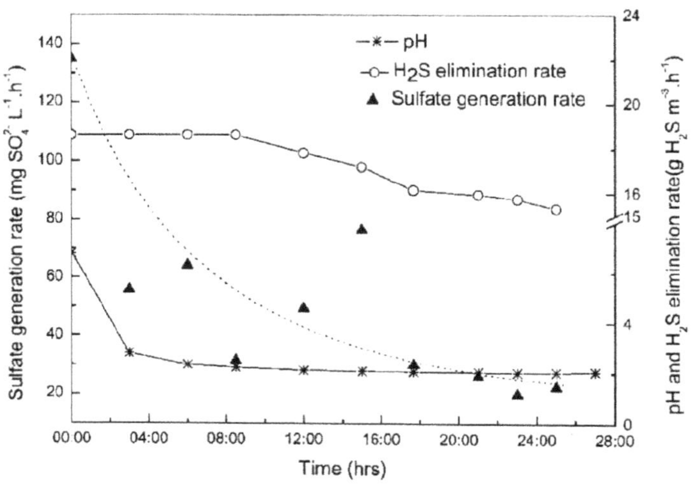

Figure 11.2 pH fluctuations as a result of hydrogen sulfide biodegradation in a bioreactor (reproduced with permission from Jin *et al.*, 2005b).

11.4.3 Fungal degradation of volatile compounds

Fungi, similarly as bacteria, are also able to degrade volatile compounds. However, most studies are dealing with volatile organic pollutants. To the best of our knowledge hardly any report has offered clear evidence about the fungal biodegradation of VIC such as hydrogen sulfide, although some authors have suggested that biodegradation of typical inorganic air pollutants might be possible (Ishikawa *et al.*, 1980; Liu *et al.*, 2016). Actually, some sulfur compounds such as hydrogen sulfide have been shown to act as fungicides (Tang *et al.*, 2014). In fixed-film bioreactors used for air pollution control, it was observed that higher pollutant removal efficiencies can be reached in presence of filamentous fungi, mainly in case of hydrophobic compounds. It has been hypothesized that this could be due to the large surface area of the filamentous biofilm layer, resulting in a better mass transfer between the polluted gas phase and the biofilm compared with near flat bacterial biofilms (Estévez *et al.*, 2005a). Fig. 11.3 shows a picture of a filamentous fungal biofilm attached on a support material in a packed bed bioreactor treating VOC polluted air. Fungi present other characteristics that may be considered relevant for the removal of gas-phase pollutants. They tolerate acidic conditions better than most bacteria and they would remain active in environments with limited moisture content (Kennes and Veiga, 2004). This would be of interest for volatile pollutants producing acidic end metabolites and for the removal of pollutants from air with reduced relative humidity levels. Similarly as for the bacteria listed in Table 11.2, the fungi in Table 11.3 are all mesophilic organisms. Identified VOC degrading thermophilic fungi are very scarce so far and basically no information is available in the literature on such topic.

11.5 Bioconversion of volatile pollutants

Although further improvements are still being done and studied, the removal of volatile pollutants from waste gases through their biodegradation in bioreactors is a quite well established technology. These bioreactors can also be used for upgrading gases, such as in the case of the removal of hydrogen sulfide and other contaminants from biogas (López *et al.*, 2012). Research is being done nowadays in order to extend the range of application of such bioprocesses, improve their efficiency, or with the goal of bioconverting volatile pollutants to useful products rather than biodegrading them. This is a quite new approach, although it has already occasionally been applied to specific cases, such as in the use of algae for the removal of carbon dioxide with the production of biodiesel (Kennes and Veiga, 2013). Another recent example is the bioconversion of waste gases from steel industries into ethanol (Abubackar *et al*, 2011). Waste gases are generated

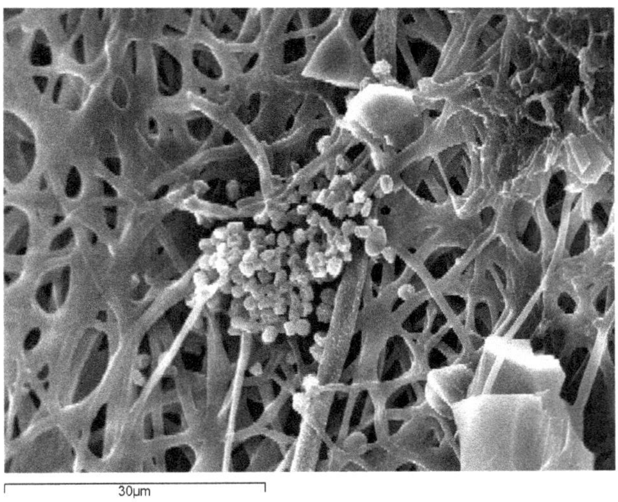

Figure 11.3 Scanning electron microscope picture of a filamentous biofilm growing in a biofilter treating toluene polluted air.

Table 11.3 Examples of fungi degrading typical common volatile air pollutants

Volatile pollutant	Fungi	References
n-Alkanes		
C_2-C_4-Alkanes	*Acremomium* sp.	Davies *et al.* (1973)
C_6-C_{19}-Alkanes	*Cladosporium resinae*	Cofone *et al.* (1973)
C_1-C_9-Alkanes	*Scedosporium* sp.	Onodera *et al.* (1989)
n-Butane	*Graphium* sp	Hardison *et al.* (1997)
Alkylbenzenes (toluene, ethylbenzene and/or xylenes)	*Cladophialophora* sp.	Prenafeta-Boldú *et al.* (2001)
	Cladosporium sphaerospermum	Weber *et al.* (1995)
	Exophiala lecanii-corni	Woertz *et al.* (2001)
	Exophiala oligosperma	Estévez *et al.* (2005b)
	Phanerochaete chrysosporium	Yadav and Reddy (1993)
Benzene	*Exophiala lecanii-corni*	Woertz *et al.* (2001)
Dimethyl (di)sulfide (DM(D)S)	*Cephalosporium* sp.	Ishikawa *et al.* (1980)
Formaldehyde	*Paecilomyces variotii*	Sakaguchi *et al.* (1975)
Hydrogen sulfide (H_2S)	*Cephalosporium* sp.	Ishikawa *et al.* (1980)
Methyl-*t*-butyl-ether (MTBE)	*Graphium* sp.	Hardison *et al.* (1997)
Perchloroethylene (PCE)	*Trametes versicolor*	Marco-Urrea *et al.* (2008)
Styrene	*Cladophialophora* sp.	Prenafeta-Boldú *et al.* (2001)
	*Exophiala jeanselmei**	Cox *et al.* (1996)
	Exophiala lecanii-corni	Woertz *et al.* (2001)
Trichloroethylene (TCE)	*Trametes versicolor*	Marco-Urrea *et al.* (2008)

*Redefined as *Exophiala oligosperma*.

in different parts of the steel plants and several of those gases are composed of three major volatile compounds, i.e. CO (carbon monoxide), CO_2 and H_2. Those volatile compounds are the same as found in syngas, obtained through gasification of biomass or waste. They can be used in different ratios by anaerobic bacteria such as clostridia to produce end-products such as ethanol or butanol, among others, according to the reactions shown hereafter for ethanol.

$$6CO + 3H_2O \rightarrow C_2H_5OH + 4CO_2$$

$$6H_2 + 2CO_2 \rightarrow C_2H_5OH + 3H_2O$$

$$6CO + 6H_2 \rightarrow 2C_2H_5OH + 2CO_2$$

Thus, this process allows to convert volatile compounds from waste gases into fuels, e.g. ethanol or butanol. Pilot plants and pre-commercial set-ups have recently been installed, with promising results suggesting that such alternative is cost-competitive compared with fossil fuels and commercial biofuels (Kennes *et al.*, 2016).

11.6 Bioreactors for the removal of volatile pollutants

The biological removal of volatile pollutants from gaseous effluents takes place in bioreactors. Different types of bioreactors are suitable for this. The most common ones, used at full-scale, are the biofilter, the biotrickling filter, the bioscrubber and bioreactors with polluted gas diffusing through a suspended-growth system (usually an activated-sludge type reactor) (Kennes and Veiga, 2001, 2013). They will briefly be described hereafter.

11.6.1 Biofilter

Two types of biofilters are commonly used in full-scale applications, namely the open-bed biofilter and the enclosed biofilter. Fig. 11.4 shows an open-bed biofilter. The enclosed biofilter works in a similar way. The only difference is that the

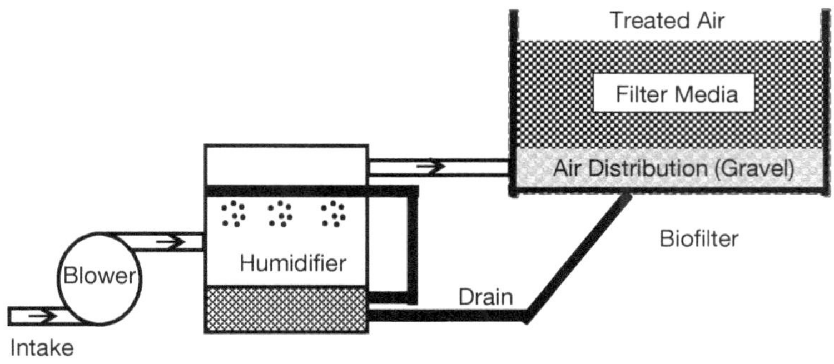

Figure 11.4 Open-bed biofilter.

open-bed biofilter is open to the atmosphere while the enclosed biofilter is not.

The main unit in the biofiltration process is the bioreactor. The latter is filled with a packing material or filter bed containing the active biomass. Some common packing materials used in biofiltration are soil, compost, peat, heather, pine bark, wood chips, or mixtures thereof, among others. Most of those packing materials contain indigenous microbial populations often able to degrade the volatile pollutants present in the waste gas. Inoculation of adapted or specialized microorganisms may be useful whenever some of the pollutants to be removed are not readily biodegraded by the indigenous microbial community of the filter bed. Natural filter beds (e.g. soil, compost, peat) may sometimes advantageously be mixed with synthetic or inert materials, such as perlite, in order to increase their porosity and ensure a better and more homogenous gas flow through that filter bed. A humidification unit in front of the bioreactor allows to increase the relative humidity of the waste gas, which is generally needed in order to ensure the presence of enough moisture required for an optimal microbial activity. Additionally, sprinklers can be installed on the edges or the upper part of the bioreactor in order to occasionally spray additional water on top of the packing material. The filter bed contains active biomass that will degrade the pollutants present in the waste gas while flowing through the system. Clean air will then leave the unit and will be released to the atmosphere after passing through the biotreatment system. The amount water present in the reactor is rather limited in biofilters and such systems are therefore usually the best option for the biodegradation of highly hydrophobic volatile compounds.

11.6.2 Biotrickling filter

Biotrickling filters look similar to absorption columns. They are also quite similar to conventional enclosed biofilters (section 'Biofilter'), except that in the present case a liquid phase is continuously supplied to the reactor and/or recycled through the system (Fig. 11.5) (Kennes and Veiga, 2001). There is thus no need for a prehumidification unit as used in conventional biofilters, as water is continuously supplied to the reactor. Compared to the biofilter, this system allows for a better and easy pH regulation through the addition of an acid or a base in the aqueous phase whenever needed. Natural filter beds typical in conventional biofilters are not suitable here. The most common packing materials are inert and some are similar as in absorption columns, e.g. plastic rings, polyurethane foam, perlite, lava rock (Kennes and Veiga, 2002). Others, such as activated carbon, have also been used. The adsorption capacity of activated carbon allows buffering load fluctuations resulting in a more stable operation and avoiding potentially inhibitory shock loads.

In a biotrickling filter, biodegradation takes place after mass transfer of the pollutant from the polluted gas phase to the liquid film and then to the biomass attached on the packing material. Biotrickling filters are generally less efficient than biofilters when dealing with highly hydrophobic compounds, in view of the presence of larger amounts of water. Suggestions on the range of compounds that can best be treated in a given reactor configuration, based

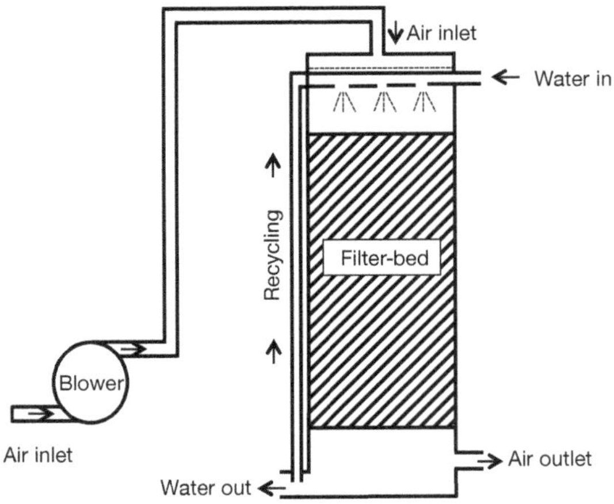

Figure 11.5 Biotrickling filter.

on their hydrophobic nature, is often expressed in terms of Henry's coefficient (Table 11.4).

11.6.3 Bioscrubber

A bioscrubber combines two separate processes, namely an absorption process followed by biodegradation in a bioreactor (Fig. 11.6) (Kennes and Veiga, 2001). The polluted gas phase flows first through an absorption unit in which the volatile compounds are transferred from that gas phase to a liquid phase, generally water. Clean gas (usually air) is then released to the atmosphere from the absorption tower, while generating at the same time a polluted liquid phase. The latter needs then to be treated, which is done by feeding the polluted liquid phase to a bioreactor in which the water soluble volatile compounds are biodegraded. Several types of bioreactors would be suitable to ensure biodegradation, but the most common one is the activated sludge reactor. In that reactor biomass grows in suspension in the liquid phase, contrary to the typical attached growth observed in biofilters and biotrickling filters. This technology is only suitable for rather hydrophilic compounds, as the pollutant must first be transferred from the gas phase to a liquid phase while being later on converted to harmless end metabolites in a suspended-growth bioreactor. In case of suspended-growth systems, no packing material is introduced in the bioreactor, although it can be used in the absorption column if a packed tower is chosen. Similarly as in the case of the biotrickling filter, a liquid phase is continuously flowing through the bioreactor, which allows for easy pH control and regulation.

Table 11.4 Typical suitable Henry's coefficients for different bioreactor configurations

Reactor type	Henry's coefficient (H) (dimensionless)
Biofilter	$H < 1$
Biotrickling filter	$H < 0.1$
Bioscrubber	$H < 0.01$

11.6.4 Gas diffusion through suspended growth bioreactors

In this type of suspended-growth bioreactor, the polluted gas phase is directly introduced, in the form of dispersed bubbles, into the bottom of the reactor where absorption and mass transfer take place between the gas phase and the liquid phase. Usually an activated sludge reactor is used, similarly as in aerobic wastewater treatment. This does then also mean that this technology is suitable both for wastewater treatment as well as for the simultaneous treatment of wastewater and waste gases. In a conventional activated sludge reactor for wastewater treatment, the polluted water phase is fed continuously to the reactor at a given flow rate, while air is introduced simultaneously into the reactor in order to provide oxygen as electron acceptor for the biodegradation process. If both wastewater and waste gas treatment are considered, then polluted air, rather than clean air, is introduced at the bottom of the bioreactor and both air and water pollutants are degraded simultaneously in the activated sludge reactor. If the system is used for waste gas treatment only, then a water phase containing nutrients is fed to the bioreactor, rather than wastewater, and polluted air is introduced simultaneously, providing both oxygen and the volatile pollutants that may act as carbon and/or energy sources.

11.6.5 Other bioreactors and hybrid processes

Besides the bioreactor configurations described above, some other systems have been tested, mainly at laboratory-scale, such as the membrane bioreactor and the rotating biological contactor (RBC) (Kennes and Veiga, 2001, 2013). The membrane bioreactor is so far not being used at full-scale. One major concern is related to the cost of the membranes. RBC have occasionally been used

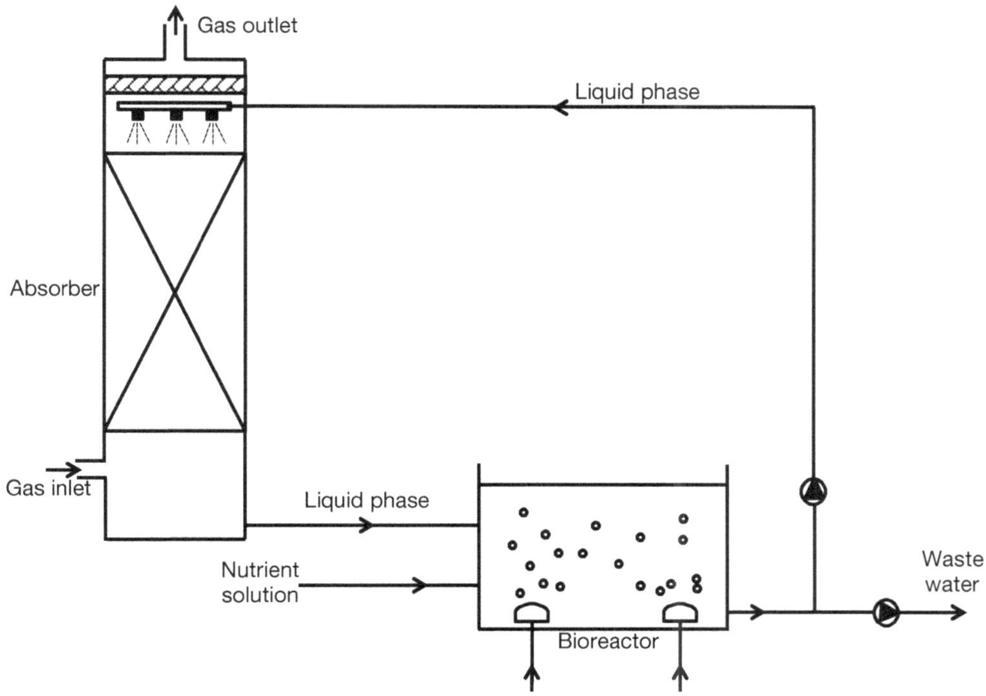

Figure 11.6 Bioscrubber.

at industrial scale, although only in a very limited number of applications.

Another alternative consists in combining bioprocesses with non-biological treatment technologies, such as also done in bioscrubbers described above, integrating absorption and biodegradation (Rene *et al.*, 2012). The combination of bioreactors and advanced oxidation processes is one example in which biodegradation may be coupled with a non biological photochemical treatment (Kennes and Veiga, 2001). Such combined processes may allow a more efficient removal of recalcitrant pollutants (Zhu *et al.*, 2015). Non-thermal plasma has been evaluated as well (Kennes and Veiga, 2001). Non-biological reactions may help in the partial degradation of pollutants that are more difficult to degrade biologically. It was already mentioned above that biodegradation can be combined with an adsorption process. If the latter is used as first stage treatment, it will allow for load equalization and will buffer any possible shock loads. If used as post-treatment, it will allow to polish the effluent and remove the non-biodegraded fraction of the inlet pollutant load (Rene *et al.*, 2012). Finally, combining two bioreactors, of the same or different configurations, may also present advantages in some cases.

11.7 Conclusions

Although the emission a volatile organic and inorganic pollutants to the atmosphere is an important environmental problem, initiatives are being taken to reduce such emissions and technologies have been developed to remove the pollutants once emitted from different sources. Bioprocesses have proven to represent a suitable alternative for the degradation of volatile pollutants from stationary sources, over a wide range of gas flow rates and up to concentrations of about 5 g/m^3 (air). Many of those compounds are partly converted to biomass besides being completely mineralized. Alternatively, recent research has shown that it would be feasible to convert volatile pollutants to useful end metabolites. Other on-going research is focusing on widening the application range of such bioprocesses. Different efficient types of bioreactors have been developed and are used in full scale applications, while others are still being optimized at lab-scale.

References

Abubackar, H.N., Veiga, M.C., and Kennes, C. (2011). Biological conversion of carbon monoxide rich syngas or waste gases to bioethanol. Biof. Bioprod. Biorefin. 5, 93–114.

Baggi, G., Bogga, M.M., and Catelani, D. (1983). Styrene catabolism by a strain of *Pseudomonas fluorescens*. System. Appl. Microbiol. 4, 141–147.

Bailón, L., Dopico, Y., Nikolausz, M., Kästner, M., Veiga, M.C., and Kennes, C. (2007). Removal of dichloromethane from waste gases using a fixed-bed biotrickling filter and a continuous stirred tank bioreactor. In: Kennes C and Veiga MC (eds.), Proc. 2nd International Conference on Biotechniques for Air Pollution Control. Universidade da Coruña (publisher), La Coruña, Spain, pp. 505–514. ISBN 978-84-9749-258-4.

Bailón, L., Nikolausz, M., Kästner, M., Veiga, M.C., and Kennes, C. (2009). Removal of dichloromethane from waste gases in one- and two-liquid phase stirred tank bioreactors and biotrickling filters. Water Res. 43, 11–20.

Beschkov, V., Sapundzhiev, T., Torz, M., Wietzes, P., and Janssen, D.B. (2008). Modeling of 1,2-dichloroethane biodegradation by *Xanthobacter autotrophicus* GJ10 under shock loading of other halogenated compounds in continuous stirred tank bioreactor. Chem. Biochem. Engin. Quater. 22, 339–348.

Chen, D.-Z., Ouyang, D.-J., Liu, H.-X., Chen, J., Zhuang, Q.-F., and Chen, J.M. (2014). Effective utilization of dichloromethane by a newly isolated strain *Methylobacterium rhodesianum* H13. Environ. Sci. Poll. Res. 21, 1010–1019.

Cofone, L., Walker, J.D., and Cooney, J.J. (1973). Utilization of hydrocarbons by *Cladosporium resinae*. J. Gen. Microbiol. 76, 243–246.

Cox, H.H.J., Faber, B.W., van Heiningen, W.N.M., Radhoe, H., Doddema, H.J., and Harder, W. (1996). Styrene metabolism in *Exophiala jeanselmei* and involvement of a cytochrome P-450-dependent styrene monooxygenase. Appl. Environ. Microbiol. 62, 1471–1474.

Davies, J.S., Wellman, A.M., and Zajic, J.E. (1973). Hyphomycetes utilizing natural gas. Can. J. Microbiol. 19, 81–85.

de Carvalho, C.C.C.R., and Fonseca, M.M.R. (2005). Degradation of hydrocarbons and alcohols at different temperatures and salinities by *Rhodococcus erythropolis* DCL14. FEMS Microbiol. Ecol. 51, 389–399.

Emanuelsson, M.A.E., Osuna, M.B., Ferreira Jorge, R.M., and Castro, P.M.L. (2009). Isolation of a *Xanthobacter* sp. degrading dichloromethane and characterization of the gene involved in the degradation. Biodegradation 20, 235–244.

Estévez, E., Veiga, M.C., and Kennes, C. (2005a). Biofiltration of waste gases with the fungi *Exophiala oligosperma* and *Paecilomyces variotii*. Appl. Microbiol. Biotechnol. 67, 563–568.

Estévez, E., Veiga, M.C., and Kennes, C. (2005b). Biodegradation of toluene by the new fungal isolates *Paecilomyces variotii* and *Exophiala oligosperma*. J. Indust. Microbiol. Biotechnol. 32, 33–37.

Firsova, J., Doronina, N., Lang, E., Spöer, C., Vuilleumier, S., and Trotsenko, Y. (2009). *Ancylobacter dichloromethanicus* sp. nov. – a new aerobic facultatively metholoytrophic bacterium utilizing dichloromethane. System. Appl. Microbiol. 32, 227–232.

Folsom, B.R., Chapman, P.J., and Pritchard, P.H. (1990). Phenol and trichloroethylene degradation by *Pseudomonas cepacia* G4: kinetics and interactions between substrates. Appl. Environ. Microbiol. 56, 1279–1285.

Garnier, P.M., Auria, R., Augur, C., and Revah, S. (1999). Cometabolic biodegradation of methyl-*t*-butyl ether by *Pseudomonas aeruginosa* grown on pentane. Appl. Microbiol. Biotechnol. 51, 498–503.

Habibi, A., and Vahabzadeh, F. (2013). Degradation of formaldehyde at high concentrations by phenol-adapted *Ralstonia eutropha* closely related to pink-pigmented facultative methylotrophs. J. Environ. Sci. Health. 48, 279–292.

Hardison, L.K., Curry, S.S., Ciuffetti, L.M., and Hyman, M.R. (1997). Metabolism of diethyl ether and cometabolism of methyl *tert*-butyl ether by a filamentous fungus, a *Graphium* sp. Appl. Environ. Microbiol. 63, 3059–3067.

Ishikawa, H., Kita, Y., and Horikoshi, K. (1980). US Patent 4225381.

Itoh, N., Yoshida, K., and Okada, K. (1996). Isolation and identification of styrene-degrading *Corynebacterium* strains, and their styrene metabolism. Biosci. Biotechnol. Biochem. 60, 1826–1830.

Jin, Y., Veiga, M.C., and Kennes, C. (2005a). Autotrophic deodorization of hydrogen sulphide in a biotrickling filter. J. Chem. Technol. Biotechnol. 80, 998–1004.

Jin, Y., Veiga, M.C., and Kennes, C. (2005b). Effects of pH, CO_2 and flow pattern on the autotrophic degradation of hydrogen sulphide in a biotrickling filter. Biotechnol. Bioeng. 92, 462–471.

Kayser, M.F., Ucurum, Z., and Vuilleumier, S. (2002). Dichloromethane metabolim and C1 utilization genes in *Methylobacterium* strains. Microbiol. 148, 1915–1922.

Kennes, C., and Veiga, M.C. (2001). Bioreactors for waste gas treatment. Kluwer Academic Publishers, Dordrecht, The Netherlands, pp. 312, ISBN: 9780792371908.

Kennes, C., and Veiga, M.C. (2002). Inert filter media for the biofiltration of waste gases – Characteristics and biomass control. Rev. Environ. Sci. Bio/Technol. 1, 201–214.

Kennes, C., and Veiga, M.C. (2004). Fungal biocatalysts in the biofiltration of VOC-polluted air. J. Biotechnol. 113, 305–319.

Kennes, C., and Veiga, M.C. (2013). Air pollution prevention and control: bioreactors and bioenergy. J. Wiley and Sons, Chichester, United Kingdom, pp. 549, ISBN: 9781119943310.

Kennes, C., Rene, E.R., and Veiga, M.C. (2009). Bioprocesses for air pollution control. J. Chem. Technol. Biotechnol. 84, 1419–1436.

Kennes, D., Abubackar, H.N., Diaz, M., Veiga, M.C., and Kennes, C. (2016). Bioethanol production from biomass: carbohydrate *vs* syngas fermentation. J. Chem. Technol. Biotechnol. 91, 304–317.

Kim, S.B., Park, C.-H., Kim, D.-J., and Jury, W.A. (2003). Kinetics of benzene biodegradation by *Pseudomonas aeruginosa*: Parameter estimation. Environ. Toxicol. Chem. 22, 1038–1045.

Kim, J., and Jeon, C. (2009). Isolation and characterization of a new benzene, toluene, and ethylbenzene degrading bacterium, *Acinetobacter* sp. B113. Current Microbiol. *58*, 70–75.

Kim, S., Bae, W., Hwang, J., and Park, J. (2010). Aerobic TCE degradation by encapsulated toluene-oxidizing bacteria, *Pseudomonas putida* and *Bacillus* spp. Water Sci. Technol. *62*, 1991–1997.

Kohler-Staub, D., and Leisinger, T. (1985). Dichloromethane dehalogenase of *Hyphomicrobium* sp. strain DM2. J. Bacteriol. *162*, 676–681.

Lee, E.-H., and Cho, K.-S. (2008). Characterization of cyclohexane and hexane degradation by *Rhodococcus* sp. EC1. Chemosphere. *71*, 1738–1744.

Liu, D., Feilberg, A., Hansen, M.J., Pedersen, C.L., and Nielsen, A.M. (2016). Modeling removal of volatile sulfur compounds in a full-scale biological air filter. J. Chem. Technol. Biotechnol. *91*, 1119–1127.

López, M.E., Rene, E.R., Veiga, M.C., and Kennes, C. (2012). Biogas technologies and cleaning techniques (chapter 9). In: Kennes C and Veiga MC (eds), Environmental chemistry for a sustainable world: remediation of air and water pollution, Springer, Dordrecht, The Netherlands, pp. 347–377.

Ma, Y.-L., Yang, B.-L., and Zhao, J.L. (2006). Removal of H_2S by *Thiobacillus denitrificans* immobilized on different matrices. Bioresour. Technol. *97*, 2041–2016.

Marco-Urrea, E., Gabarrell, X., Caminal, G., Vicent, T., and Reddy, C.A. (2008). Aerobic degradation by white-rot fungi of trichloroethylene (TCE) and mixtures of TCE and perchloroethylene (PCE). J. Chem. Technol. Biotechnol. *83*, 1190–1196.

Mukherjee, A.K., and Bordoloi, N.K. (2012). Biodegradation of benzene, toluene and xylene (BTX) in liquid culture and in soil by *Bacillus subtilis* and *Pseudomonas aeruginosa*. Environ. Sci. Poll. Res. *19*, 3380–3388.

O'Connor, K., Buckley, C.M., Hartmans, S., and Dodson, A.D.W. (1995). Possible regulatory role for monoaromatic carbon sources in styrene degradation by *Pseudomonas putida* CA-3. Appl. Environ. Microbiol. *61*, 544–548.

Oldenhuis, R., Oedzes, J.Y., van der Waarde, J.J., and Janssen, D.B. (1991). Kinetics of chlorinated hydrocarbon degradation by *Methylosinus trichosporium* OB3b and toxicity of trichloroethylene. Appl. Environ. Microbiol. *57*, 7–14.

Onodera, M., Endo, Y., and Ogasawara, N. (1989). Utilization of short-chain hydrocarbons and accumulation of methylketones by a gaseous hydrocarbon assimilating mold, *Scedosoprium* sp. A-4. Agric. Biol. Chem. *53*, 1431–1432.

Pol, A., van der Drift, C., and op den Camp, H.J.M. (2007). Isolation of a carbon disulfide utilizing *Thiomonas* sp. and its application in a biotrickling filter. Appl. Microbiol. Biotechnol. *74*, 439–446.

Prenafeta-Boldú, F.X., Kuhn, A., Luykx, D.M.D.A., Anke, H., van Groenestijn, J.W., and de Bont, J.A.M. (2001). Isolation and characterization of fungi growing on volatile aromatic hydrocarbons as their sole carbon and energy source. Mycol. Res. *105*, 477–484.

Rene, E.R., Veiga, M.C., and Kennes, C. (2012). Combined biological and physico-chemical waste-gas cleaning techniques. J. Environ. Sci. Health. *47*, 920–939.

Rodrigues, E.M., Kalks, K.H.M., and Tótola, M.R. (2015). Prospect, isolation and characterization of microorganisms for potential use in cases of oil bioremediation along the coast of Trindade island, Brazil. J. Environ. Manage. *156*, 15–22.

Sakaguchi, K., Kurane, R., and Murata, M. (1975). Assimilation of formaldehyde and other C1-compounds by *Gliocladium deliquescens* and *Paecilomyces variotii*. Agr. Biol. Chem. *39*, 1695–1702.

Shim, H., and Yang, S.T. (1999). Biodegradation of benzene, toluene, ethylbenzene, and o-xylene by a coculture of *Pseudomonas putida* and *Pseudomonas fluorescens* immobilized in a fibrous-bed reactor. J. Biotechnol. *67*, 99–112.

Smet, E., Van Langenhove, H., and Verstraete, W. (1996). Long-term stability of a biofilter treating dimethyl sulphide. Appl. Microbiol. Biotechnol. *46*, 191–196.

Smith, N.A., and Kelly, D.P. (1988). Oxidation of carbon disulphide as the sole source of energy for the autotrophic growth of *Thiobacillus thioparus* strain TK-m. J. Gen. Microbiol. *134*, 3041–3048.

Tanagawa, T., and Mikami, E. (1989). Removal of methanethiol, dimethyl sulfide, dimethyl disulfide and hydrogen sulfide from contaminated air by *Thiobacillus thioparus* TK-m. Appl. Environ. Microbiol. *55*, 555–558.

Tang, J., Hu, K.-D., Hu, L.-Y., Liu, Y.-S., and Zhang, H. (2014). Hydrogen sulfide acts as a fungicide to alleviate senescens and decay in fresh-cut sweetpotato. HortScience. *49*, 938–943.

Toda, H., Imae, R., Komio, T., and Itoh, N. (2012). Expression and characterization of styrene monooxygenases of *Rhodococcus* sp. ST-5 and ST-10 for synthesizing enantiopure (S)-epoxides. Appl. Microbiol. Biotechnol. *96*, 407–418.

Weber, F.J., Hage, K.C., and de Bont, J.A.M. (1995). Growth of the fungus *Cladosporium sphaerospermum* with toluene as the sole carbon and energy source. Appl. Environ. Microbiol. *61*, 3562–3566.

Willers, A., Dressler, C., and Kennes, C. (2013). Biotrickling filtration of waste gases from the viscose industry (chapter 19). In Air pollution prevention and control: bioreactors and bioenergy, Kennes, C., and Veiga, M.C., eds. (J. Wiley and Sons, Chichester, UK) pp. 467–484.

Woertz, J.R., Kinney, K.A., McIntosh, N.D.P., and Szaniszlo, P.J. (2001). Removal of toluene in a vapor-phase bioreactor containing a strain of the dimorphic black yeast *Exophiala lecanii-corni*. Biotechnol. Bioeng. *75*, 550–558.

Wu, S.J., Zhang, L.L., Wang, J.D., and Chen, J.M. (2007). *Bacillus circulans* WZ-12 – A newly discovered aerobic dichloromethane-degrading methylotrophic bacterium. Appl. Microbiol. Biotechnol. *76*, 1289–1296.

Xiao, B., Chen, M., Tan, Z., Liu, Q., Zhou, H., and Li, X. (2015). Isolation and characterization of *Bacillus* sp. BZ-001H capable of biodegrading formaldehyde at high concentration. Environ. Engin. Sci. 32, 824–830.

Yadav, J.S., and Reddy, C.A. (1993). Degradation of benzene, toluene, ethylbenzene, and xylenes (BTEX) by the lignin-degrading basidiomycete *Phanerochaete chrysosporium*. Appl. Environ. Microbiol. *59*, 756–762.

Yu, J.M., Cai, W., Cheng, Z., and Chen, J.M. (2014). Degradation of dichloromethane by an isolated

strain *Pandoraea pnomenusa* and its performance in a biotrickling filter. J. Environ. Sci. *26*, 1108–1117.
Zhao, H., Geng, Y., Fan, J., Tao, K., and Hou, T. (2013). Isolation and identification of *Paracoccus* sp. FD3 and evaluation of its formaldehyde degrading kinetics. Biotechnol. Bioproc. Engin. *18*, 300–305.
Zhu, R., Kennes, C., Cheng, Z., Lu, L., Yu, J.M., and Chen, J.M. (2015). Styrene removal in a biotrickling filter and a combined UV-biotrickling filter: Steady- and transient-state performance and microbial analysis. Chem. Engin. J. *275*, 168–178.

Heavy Metal Removal by Microbial Cells

12

Mirosława Słaba, Katarzyna Hrynkiewicz and Geoffrey M. Gadd

Contents

Abstract

Environmental pollution with heavy metals poses a real ecological and human-health threat due to their high toxicity to living organisms. Metals cannot be degraded. Hence, microbial removal by binding to biomass, leaching, enzymatic transformations, and changes in metal solubility and, as a consequence, in toxicity are major strategies for the remediation of metal-contaminated environments. For this reason, more knowledge on the mechanisms responsible for microbial tolerance and transformations of toxic metals are important for possible applications in environmental biotechnology. This chapter includes the results of studies on passive metal removal through biosorption and metal uptake and accumulation involving the metabolic activity of microbial cells. Selected aspects of metal–microbe interactions have been presented.

Proteomic studies that were carried out to identify microbial responses to heavy metal toxicity are also described. In the second part of the review, the potential of different groups of microorganisms for remediation and enhanced phytoremediation of metal-polluted soils is discussed. Attention has also been focused on the role of microorganisms in plant protection against metal stress as well as the mechanisms related to plant–microbe associations under unfavourable soil conditions.

12.1 Introduction

Global pollution with heavy metals, which can result from natural processes (e.g. forest fires, volcanic eruptions, rock erosion) and primarily anthropogenic activity, is a serious environmental problem. Natural sources can contribute to local heavy-metal contamination of the environment (e.g. serpentine soil) (Gadd, 2010). The main sources of metal pollution are burning processes and industrial operations such as mining, metal processing, electroplating, battery, dye, textile, and fertilizer production, and leather tanning. Metallic elements with atomic weights from 63.5 to 200.6 and a specific gravity above 5.0 are often termed heavy metals and are reported as being a priority pollutant, due to their high toxicity and environmental mobility (Fu and Wang, 2011). Additionally, they accumulate in cells, tissues, and organs of living organisms, and can generate a serious risk to humans. For example, chronic exposure to cadmium, chromium (VI), and nickel can result in the development of certain cancers (Beyersmann, 2002; Arita and Costa, 2009). In the case of chromium, its biological effects are strongly dependent on the valence state. Cr(VI) is a hundred times more toxic and a thousand times more mutagenic than Cr(III) (Arévalo-Rangel *et al.*, 2013). Another health risk is related to the harmful effects of these pollutants on the hormonal systems of animals including humans. Mercury, cadmium, nickel, and lead are known endocrine-disrupting compounds (EDCs) (Takiquchi and Yoshihara, 2006). Cadmium and nickel combine endocrine disruption and carcinogenesis. They have been hypothesized to play a role in breast cancer development by binding to oestrogen receptors mimicking the actions of oestrogen (Aquino *et al.*, 2012).

In recent years, biological methods for the removal of hazardous metals have been intensively explored due to the environmentally friendly character and purported low cost.

Microorganisms, including bacteria, actinomycetes, filamentous fungi, yeasts, and microalgae, can adapt and grow even at high concentrations of metals (Gadd, 2010; Mukherjee *et al.*, 2010; Słaba and Długoński, 2011). They also interact with metals emitted by anthropogenic sources and with metal-containing minerals, altering their solubility and toxicity. Biosorption and bioaccumulation by microbial cells causes a decrease in environmental mobility. Bacteria and fungi play the most important roles in the bioremediation of metal polluted soils and they enhance phytoremediation efficiency.

12.2 Passive heavy metal uptake via biosorption

Microorganisms are good biosorbents for heavy metal removal or recovery. Various functional groups in the cell wall or external polymers of different groups of microorganisms are responsible for effective metal biosorption (Table 12.1).

Active groups (mainly carboxyl groups) in the cell wall polymers of Gram-positive bacteria such as peptidoglycan and teichoic and teichuronic acids play an essential role in surface metal binding, but the heavily cross-linked peptidoglycan layer may result in a lower level of metal sorption in comparison to Gram-negative cells (Joo *et al.*, 2010). Gram-negative bacteria are described as good metal sorbents due to better exposure of their lipopolysaccharides, phospholipids, and proteins on the cell wall surface (Aryal and Liakopou-Kyrakides, 2015). Carboxyl and sulfate groups are predominant in cyanobacteria, and metal cations are effectively adsorbed by the negatively charged archaeal cell wall.

Algae represent a non-homogeneous group ranging from unicellular to macroscopic forms. They also differ in cell wall structure and, as a consequence, different functional groups from various cell wall polymers are the most active. Algal wall polymers like mannan, alginic acid, and xylans, as well as proteins, supply hydroxyl, phosphate, sulfate, and amino groups (Mehta and Gaur, 2005; Wang and Chen, 2009; Fomina and Gadd, 2014). Within this complex group of organisms, brown algae seem to be the best biosorbents, due to high

Table 12.1 Functional groups in the cell wall or external polymers of selected microorganisms originating from differ systematic groups, the most active in metal sorption process

Group of microorganisms	Characteristic polymers of cell wall, outer membrane or external surface	The most active chemical groups in the metal biosorption	References
Cyanobacteria	Galacturonic acid Glucuronic acid Exopolymers	Carboxyl Sulfate Phosphate	Parker *et al*. (2000) Michalak *et al*. (2013)
Gram-positive bacteria	Peptidoglycans Teichoic acids Teichuronic acids	Carboxyl, hydroxyl, amide Phosphate Carboxyl	Joo *et al*. (2010) Fomina and Gadd (2014)
Gram-negative bacteria	Lipopolysaccharides (LPS) Phospholipids (PLs) Proteins	Carboxyl Phosphate Amide	Aryal and Liakopou-Kyrakides (2015)
Filamentous fungi	Chitin, chitosan, glucans Some with melanin pigment	Carboxyl Hydroxyl Amide	Gadd (1993, 2007, 2009)
Yeast	Mannan Exopolymers Proteins and amino acids Some with melanin pigment	Carboxyl Amide Hydroxyl	Breierová *et al*. (2002)
Microalgae	Polysaccharides Proteins Lipids	Phosphate Carboxyl Amide	Brinza *et al*. (2007); Monteiro *et al*. (2012) Kumar *et al*. (2015)

content of alginate containing negative-charged active groups (Wang and Chen, 2009). Microalgae from the genera *Chlorella*, *Scenedesmus* and *Spirulina* have been often reported as effective biosorbents for metal uptake (Brinza *et al*., 2007; Kumar *et al*. 2015). The cell walls of fungi are composed of polysaccharides such as glucans, chitin, and chitosan, which possess a strong ability to bind metal cations (Gadd 1993, 2007, 2009).

The main purported advantages of biosorption are low cost. Biosorbents are more suitable for the treatment of dilute heavy-metal-containing wastewaters than traditional clean-up techniques such as chemical precipitation (mainly hydroxides and sulfides), ion exchange, and activated carbon adsorption (Fu and Wang, 2011). Non-living or waste biomass from the fermentation industry may be useful for biosorption and frequently exhibit a higher affinity for metal ions than living cells, possibly due to the lack of protons, that are produced by active microorganisms competing with metal cations (Acheampong *et al*., 2010). Metal biosorption can also be performed by living microbial cells through passive binding or adsorption of metal ions onto the cell surfaces. However, metabolic activity can indirectly affect the process. The biosorption process is fast, taking from a few minutes to a few hours (fastest when the ion exchange mechanism is predominant). The fast saturation of binding sites limits sorption capacity in comparison to intracellular accumulation, but this disadvantage can be overcome by desorption cycles. Since microbial morphology and cell wall composition are variable, the main problem is preparation of conditions for the use of biomass as an efficient metal sorbent. In the case of batch experiments, separation of metal-loaded biomass can be difficult.

12.2.1 Definition and mechanism

Biosorption of metals can be defined as the ability of materials of biological origin (biosorbents) to bind metal ions from solution. This simple physicochemical process involves different mechanisms such as ion exchange, complexation, or physical adsorption. If changes in metal solubility occur, adsorption can result in the formation of surface-bound precipitates (Gadd, 2009; Fomina and Gadd, 2014). It has been found that one mechanism of metal uptake usually dominates over others (Słaba and Długoński, 2011). Most studies on metal biosorption concern a situation when metals occur as divalent cations being bound by anionic active groups in the biosorbents. Negatively charged metal species are bound to positively charged functional groups in the biosorbent (Saha and Orvig, 2010). The mechanism of Cr(VI) biosorption is more complex and may differ from the mechanisms for uptake of other metals (Park *et al.*, 2008; Saha and Orvig, 2010). It involves adsorption of Cr(VI) to positively charged functional groups (mainly at low pH), followed by passive reduction of Cr(VI) to Cr(III), after which a proportion of reduced Cr(III) is adsorbed by the biomass (Park *et al.*, 2008). According to a third mechanism, after partial reduction of Cr(VI), reduced Cr(III) as a cation and the remaining Cr(VI) in the form of anionic species are adsorbed on the cell surface. A fourth possibility leads to a partial reduction of Cr(VI) and anionic adsorption of Cr(VI) residues, whereas Cr(III) remains in solution (Saha and Orvig, 2010).

The metal sorption mechanism can change even in the same microorganism, depending on the nature of the biomass, e.g. binding of strontium (Sr) by dead brewing yeast proceeds mainly via ionic bonding (a significant role for carboxyl and phosphate groups), whereas Sr uptake by living yeast cells exhibits a high role of covalent bonding, probably due to the engagement of amine and sulfhydryl groups, which are active on the cell surface of living cells but denatured in dead biomass (Avery and Tobin, 1993).

12.2.2 Affecting factors

The pH is of great importance in biosorption. Generally, sorption of metal cations increases with an increase in pH. Competition between metal cations and protons for anionic binding sites on the biosorbent is responsible for reduced sorption at low pH. In the same way, metals forming anionic groups such as CrO_4^{2-} $Au(CN_2)^-$, and SO_4^{2-} bind better at lower pH values. Metal ions with a tendency to form covalent bonds with N- and S-containing ligands, like Ag^+, Hg^{2+}, and $AuCl^-$, show a lack of pH dependence or a decrease in binding capacity when the pH increases (Gadd, 2009; Mukhopadhyay *et al.*, 2011).

Competition between metal and other cations in the solution can also disturb the biosorption process. In some cases, supplied cations (Ca^{2+}, Mg^{2+}) can stimulate cation biosorption via pH buffering (Gadd, 2009). Organic pollutants coexisting in the sorption system may inhibit this process (Park *et al.*, 2010). High salinity (chlorides) affects biosorption in several different ways, e.g. through the formation of complexes of $CdCl_3$ or precipitation of $PbCl_2$ (Gadd, 2009). Similarly, anions such as CO_3^{2-} and PO_4^{2-} cause metal precipitation. Other parameters influencing biosorption, e.g. biomass loading, are metal concentration, contact time, and temperature are also well recognized (Gadd, 2009; Wang and Chen, 2009; Aryal and Liakopoulou-Kyriakides, 2015). Recent studies also have included characterization of biosorbent surface morphology, surface area, and particle size (Acheampong *et al.*, 2010).

12.2.3 Desorption and biosorbent regeneration

Desorption and biosorbent regeneration are essential aspects of biosorbent use. Desorption efficiency requires selection of a good desorbent which may strongly depend on the type of biosorbent and the mechanism of biosorption. Other factors such as metal properties, metal-containing materials, and the applied techniques (immobilized, packed-column, or batch studies) also affect metal desorption. Acids, alkalis, salts, chelating agents, and buffer solutions have been used as efficient desorbents in many studies. Desorption of metal ions in an acidic environment appears to be rapid and higher than in neutral and basic environments (Srivastava and Goyal, 2010), whereas EDTA has been recommended as a non-destructive eluent (Lata *et al.*, 2015). Generally, there is no universally effective desorbent. Biosorbent reuse is one of the advantages of the biosorption process. Sorbent regeneration with common salt and mineral acid solutions such as $CaCl_2$, KCl and KNO_3 has also been described (Njikam and Schiewer, 2012; Aryal and Liakopoulou-Kyriakides, 2015).

12.2.4 Biosorbent modification

Surface modification can affect biosorption capacity and many studies have investigated the physical and chemical improvement of biomass. Physical procedures such as autoclaving, boiling, drying, lyophilization, and mechanical disruption can affect binding properties (Gabr *et al.*, 2008; Won *et al.*, 2014).

The most popular modifications are chemical pretreatments with acids, alkalis, organic solvents, or other chemicals. The improvement in binding ability are obtained through addition or modification of functional groups on the biomass. Increased sorption properties can also be achieved through chemical elimination of some inhibitory groups that may be present in raw biomass (Vijayaraghhavan and Yun, 2008). Many reports describe improved sorption of metal cations after washing with inorganic acids (Park *et al.*, 2010). However, a disadvantage of the chemical treatment is that biomass may lose structural integrity (Yan and Viraraghavan, 2001). Chemical modification of fungal biomass by an alkali often leads to higher biosorption capacity due to chitosan formation via chitin deacetylation. Glucan–chitosan complexes possess a higher metal affinity (Gadd, 2009). In the case of algae, the best results were obtained after $CaCl_2$ pretreatment (Mehta and Gaur, 2005). When living microbial cells are used as a metal biosorbent, optimization of culture conditions can enhance biosorption capacity. Biosorbent modifications increase the cost, and in the case of chemical reagents, may cause environmental secondary contamination. Genetic modifications based on expression of hybrid proteins on the cell surface have also been investigated (Kuroda and Ueda, 2010), but only low-cost biomass modifications would have real importance in any practical applications (Park at al., 2010).

12.3 Heavy metal immobilization by living microbial cells – mechanisms and applications

12.3.1 Accumulation

Microorganisms have been documented to efficiently remove soluble metal forms through accumulation, precipitation, and transformation. In contrast to biosorption, accumulation is a more complex phenomenon, involving cellular metabolism and active transport. This process proceeds slower than biosorption but may lead to higher metal loading. For accumulation, living microorganisms are required which generates costs of media and cultivation. Additionally, metal recovery needs disruptive methods. Metals may be located in different organelles (eukaryotes) or precipitated in the cytoplasm. Metal-binding cysteine-rich peptides or proteins contribute to detoxification of heavy metal ions. Metallothioneins (MTs) are proteins of low molecular weight (<14 kDa) detected in animals, plants, algae, bacteria and fungi, e.g. Cu^{I}-specific MT in *Saccharomyces cerevisiae* (Loebus *et al.* 2013). Phytochelatins (PCs) play a major role in defence against heavy metals in plants and are synthesized *de novo* from a glutathione precursor by phytochelatin synthase (PCS, glutathione-glutamylcysteinyl transferase), which is induced by an elevated concentration of metal ions. In fungi exposed to metal ions (especially Cd and Cu), PC–metal complexes were found in the vacuoles (Gadd, 1993; Clemens *et al.*, 1999; Shine *et al.*, 2015). Some fungi, e.g. *Heliscus lugdunensis* and *S. cerevisiae*, are not capable of PC synthesis (PC1) due to lack of phytochelatin synthase. Metals, especially Cd, induce a defence response manifested by the synthesis of another class of phytochelatins (PC2), formed by a vacuolar carboxypeptidase (Wunschmann *et al.*, 2007; Krauss *et al.* 2011; Loebus *et al.* 2013).

Elevated metal accumulation can be also caused by the simultaneous toxic action of metal ions and organic xenobiotics. Słaba *et al.* (2009) observed that the presence of alachlor, a chloroacetanilide herbicide, stimulated zinc uptake by mycelium in stationary *Paecilomyces marquandii* cultures. Enhanced zinc uptake in the presence of alachlor was correlated with changes in the cell wall and membrane (Słaba *et al.*, 2009).

12.3.2 Precipitation/transformation

Metal precipitation and transformation is a widespread microbial phenomenon leading to a decrease in metal toxicity. This may occur as a consequence of dissimilatory metal reduction. For instance, Cr(VI) reduction by chromate reductase results in formation of a Cr(III) precipitate. A common mechanism of metal precipitation is sulfide production by sulfate-reducing bacteria (SRB), e.g. *Desulfovibrio* and *Desulfotomaculum* spp. Microbial reduction of a metal to a lower redox state may result in decreased metal toxicity and mobility (Gadd, 2010). In many chromium-reducing bacteria (e.g. *Enterobacter*, *Pseudomonas*, *Micrococcus*, *Bacillus*, *Escherichia*, and *Achromobacter* spp.) or fungi (e.g. *Aspergillus*, *Penicillium*, *Trichoderma*, and *Paecilomyces* spp.) a specific chromate reductase has been identified (Dhal *et al.*, 2013; Rahman and Singh, 2014).

A common fungal strategy, which may lead to a decrease in toxicity is extracellular precipitation in the form of various minerals (e.g. carbonates or oxalates) (Słaba and Długoński, 2011; Gadd *et al.*, 2012; Li *et al.*, 2014).

Biologically controlled mineralization (BCM)

occurs when microorganisms exercise a great degree of process control and metabolic activity directs the nucleation, growth, morphology, and final location of minerals, whereas biologically induced mineralization (BIM) is when microorganisms alter their surroundings giving conditions favourable for extracellular precipitation (Gadd, 2010; Li *et al.*, 2014). Microbiologically induced pH increase can result in metal precipitation. For example, urea-hydrolysing bacteria and fungi cause extensive biomineralization of metal carbonates on cell surfaces. In this way urease-positive fungi, e.g. *Neurospora crassa*, may play a role in metal remediation or biorecovery (Li *et al.*, 2014; Li, Q. *et al.*, 2015).

Several fungi are often capable of transforming lead into pyromorphite, $Pb_5(PO_4)_3Cl$, a very stable lead mineral and in this way, they can contribute to Pb immobilization and alleviation of potent toxicity (Rhee *et al.*, 2012; 2014a,b; Jarosławiecka and Piotrowska-Seget, 2014).

An additional advantage of metal biomineralization is the ability to produce nanocrystals. Formation of nanocrystals by living biomass of *Aspergillus* and *Fusarium* spp. has been described (Sanyal *et al.*, 2005; Velmurugan *et al.*, 2010). The fact that fungal proteins play a crucial role in the morphology and size of synthesized crystals is important for further applications (Velmurugan *et al.*, 2010).

12.4 Metal–microbe interactions

12.4.1 Metal-induced oxidative stress

Induction of oxidative stress is one of the possible mechanisms of metal ion toxicity. Metals that exhibit two or more oxidative states such as Fe, Cu, and Cr, directly produce reactive oxygen species (ROS) via Fenton reactions. Redox-inert metals, such as cadmium, stimulate oxidative stress by inhibition of antioxidant enzyme (by binding their thiol groups in the active centre). On the other hand, cadmium can replace free reactive ions of Fe or Cu from cellular polymers, which leads to the induction of a direct stress mechanism (Koedrith and Seo, 2011). Additionally, metals cause oxidative stress by depletion of the pool of cellular non-enzymatic antioxidants. The consequence of metal-catalysed oxidative imbalance is damage to proteins, lipids, and nucleic acids. An increase in lipid peroxidation is often observed as a result of free radical and ROS production under the influence of Cd, Cu, As, Ni, and Pb in microbial cells (Howlett and Avery, 1997; Mukherjee *et al.*, 2010; Paraszkiewicz *et al.*, 2010; Słaba *et al.*, 2013a). A correlation between metal-induced lipid peroxidation, increased membrane permeability, and membrane damage has also been reported. Among metals, Cu especially causes strong oxidative membrane damage in fungi (Avery *et al.*, 1996; Azavedo *et al.*, 2007; Słaba *et al.*, 2013a,b). Polyunsaturated fatty acids of lipids are especially susceptible to peroxidation. Ni-induced damage of lipid double bonds resulting in elevated saturation of fatty acid in fungal cultures has been documented (Paraszkiewicz *et al.*, 2010). Toxic actions of metals can be manifested by interruption to cellular lipid synthesis. Phospholipids (PLs) are the main structural lipids in the cell membrane, playing an essential role in correct fluidity maintenance (de Kroon *et al.*, 2013). Changes in PLs composition and its saturation in metal-stressed microorganisms have often been observed (Čertík *et al.*, 2005; Słaba *et al.*, 2013b). Enhanced hydrolysis of phosphatidylcholine (PC) to phosphatidic acid (PA) by phospholipase D is postulated as a result of Pb toxicity in the ubiquitous soil fungus *Paecilomyces marquandii* (Słaba *et al.*, 2013b). The same mechanism causing a decrease in the PC level in stressed tissues of plants has also been documented (Welti *et al.*, 2002).

Antioxidative enzymes constitute the first line of cellular defence against ROS. Microbial cells exposed to toxic metal ions often exhibit a high activity of superoxide dismutase (SOD), converting the superoxide anion to H_2O_2 and O_2, and catalase (CAT), which scavenges hydrogen peroxide and decomposes it to H_2O and O_2 (Behera *et al.*, 2014).

Heavy metal stress may cause an increase only in the SOD activity, whereas generated ROS inhibit CAT activity (Cyrne *et al.*, 2003; Todorova *et al.*, 2008). In tolerant fungi *S. cerevisae* and *A. niger* an alternative mechanism of protection against H_2O_2 is possible, with participation of glutathione (Cyrne *et al.*, 2003; Todorova *et al.*, 2008).

Enhanced production of low-molecular-weight antioxidants such as glutathione or ascorbic acid also help microorganisms to neutralize metal-induced oxidative stress. Metal–glutathione conjugates in cultures of *Saccharomyces cerevisiae* exposed to Pb have been documented (Van der Heggen, 2010). Addition of an exogenous antioxidants have been shown to protect aquatic fungi from Zn and Cu (Azavedo *et al.*, 2007). Similarly, ascorbic acid supplementation alleviates Pb toxicity to the filamentous fungus *Paecilomyces marquandii* (Słaba *et al.*, 2013b).

12.4.2 Microbial resistance to heavy metals

Despite the high toxicity of heavy metals, resistant microorganisms have been isolated from metal-contaminated soils, industrial wastes, effluents, and sediments. Among isolated strains, the following bacterial genera have often been identified: *Pseudomonas, Enterobacter, Bacillus,* and *Arthrobacter* (Gupta *et al.*, 2014; Singh *et al.*, 2014). The most frequent microfungi from heavy-metal-polluted environments were from the genera *Aspergillus, Penicillium, Fusarium, Trichoderma,* and *Paecilomyces* (Massaccesi *et al.*, 2002; Ezzouhri *et al.*, 2009; Sun *et al.*, 2012; Iram *et al.*, 2013; Gadd *et al.*, 2014). Microorganisms have developed different strategies to survive heavy-metal contamination, including production of extracellular polymers, intracellular metal sequestration, changes in metal transport, enzymatic conversion into less toxic forms and specific efflux systems (Fig. 12.1). An increased metal concentration may cause changes in the morphology and physiology of metal-stressed fungi. In the presence of metals, hyphae may become shorter and denser. Aggregation of hyphae is recognized as a defensive response to toxic metals and is aimed at reduction of exposed surfaces (Gadd, 2007). Negative chemotropism of hyphae in response to Cu may be of similar importance (Fomina *et al.*, 2003).

It is well documented that toxic metals can enhance cell wall pigmentation in fungi. Fungal melanins, rich in hydroxyl groups, have high

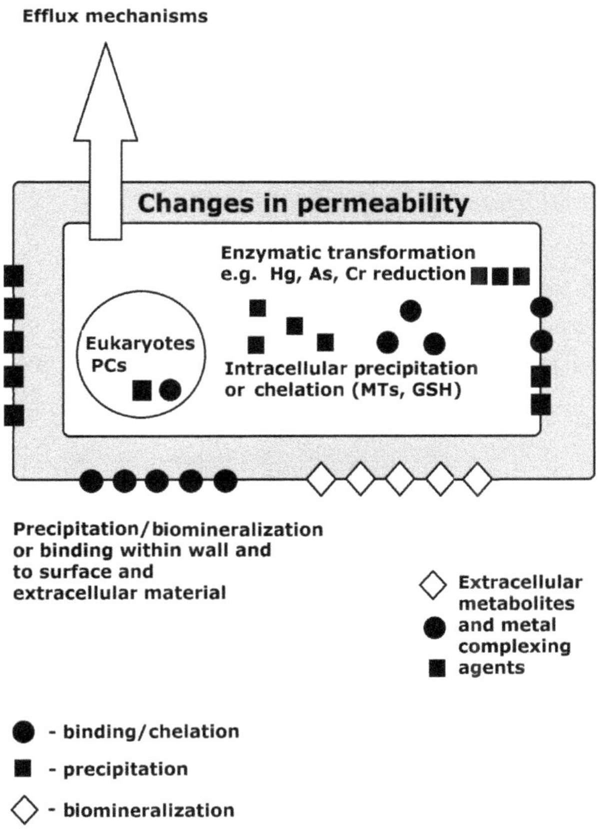

Figure 12.1 Scheme illustrating overall metal location in/around microbial cells and metal detoxification strategies (adapted from Gadd, 2010). PCs, phytochelatins; MTs, metallothioneins; GSH, glutathione.

biosorptive capacity and protect pigment-producing microfungi, e.g. *Cladosporium*, *Alternaria*, and *Aureobasidium* spp., against metal toxicity (Fomina and Gadd, 2002). Some metals (e.g. Zn, Cu, Cr, and Ni) induce changes in colony colouration, probably as a stress response. Yellow and red pigments are often accumulated in Zn and Cu-stressed mycelia from *Penicillium* and *Fusarium* sp. (Griffith and al., 2007; Ezzouhri *et al.*, 2009). Blue pigment production in the bacterium *Vogesella indigofera* was correlated with increasing Cr(VI) concentration (Cheung and Gu, 2002).

Another non-specific defence mechanisms of microorganisms is enhanced production of exopolysaccharides (EPS). The crucial role of exopolymers for survival in a metal-polluted environment was shown for *Curvularia lunata* (Paraszkiewicz *et al.*, 2007). This steroid-transforming fungus exhibited a low metal tolerance when grown on a solid medium preventing emulsifier production, whereas tolerance to Cd, Zn, and Pb at high concentrations (up to 15 mM) was observed in rich liquid cultures that stimulated exopolymer formation. Pullulan from *Aureobasidium pullulans* also mitigates the toxic effect of metal ions (Certik *et al.*, 2005). Enhanced production of intracellular polymers can also enhance microbial tolerance to metals. Lipid synthesis in *Chlorella minutissima* was induced by Cd and Cu. Based on this phenomenon, an integrated biotechnology of biofuel production involving intracellular metal accumulation and extracellular biosorption connected with enhanced lipid production by this oleaginous microalga was proposed (Yang *et al.*, 2015).

Toxic metals inhibit enzymatic activities. Metal-tolerant microorganisms are able to alter enzymatic activities in the presence of metal ions. Extracellular enzymes from *Streptomyces subrutilus* P5 with superoxide dismutase (SOD) activity under Pb exposure have been described (So *et al.*, 2001). The extracellular precipitation of a protein as a result of Pb chelation seems to be a major mechanism of Pb tolerance, but the production of SOD was not induced by metal ions. In a metal-tolerant *Variovorax* strain, catechol 2,3-dioxygenase activity increased in the presence of 0.1 mM Cd (Hupert-Kocurek *et al.*, 2013). This enabled bacterial phenol utilization in the environment polluted with inorganic and organic chemicals, whereas Cd significantly reduced the activities of dioxygenases and catechol dioxygenase in metal-sensitive bacteria (Guzik *et al.*, 2012).

12.4.3 Proteomic analysis of metal resistance

A number of reports apply proteomics to identify proteins produced by microorganisms in response to heavy-metal exposure increases (Luque-Garcia *et al.*, 2011). Among them, Cd action towards microbial cells has been best documented (Vido *et al.*, 2001; Lacerda *et al.*, 2007; Jain and Bhatt, 2013; Orell *et al.*, 2013; Siripornadulsil *et al.*, 2014).

One of the first proteomic studies addressed Cd responses in *S. cerevisiae* (Vido *et al.*, 2001). Cd strongly affected the yeast cells and induction of nine enzymes of the biosynthetic pathway of sulfur amino acids was detected. An essential role for yeast Cd tolerance was played by thioredoxin and thioredoxin reductase. Mutants without the YAPS1 genes encoding these proteins were hypersensitive to Cd. A *Klebsiella pneumoniae* strain isolated from a metal-contaminated river overexpressed two proteins, which were identified as L-isoaspartate carboxymethyltransferase type II and DNA gyrase (Bar *et al.*, 2007). DNA gyrase can alter DNA topology in response to environmental stress. The other overexpressed enzyme is responsible for transferring the methyl group from S-adenosylmethionine to the carboxyl group of isoaspartyl D- and L-isomers and is probably involved in the repair or degradation of damaged proteins.

Examination of the proteome of the plant growth-promoting bacterium (PGPB) *Pseudomonas putida under Ni stress* revealed proteins responsible for adaptation to both general stress and antioxidative stress as well as a specific resistance mechanism involving heavy-metal efflux (Cheng *et al.*, 2009). Proteomic analysis of metal-tolerant *Sulfolobus metallicus*, a thermoacidophilic archaeon used for thermophilic Cu leaching, revealed overproduction of proteins related to the synthesis of amino acids, the production and conversion of energy, and a stress response (Orell *et al.*, 2013). Generally, sublethal toxic actions of heavy metals causes overexpression of more proteins in comparison with down-regulation (Table 12.2).

Table 12.2 Selected proteomic approaches for documenting heavy metal toxicity and microbial survival strategy

Microorganism	Metal	Amount of detected proteins with changed expression	The most important proteins/enzymes for microbe survival in heavy metal polluted environment	References
Saccharomyces cerevisiae (yeast)	Cd	54 up-regulated 43 down-regulated	thioredoxin and thioredoxin reductase	Vido *et al.* (2001)
Schizosaccharomyces pombe (yeast)	Cd	106 up-regulated 55 down-regulated	27 proteins important in stress response and repair proteins, including thioredoxin and glutathione S- transferase (|GST)	Bae and Chen (2004)
Klebsiella pneumoniae (Gram-negative bacterium)	Co Pb	Not listed	L-isoaspartate carboxymethyltransferase type II and DNA gyrase	Bar *et al.* (2007)
Pseudomonas fluorescens ATCC 948 (Gram-negative bacterium)	Co Cu	Not listed	Co – induction of outer membrane proteins Cu – overexpression of enolase	Sharma *et al.* (2006)
Cuprividus taiwanensis (Gram-negative nitrogen fixing bacterium)	Cd	982 differentially expressed	overexpression of chaperone GroEL down-regulated protein flagellin, up-regulation of enzymes involved in response to oxidative stress	Siripornadulsil *et al.* (2014)
Sulfolobus metallicus (thermoacidophilic archaeon)	Cu	30 up-regulated 23 down-regulated	duplicated version of the *cop* gene cluster encoding a Cu-exporting ATPase	Orelll *et al.* (2013)

Proteomics and related omics techniques are very useful tools that may help in understanding the defensive responses of metal tolerant microorganisms and to choose the best candidates for bioremediation. Microbial proteome analysis has revealed universal stress-avoiding strategies such as stress response proteins, antioxidant enzymes, proteins responsible for energy production and conversion, DNA protection and repair, and recognition and elimination of abnormal proteins (Siripornadulsil *et al.*, 2014). Two-dimensional gel electrophoresis showed the presence of an ATP-dependent protease (induced stress protein) and 20S proteasome contributing to the degradation of oxidized proteins in psychrotolerant *Pseudomonas putida* (Jain and Bhatt, 2013). Specific cellular responses of microorganisms to metal toxicity have also been detected in proteomic studies (Bell *et al.*, 2014). Nitrate reductase expression in the proteome of alkalophilic *Pseudomonas monteilii* after Cd exposure has been documented. Nitrate reductase may have chromate-reducing activity, but in the case of Cd with stable valency, its role needs clarification (Jain and Bhatt, 2013). Cd toxicity via inhibition of flagella-mediated motility in *Cupriavidus taiwanensis* has been described (Siripornadulsil *et al.*, 2014).

12.5 Microbial bioremediation of metal-contaminated environments

Soil remediation is the process of correcting or cleansing soil of hazardous contaminants or pollutants, e.g. heavy metals. These procedures can involve physical, chemical, and biological methods (Tan, 2009). Among the many physicochemical methods that are utilized for reclamation of degraded soils, vitrification, chemical immobilization with organic or inorganic compounds, and electrokinetic techniques have been applied (Wuana and Okieimen, 2011). However, conventional methods are usually very expensive and in general adversely affect the texture and organic chemical compounds in the soil, which negatively affects soil fertility (Rajkumar *et al.*, 2010). Therefore, biological methods of remediation are often considered as alternative cheap and ecologically friendly technologies that can be used under natural conditions with high efficiency.

Biological remediation, known as bioremediation, is a process of cleaning the environment with the assistance of living organisms and/or their products that are capable of breaking down or immobilizing pollutants (Abhilash *et al.*, 2009; Tan, 2009). Bioremediation that revitalizes degraded or contaminated land is an attractive strategy to meet the food and energy needs of the growing world population (Abhilash *et al.*, 2012). Some examples of bioremediation are phytoremediation, bioventing, bioleaching, land forming, bioreactors, composting, bioaugmentation, rhizofiltration, and biostimulation. In these biotechnologies, microbes play a key role. Microorganisms used in bioremediation are known as bioremediators. These may include bacteria and fungi, including yeasts. In organic bioremediation, microorganisms may use the contaminants as nutrient and energy sources (Tang *et al.*, 2007). However, metals cannot be degraded (unlike organic pollutants), but only removed or transformed into their less toxic forms, e.g. by changing their redox state, and as a consequence, their chemical properties (Tabak *et al.*, 2005). This is why metal-related bioremediation strategies are aimed only at increasing solubility (biomobilization) increasing their stability (bioimmobilization) in a solid matrix (Beolchini *et al.*, 2011).

12.5.1 Metal removal by microbial consortia

In natural environments, microorganisms commonly exist as communities of multiple species that are capable of performing more varied and complex tasks and of enduring more changeable environments than monocultures (Brune and Bayer, 2012). Natural microbial consortia are known to facilitate a wide range of complex tasks such as inter-species biofilm formation, which allows microorganisms to persist in inhospitable environments (Keller and Surette, 2006). This is possible because members of the consortium communicate with one another, e.g. by trading metabolites or by exchanging special molecular signals. Moreover, each population or individual detects and responds to the presence of others in the consortium (Keller and Surette, 2006). This is why the overall output of the consortium depends on a combination of tasks performed by constituent individuals or subpopulations (Brenner *et al.*, 2008).

Natural consortia play crucial roles in biomining and the bioremediation of heavy-metal-contaminated environments. In biomining, acidophilic microbes facilitate the recovery of metals from sulfide minerals in the processes of bioleaching (solubilization of metals of interest such as Co, Cu, and Ni from sulfide minerals) and bio-oxidation (enrichment of metals, particularly Au, by mobilization and thus removal of interfering metal sulfides from ores bearing the precious metals) (Brune and Bayer, 2012). The microbes that are found in these environments are (extreme) acidophiles growing at pH 3 or lower and span a wide range of different phyla: mostly bacterial and archaeal domains, although unicellular eukaryotes have also been reported (Brune and Bayer, 2012). Boelchini *et al.* (2011) described experiments on Fe/S-oxidizing strains that showed a strong correlation between elemental S, bacterial abundance and metal solubilization. The efficiency of metal removal using autotrophic Fe/S-oxidizing and heterotrophic acidophilic bacteria reaches >90% for Cu, Cd, Hg and Zn. Such experiments confirm that application of bioleaching technologies based on mutual interactions among bacterial strains having different metabolic pathways and requirements may open up new opportunities for remediation (Boelchini *et al.*, 2011).

Natural communities of microorganisms in various habitats have an extensive physiological versatility and are able to metabolize a great variety of organic molecules. However, Singh *et al.* (2014) showed that individual microorganisms cannot degrade the most hazardous compounds because complete decomposition depends on sequential degradation by a consortium of microorganisms and involves synergistic and co–metabolic interactions (Singh *et al.*, 2014). The most complicated microbial consortia form biofilms, which are composed of a large number of cell phenotypes with complex interactions and self-organize into three-dimensional structures. In general, biofilms are assemblages of one or more populations that are attached to an abiotic or biotic surface through extracellular polymeric substances. A biofilm can be formed by a single bacterial species although they can also consist of many species of bacteria, fungi, algae, and protozoa (Singh *et al.*, 2006). In addition to water (approximately 97%) and microbial cells, the biofilm matrix is a complex of secreted polymers, absorbed nutrients and metabolites, cell lysis products, and even particulate material and detritus from the immediate environment (Flemming, 1995). Therefore, in addition to peptidoglycans, lipids, phospholipids, and other cell components, all major classes of macromolecules (i.e. proteins, polysaccharides, DNA, and RNA) can be found within a biofilm environment (Sutherland, 2001; Singh *et al.*, 2006). Recently, biofilm communities have been applied for bioremediation. Biofilm-mediated bioremediation may present a proficient and safer alternative method of bioremediation because cells in a biofilm have a better chance of adaptation and survival (especially during periods of stress) since they are protected within the matrix. Aggregation of bacteria into complex microbial communities gives some special advantages that are not available to discrete bacteria: e.g. organisms within a biofilm can withstand shear forces, nutrient deprivation, pH changes, and antibiotics because of the protection provided by the EPS (Davey and O'Toole, 2000; Jefferson, 2004). Moreover, transformation, transduction, and conjugation in bacterial biofilms results in gene transfer between biofilm organisms. White and Gadd (2000) reported that sulfate-reducing bacterial biofilms grown in continuous culture and exposed to a medium containing 20 or 200 μM Cu accumulate Cu. The mechanism of action is linked to precipitation of Cu sulfides at the biofilm surface or in the liquid phase followed by entrapment of precipitated Cu sulfide by the exopolymer-enhanced biofilm. Labrenz *et al.* (2000) observed formation of sphalerite (ZnS) by members of the aerotolerant Desulfobacteriaceae in a natural biofilm. In this process, Zn was concentrated and metal sulfides were then precipitated by sulfate-reducing bacteria in the second phase of a combined sulfur oxidation–reduction biotreatment method.

12.5.2 Microbial modifications for increasing the potential of remediation

Application of genetically engineered (GE) microorganisms could be a factor in successful *in situ* bioremediation of environments contaminated with various pollutants. However, several critical research questions pertaining to the development and implementation of GE bacteria for enhanced bioremediation have to be formulated.

To date, only in a few reports relate to GE bacteria in heavy-metal remediation in combination with plants have been published. Sriprang *et al.* (2003) reported the ability of recombinant *Mesorhizobium huakuii* subsp. *rengei* strain B3 to produce PCs and to accumulate Cd^{2+}. Such bacterial strains may be useful for the development of a novel plant–bacterium remediation system for removal of heavy metals from rice fields when genetically engineered *M. huakuii* subsp. *rengei* strain B3 establishes a symbiotic relation with *Astragalus sinicus* (a nitrogen-fixing legume). Lee *et al.* (2001) described *Pseudomonas putida* 06909, a robust and versatile antifungal rhizosphere bacterium, that is engineered to produce a metal-binding peptide (MBP)-EC20, which has high affinity for Cd. *P. putida* 06909 was moderately Cd resistant due to the presence of an efflux pump (Lee *et al.*, 2001). Production of the MBP-EC20 peptide in *P. putida* 06909 not only enabled enhanced Cd binding but also protected the engineered strain from metal toxicity. The above results demonstrate that a combination of enhanced microbial uptake in a plant–bacterium symbiosis could be a promising strategy for heavy-metal clean-up. Increased resistance to metals by the engineered rhizobacteria at higher concentrations is particularly important because it gives a competitive advantage to the engineered bacterial strains in contaminated environments. This advantage could be crucial for sustaining the growth of the engineered strain in the presence of a native bacterial population.

Recently, interest arose in modification of microbial consortia. Engineers have exploited components of bacterial quorum sensing (QS) and successfully constructed microbial consortia by implementing cell–cell communication and differentiation of functions in traditional laboratory microbes (Brenner *et al.*, 2009). However, to fully exploit the potential of engineered consortia, stable engineering of organisms that are currently recalcitrant to genetic manipulation must be achieved.

The use of genetic engineering to create organisms specifically designed for bioremediation holds great promise, although many challenges have to be addressed concerning the release of genetically engineered bacteria into nature.

12.5.3 The role of microorganisms in metal phytoremediation

Bioremediation of heavy-metal-polluted soil, water, and sediments on the basis of plant uptake is known as phytoremediation (Peng *et al.*, 2009). The most important limitations of this technology include the time required for clean-up, the amount of produced biomass, the level of contamination, climatic conditions, and uncertain efficiency (cost/benefit ratio). This is why the success of this method depends on the selection of suitable plant species that can accumulate heavy metals and produce large amounts of biomass via established crop production and management practices (Rodriguez *et al.*, 2005; Tangahu *et al.*, 2011). The best potential for phytoremediation are hyperaccumulators, e.g. some representatives of the family Brassicaceae that accumulate large amounts of heavy metals in the aerial organs, without suffering phytotoxic effects. It is assumed that hyperaccumulators have a >1 shoot:root metal-accumulation ratio, which they achieve by (i) overexpression of transport systems for improved sequestration, (ii) tissue-specific protein expression, and (iii) production of high concentrations of metal chelators (Iqbal *et al.*, 2015). An important function in the process of phytoremediation is also performed by non-hyperaccumulator woody plant species, such as willow (*Salix* sp.) and poplar (*Populus* sp.). These woody species possess enormous plant biomass, and the amount of heavy metals taken up from the soil in such large biomass would be significant (Hrynkiewicz *et al.*, 2010; de Souza *et al.*, 2012). Nonetheless, the effectiveness of plant-based remediation can be limited by (i) the toxicity of the targeted pollutants and (ii) a variety of environmental extremes such as reduction in vegetation and soil biotic diversity, loss of soil organic matter, extremes in pH, and modifications of soil texture and bulk density that impact plant–soil–water relationship and nutrient availability (Abhilash *et al.*, 2012). A crucial role in the proper development of plants under unfavourable soil conditions is played by plant-associated microorganisms, which can increase the tolerance of plants to abiotic stress, stimulate plant growth, and therefore contribute to accelerated remediation of disturbed soils (Hrynkiewicz and Baum, 2012, 2013). The existence of positive plant–microbial interactions in disturbed soils is unquestionable, but the mechanisms are often poorly understood. Because microbes can promote plant growth in heavy-metal-polluted areas in various ways (directly and indirectly), they should be analysed in relation to particular taxonomic groups, their metabolic potential (e.g. siderophore synthesis and biosorption), and under the influence of different rhizosphere soil conditions, including organic-matter content, pH, temperature, nutrients, and pollutant levels (Glick, 2010; Hrynkiewicz *et al.*, 2012) (Fig. 12.2).

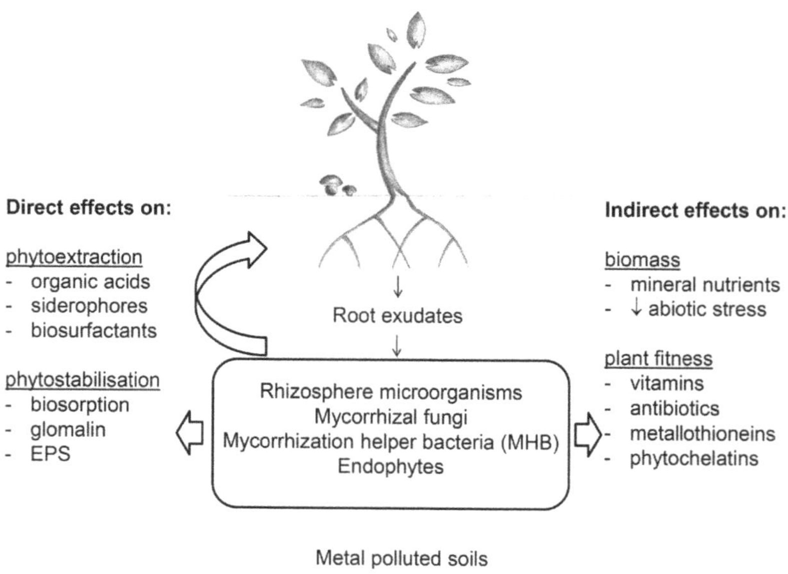

Figure 12.2 Direct and indirect effect of plant-associated microorganisms on phytoremediation in metal polluted soil. EPS, extracellular polymeric substances; PGPB, plant growth promoting bacteria, PGPF, plant growth promoting fungi; MHB, mycorrhization helper bacteria.

12.5.3.1 Rhizosphere microorganisms

Among the rhizosphere microorganisms involved in plant interactions with the metal-contaminated soil, those deserving special attention are plant growth-promoting bacteria (PGPB) and plant growth-promoting fungi (PGPF) (Hrynkiewicz *et al.*, 2009, 2010, 2012; Elsharkawy *et al.*, 2012).

12.5.3.2 Endophytic microorganisms

Microorganisms colonizing the internal tissues of plants without causing symptomatic infections or having negative effects on their host are defined as endophytes (Schulz and Boyle, 2006). In general, the beneficial effects of endophytes are greater than those of many rhizobacteria (Pillay and Nowak, 1997) and might be enhanced when the plant is growing under either biotic or abiotic stress conditions (Hardoim *et al.*, 2008). Endophytic microorganisms may serve as highly efficient biosorbents for heavy-metal metal accumulators because they have the advantage of being relatively protected from the competitive, high-stress environment of the soil (Sturz *et al.*, 2000; Xiao *et al.*, 2010).

12.5.3.3 Mycorrhizal fungi

Mycorrhiza formation can efficiently ameliorate abiotic stress on the host plant (Jentschke and Godbold, 2000; Hrynkiewicz *et al.*, 2008; Hrynkiewicz and Baum, 2012). Mycorrhizal fungi have been used in several remediation studies involving heavy metals, and the results showed that mycorrhizas use different mechanisms for the remediation of heavy-metal-polluted soils. For instance, although some studies have shown enhanced phytoextraction through accumulation of heavy metals in plants (Joner and Leyval, 2001; Jamal *et al.*, 2002; Marques *et al.*, 2006), others reported enhanced phytostabilization through metal immobilization and reduced metal concentration in the plants (Chibuike and Obiora, 2014). Increased exclusion processes of mycorrhizal plants can be related e.g. to absorption of metals by the hyphal sheath, reduced access to the apoplast because of hydrophobicity of the fungal sheath, chelation by fungal exudates, and sorption to the external mycelium (Hall, 2002). Exclusion of metals by mycorrhizal fungi can decrease phytoextraction of metals by willows. Nonetheless, this effect can be compensated by the significant promotion of plant growth due to the mycorrhizas (Baum *et al.*, 2006).

12.3.5.4 Mycorrhiza helper bacteria (MHB)

Symbiosis between host plants and mycorrhizal fungi can be enhanced by the associated bacteria that can support mycorrhiza formation and function (Hrynkiewicz *et al.*, 2010). The associated bacteria can also increase the efficiency of mycorrhizal fungi for phytoremediation (Zimmer *et al.*, 2009). This effect can be caused by enzymatic mobilization of nutrients from organic matter and by synthesis of plant hormones or vitamins (Belimov *et al.*, 2001; Hrynkiewicz *et al.*, 2010). Dual inoculation was confirmed to be more effective at the promotion of willow growth on contaminated soils than single inoculation. The possible usefulness of dual inoculation with ectomycorrhizal (EM) fungi and associated bacteria for promoting willow growth is in agreement with the potential of other dual inoculation methods for willows (Kuffner *et al.*, 2008; Zimmer *et al.*, 2009). Furthermore, dual inoculation of *Tripolium repens* with arbuscular mycorrhizal fungi and *B. cereus* (Azcón *et al.*, 2010) and of *Brassica juncea* with other *Bacillus* spp. (*B. megaterium* HKP-1 and *B. mucilaginosus* HKK-1) and *Azotobacter chroococcum* HKN5 (Wu *et al.*, 2006) also promoted plant growth on polluted soils. These findings point to significant benefits of the combined use of diverse microbial activities as a result of dual inoculation.

12.5.4 Direct and indirect effects of microorganisms on phytoremediation

Microorganisms can be key factors in heavy-metal mobilization and have direct effects on the process of phytoremediation (Fig. 12.2). The influence of bacterial and fungal activity on metal mobility and the use of these effects for bioremediation have been reviewed by Gadd (2010). Adsorbed, precipitated, and occluded trace elements can be solubilized by acidification, chelation, or ligand-induced dissolution (Sessitsch *et al.*, 2013). Soil microbiota may increase solubility of heavy metal ions by secretion of various organic ligands, decomposition of organic matter, and release of metabolites and siderophores.

Low-molecular-weight organic acids of microbial origin, such as gluconic acid, 2-oxoglutaric acid, oxalic acid, citric acid, malic acid, and succinic acid, may sorb or desorb metal ions depending on their chemical form and pH conditions (Gadd, 1999, 2001, 2004, 2010). Bacteria producing trace element-chelating organic acids, such as citric, oxalic, or acetic acid, have been shown to mobilize various trace elements in soil (Li, H. *et al.*, 2015). The authors demonstrated that the release of carboxylic acids from bacterial cells is stimulated by the presence of trace elements (Li, H. *et al.*, 2015). Ma *et al.* (2011) showed that 5-ketoglutaric acid produced by *Gluconobacter diazotrophicus* releases Zn from insoluble compounds, such as ZnO, ZnCO, and Zn(PO). Moreover, acid-producing rhizosphere bacteria have a plant growth-promoting capacity because they can release phosphorus from insoluble phosphates, and are often referred to as phosphate solubilizers (Sessitsch *et al.*, 2013).

Siderophores are another important group of compounds that alter the mobility of metals. These low-molecular-weight organic chelators bind Fe(III) in Fe-deficient environments. However, siderophores can bind metals other than Fe and thus enhance their bioavailability in the rhizosphere of plants. They can form stable complexes with other metals that are of environmental concern, such as Al, Cd, Cu, Ga, In, Pb, and Zn, as well as with radionuclides, including U and Np (Rajkumar *et al.*, 2010). Braud *et al.* (2009) reported that inoculation of soils with pyoverdin- and pyochelin-synthesizing *P. aeruginosa* significantly increases the concentrations of bioavailable Cr and Pb as compared with those seen for uninoculated controls. Moreover, they observed a significant increase in accumulation of Cr and Pb in maize shoots after inoculation with this bacterium. It should be noted that the efficiency of microbial siderophores for mobilization of heavy metals can depend on several factors, e.g. the binding form of the heavy metals, the charge of the siderophores, pH of the soil, and its mineral composition and organic matter content (Rajkumar *et al.*, 2010).

Biosurfactants are additional important metabolites produced by plant-associated microorganisms that have the potential to improve metal mobilization and phytoremediation (Ullah *et al.*, 2015). They accelerate metal bioavailability by decreasing the tight binding between metals and soil particles and preferentially bind toxic metals with a stronger affinity than normal soil metal cations. The biosurfactants produced by microbes form complexes with heavy metals at the soil interface and desorb metals from the soil matrix, thus increasing metal solubility and bioavailability (Gadd, 2010; Rajkumar *et al.*, 2010). Juwarkar *et al.* (2007) verified metal mobilization through biosurfactants produced by *P. aeruginosa* BS2, which facilitated solubilization of Pb and Cd. Bacterial biosurfactants also increase the phytoavailability of metals, including Cr, because the bacterial biosurfactant helps to release metals that are strongly bound to soils (Ahemad, 2015).

Microorganisms, due to high surface and contact area, can easily interact with metals. They can be specific to one or a few metals and bind high concentrations of heavy metals via a variety of mechanisms – passively (metabolism-independent) or actively (metabolism-dependent) – and can remove heavy metals through bioaccumulation or biosorption. Differences and mechanisms responsible for the above processes have been described previously: see subsections on 'Passive heavy metal uptake via biosorption' and 'Heavy metal immobilization by living microbial cells – mechanisms and applications'. Hrynkiewicz *et al.* (2015) confirmed that the rhizosphere is a promising source of highly Cd-tolerant and Cd-accumulating bacterial strains. They found that, within bacterial cells, Cd^{2+} mainly accumulates in the interior of the cells, and in smaller quantities, in the cell wall. They suggested that strains from the genera *Massilia, Pseudomonas,* and *Bacillus* were the most efficient for bioremediation. Deng *et al.* (2007) stated that the green marine algae *Cladophora fascicularis* can be used as an efficient biosorbent for removal of Pb(II) from wastewater.

Some bacteria secrete exopolysaccharides (EPS) or capsular materials that bind large amounts of heavy metals, e.g. Cd^{2+} (Ron *et al.*, 1992). EPS synthesized by *Arthrobacter viscosus* accumulate 2.3-fold more Cd^{2+} than an equivalent weight of intact cells and have a 13.7-fold sorptive capacity over *Arthrobacter globiformis* cells which do not produce EPS (Scott and Palmer, 1990).

Arbuscular mycorrhizal fungi produce glomalin, a glycoprotein that contributes to the hydrophobicity of soil particles and participates in the initiation of soil aggregates (Barea *et al.*, 2002). González-Chávez *et al.* (2004) for the first time described glomalins as reducing agents involved in trace element phytotoxicity. They were later classified as probable heat shock proteins (HSPs), a class of highly conserved proteins (Rillig *et al.*, 2007). Some studies have confirmed the role of glomalins in reduction of the toxicity of some trace elements, such as Cu, Cd, Zn, Fe, Pb, and Mn (González-Chávez *et al.*, 2004; Cornejo *et al.*, 2008; González-Chávez *et al.*, 2009). González-Chávez *et al.* (2004) confirmed the ability of these proteins to extract trace elements from polluted soils, and found that glomalins that are recovered from contaminated soils contain 1.6–4.3 mg Cu, 0.02–0.08 mg Cd and 0.62–1.12 mg Pb per gram of glomalin. The data on cell wall binding of glomalin suggest that this protein has additional functions other than cytoplasmic (related to heat shock proteins) and thereby may influence the biotic and abiotic interactions in the soil.

Microorganisms can affect phytoremediation indirectly by stimulation of plant growth, e.g. by increasing levels of mineral nutrients and production of phytohormones (IAA), as well as increase plant tolerance e.g. by decreasing abiotic stress (H_2O_2), increasing expression of plant proteins (MTs and PCs), and synthesis of vitamins and antibiotics.

Microorganisms known as PGPB or PGPF may be free-living, symbiotic, or endophytic. Some key genera of PGPB include *Serratia, Bacillus, Pseudomonas, Burkholderia, Enterobacter, Erwinia, Klebsiella, Beijerinckia, Flavobacterium* and *Gluconacetobacter* (Ullah *et al.*, 2015). PGPB can improve plant growth by atmospheric nitrogen fixation (diazotrophs), phytohormone production (e.g. IAA), specific enzymatic activity, and protection of the plant from diseases via production of antibiotics or other anti-pathogen substances such as siderophores and chelating agents (Hrynkiewicz *et al.*, 2010; Ullah *et al.*, 2015). In addition, some PGPB can decrease the level of ethylene in plants, and this change increases plant growth. This phenomenon is attributed to ACC deaminase, which hydrolyses ACC, the biosynthetic precursor of ethylene in plants, into ammonia and α-ketobutyrate (Ullah *et al.*, 2015).

Mycorrhizal fungi are known to enhance the nutrient supply of their host plants, especially under nutrient-deficient conditions by very efficient soil exploitation and by excretion of significant quantities of phosphatases (Tibbett and Sanders, 2002). Mechanisms that ectomycorrhizal fungi use for adaptation to N and P shortages under unfavourable soil conditions, e.g. by the utilization of seed protein N, were demonstrated by Tibbett *et al.* (1998). Ectomycorrhizal fungal species vary in their ability to acquire specific nutrients from soil (Erland and Taylor, 2002), and this differential efficiency can be considered niche partitioning. These organisms may use the same substrate but extract different components. This idea may support the theory of high ectomycorrhizal fungal diversity in nutrient-deficient soils (Hrynkiewicz and Baum, 2012).

Plants can regulate tolerance to high concentrations of heavy metals by the expression of MTs and PCs (DalCorso *et al.*, 2008). MTs are cysteine-rich proteins of low molecular weight that are responsible for intracellular ion homeostasis and contribute to detoxification of heavy metal ions (Zhu *et al.*, 2010). The impact of MTs on heavy-metal tolerance was explored in several plants such as *Vicia faba* (Foley *et al.*, 1997), *Bruguiera gymnorrhiza* (Huang and Wang, 2009), *Oryza sativa* (Chun-Juan *et al.*, 2010), and *Piper nigrum* (Ozkuthu *et al.*, 2006). MTs can affect the metal concentrations in the plant biomass, and hence, the efficiency of metal extraction. For instance, a MT that is inserted into transgenic tobacco increases Cd concentrations in the plant shoots (Janoušková *et al.*, 2005). It was demonstrated that the expression of MTs in plants can be significantly affected by ectomycorrhiza formation (Voiblet *et al.*, 2001). Therefore, increased tolerance to abiotic stress in ectomycorrhizal plants might be promoted by increased expression of MTs. Bacterial strains, e.g. the plant pathogen *Xanthomonas campestris*, can also affect the expression of MTs in their host plants depending on biotic stress (Obertello *et al.*, 2007). Hrynkiewicz *et al.* (2012) confirmed that microorganisms in the rhizosphere control expression of MTs in willows growing on a heavy-metal-contaminated soil. However, there seems to be an optimal level of expression of MTs in willows for the tolerance of increased concentrations of heavy metals in the soil, above which growth inhibition may occur. During experiments, those researchers found that the strongest expression of MT1 after a single inoculation with *B. cereus* B1 was accompanied with a rather small increase in Cd uptake into willow stems and significant inhibition of the growth of *S. viminalis*. The above results suggest that increased expression of MT1 can be caused by biotic stress through invasive root colonization by microorganisms, not only by abiotic stress via high concentrations of a heavy metal (Hrynkiewicz *et al.*, 2012).

The plant enzyme called PC synthase (PCS), also known as glutathione-glutamylcysteinyl transferase (EC 2.3.2.15) is involved in the synthesis of PCs, which are metal-binding cysteine-rich peptides. This enzyme plays a major role in the detoxification of heavy metals in plants (Li, H. *et al.*, 2015). Garg and Aggarwal (2011) reported that exposure of mycorrhizal and non-mycorrhizal genotypes of *Cajanus cajan* (L.) Millsp. to Cd and Pb significantly increases the levels of PCs in a concentration- and genotype-dependent manner; these effects directly correlate with the intensity of mycorrhizal infection (MI). Li, H. *et al.* (2015) found that *Escherichia coli* with overexpressed PcPCS1 shows enhanced tolerance to Cd, Cu, Na and Hg.

12.6 Conclusions

Environmental pollution with heavy metals is still a major challenge in view of their negative effect on all living organisms and the problems associated with reclamation. In the above-mentioned chapter, we indicate a key role of microorganisms in this process and resume mechanisms potentially involved in the removal of heavy metals from the environment. Since biotransformation of metals have great significance in environmental biotechnology, e.g. for metal removal, recovery, detoxification, and bioremediation, we present their potential application in natural environment. We stress individual properties of microorganisms, which are often strain and environment specific as well as emphasizing the importance of plant–microbial interactions in the enhancement of phytoremediation. Additionally, some metal–microbe interactions, which may result in the production of novel biomaterials including nanoparticles forms, were described. A proteomic approach to heavy metal resistance was also introduced as a useful tool in selection of strains for bioremediation.

References

Abhilash, P.C., Jamil, S., and Singh, N. (2009). Transgenic plants for enhanced biodegradation and phytoremediation of organic xenobiotics. Biotechnol. Adv. *27*, 474–488.

Abhilash, P.C., Powell, J.R., Singh, B.S., and Singh, B.K. (2012). Plant–microbe interactions: novel applications for exploitation in multipurpose remediation technologies. Trends Biotechnol. *30*, 416–420.

Acheampong, M.A., Meulepas, R.J., and Lens, P.N.L. (2010). Removal of heavy metals and cyanide from gold mine wastewater. J. Chem. Technol. Biotechnol, *85*, 590–613.

Ahemad, M. (2015). Enhancing phytoremediation of chromium-stressed soils through plant-growth-promoting bacteria. J. Gen. Engin. Biotechnol. *13*, 51–58.

Aquino, N.B., Sevigny, M.B., Sabangan, J., and Louie, M.C. (2012). Role of cadmium and nickel in estrogen receptor signaling and breast cancer: metalloestrogenes or not? J. Environ. Sci. Health C Environ. Carcinog. Ecotoxicol. Rev. *30*, 189–224.

Arévalo-Rangel, D.L., Cárdenas-González, J.F., Martínez-Juárez, V.M., and Acosta-Rodríguez, I. (2013). Hexavalent chromate reductase activity in cell free extracts of *Penicillium* sp.. Bioinorganic Chem. Appl. *2013*, Article ID 909412, 1- 6.

Arita, A., and Costa, M. (2009). Epigenetics in metal carcinogenesis: nickel, arsenic, chromium and cadmium. Metallomics *1*, 222–228.

Aryal, M., and Liakopoulou-Kyriakides, M. (2015). Bioremoval of heavy metals by bacterial biomass. Environ. Monit. Assess. *187*, 4173.

Avery, S.V., and Tobin, J.M. (1993). Mechanism of adsorption of hard and soft metal ions to *Saccharomyces cerevisiae* and influence of hard and soft anions. Appl. Environ. Microbiol. *59*, 2851–2856.

Avery, S.V., Howlett, N.G., and Radice, S. (1996). Copper toxicity towards *Saccharomyces cerevisiae*: dependence on plasma membrane fatty acid composition. Appl. Environ. Microb. *62*, 3960–3966.

Azcón, R., Perálvarez, M.C., Roldán, A., and Barea, J.M. (2010). Arbuscular mycorrhizal fungi, *Bacillus cereus*, and *Candida parapsilosis* from a multicontaminated soil alleviate metal toxicity in plants. Microb. Ecol. *59*, 668–677.

Azevedo, M.M., Carvalho, A., Pascoal, C., Rodrigues, F., and Fernanda, C. (2007). Responses of antioxidant defenses to Cu and Zn stress in two aquatic fungi. Sci. Total Environ. *377*, 233–243.

Bae, W., and Chen, X. (2004). Proteomic study for the cellular responses to Cd^{2+} in *Schizosaccharomyces pombe* through amino acid-coded mass tagging and liquid chromatography tandem mass spectrometry. Mol. Cell. Proteom. *3.6*, 596–607.

Bar, C., Patil, R., Doshi, J., Kulkarni, M.J., and Gade, W.N. (2007). Characterization of the proteins of bacterial strain isolated from contaminated site involved in heavy metal resistance-a proteomic approach. J. Biotechnol. *128*, 444–451.

Barea, J.M., Azcón, R., and Azcón-Aguilar, C. (2002). Mycorrhizosphere interactions to improve plant fitness and soil quality. Antonie van Leeuwenhoek *81*, 343–351.

Baum, C., Hrynkiewicz, K., Leinweber, P., and Meißner, R. (2006). Heavy-metal mobilization and uptake by mycorrhizal and nonmycorrhizal willows (*Salix×dasyclados*). J. Plant Nutr. Soil Sci. *169*, 516–522.

Behera, M., Dandapat, J., and Rath, C.C. (2014). Effect of heavy metals on growth response and antioxidant defense protection in *Bacillus cereus*. J. Basic Microbiol. *54*, 1201–1209.

Belimov, A.A., Safronova, V.I., Sergeyeva, T.A., Egorova, T.N., Matveyeva, V.A., Tsyganov, V.E., Borisov, A.Y., and Tikhonovich, I.A. (2001). Characterization of plant growth-promoting rhizobacteria isolated from polluted soils and containing 1-aminocyclopropane-1-carboxylate deaminase. Can. J. Microbiol. *47*, 642–652.

Belimov, A.A., Safronova, V.I., Sergeyeva, T.A., Egorova, T.N., Matveyeva, V.A., Bell, T.H., Joly, S., Pitre, F.E., and Yergeau, E. (2014). Increasing phytoremediation efficiency and reliability using novel omics approaches. Trends Biotechnol. *32*, 271–280.

Bell, T.H., Joly, S., Pitre, F.E. and Yergeau, E. (2014). Increasing phytoremediation efficiency and reliability using novel omics approaches. Trends Biotechnol. *32*, 271–280.

Beolchini, F., Fonti, V., Dell'Anno, A., and Veglio, F. (2011). Bio-mobilization of heavy metals from contaminated sediments by acidophilic microbial consortia. Chem. Engin. Trans. *24*, 883–888.

Beyersmann, D. (2002). Effects of carcinogenic metals on gene expression. Toxicol. Lett. *127*, 63–68.

Braud, A., Jézéquel, K., Bazot, S., and Lebeau, T. (2009). Enhanced phytoextraction of an agricultural Cr-, Hg- and Pb-contaminated soil by bioaugmentation with siderophore producing bacteria. Chemosphere *74*, 280–286.

Breierová, E., Vajczíková, I., Sasinková, V., Stratilová, E., Fišerac, M., Gregorc, T., and Šajbidord, J. (2002). Biosorption of cadmium ions by different yeast species. Z. Naturforschung C *57*, 634–639.

Brenner, K., Lingchong, Y., and Arnold, F.H. (2008). Engineering microbial consortia: a new frontier in synthetic biology. Trends Biotechnol. *26*, 483–489.

Brenner, K., You, L., and Arnold, F.H. (2009). Response to Goldman and Brown: Making sense of microbial consortia using ecology and evolution. Trends Biotechnol. *27*, 4.

Brinza, L., Dring, M.J., and Gavrilescu, M. (2007). Marine micro and macro algal species as biosorbent for heavy metals. Environ. Eng. Manag. J. *6*, 237–251.

Brune, K.D.. and Bayer, T.S. (2012). Engineering microbial consortia to enhance biomining and bioremediation. Front. Microbiol. *3*, 1–6.

Čertík, M., Breierová, E., and Juršíková, P. (2005). Effect of cadmium on lipid composition of *Aureobasidium pullulans* grown with added extracellular polysaccharides. Int. Biodeter. Biodegr. *55*, 195–202.

Cheng, Z., Wei, Y.Y.C., Sung, W.W.L., Glick, B.R., and McConkey, J. (2009). Proteomic analysis of the response of the plant growth-promoting bacterium *Pseudomonas putida* UW4 to nickel stress. Proteome Sci. *7*, 18.

Cheung, K.H., and Gu, J.D. (2002). Bacterial colour response to hexavalent Cr^{+6}. J. Microbiol. *40*, 234–236.

Chibuike, G.U., and Obiora, S.C. (2014). Heavy metal polluted soils: effect on plants and bioremediation methods. Appl. Environ. Soil Sci. Vol. 2014 Article ID 752708, 12 pages.

Chun-Juan, D., Wang, Y., Shi-Shi, Y., and Liu, J.Y. (2010). Characterization of a novel rice metallothionein gene promoter: its tissue specificity and heavy metal responsiveness. J. Integrat. Plant Biol. *52*, 914–924.

Clemens, S., Kim, E.J., Neumann, D., and Schroeder, J.I. (1999). Tolerance to toxic metals by a gene family of phytochelatin synthases from plants and yeast. EMBO J. *18*, 3325–3333.

Cornejo, P., Pérez-Tienda, J., Meier, S., Valderas, A., Borie, F., Azcón-Aguilar, C., and Ferrol, N. (2013). Copper compartmentalization in spores as a survival strategy of arbuscular mycorrhizal fungi in Cu-polluted environments. Soil Biol. Biochem. *57*, 925–928.

Cyrne, L., Martins, L., Fernandes, L., and Marinho, H.S. (2003). Regulation of antioxidant enzymes gene expression in the yeast *Saccharomyces cerevisiae* during stationary phase. Free Radic. Biol. Med. *34*, 385–393.

DalCorso, G., Farinati, S., Maistri, S., and Furini, A. (2008). How plants cope with cadmium: staking all on metabolism and gene expression. J. Integrat. Plant Biol. *50*, 1268–1280.

Davey, M.E., and O'Toole, G.A. (2000). Microbial biofilms: from ecology to molecular genetics. Microbiol. Mol. Biol. Rev. *64*, 847–867.

de Kroon, A.I.P.M., Rijken, P.J., and De Smet, C.H. (2013). Checks and balances in phospholipid class and acyl chain homeostasis, the yeast perspective. Prog. Lipid Res. *52*, 374–394.

De Souza, C.E.T., Guilherme, L.R., de Melo, E.E., Ribeiro, B.T., Dos Santos, B.I.E., da Costa, S.E., Faquin, V., and Hale, B.A. (2012). Assessing the tolerance of castor bean to Cd and Pb for phytoremediation purposes. Biol. Trace Elem. Res. *145*, 93–100.

Deng, L., Su, Y., Su, H., Wang, X., and Zhu, X. (2007). Sorption and desorption of lead (II) from wastewater by green algae *Cladophora fascicularis*. J. Hazard. Mater. *143*, 220–225.

Dhal, B., Thatoi, H.N., Das, N.N., and Pandey, B.D. (2013). Chemical and microbial remediation of hexavalent chromium from contaminated soil and mining/metallurgical solid waste: a review. J. Hazard. Mater. *250–251*, 272–291.

Elsharkawy, M.M., Shimizu, M., Takahashi, H., and Hyakumachi, M. (2012). The plant growth-promoting fungus *Fusarium equiseti* and the arbuscular mycorrhizal fungus *Glomus mosseae* induce systemic resistance against *Cucumber mosaic* virus in cucumber plants. Plant Soil *361*, 397–409.

Erland, S., and Taylor, F.S. (2002). Diversity of ecto-mycorrhizal fungal communities in relation to the abiotic environment. In Mycorrhizal Ecology,: M.G.A. van der Heijden MGA and, I.R. Sanders (eds) Mycorrhizal Ecology. (Springer, Berlin), pp. 163–200.

Ezzouhri, L., Castro, E., Moya, M., Espinola, F., and Lairini, K. (2009). Heavy metal tolerance of filamentous fungi isolated from polluted sites in Tangier, Morocco. Afr. J. Microbiol. Res. 3, 35–48.

Flemming, H.C. (1995). Sorption sites in biofilms. Water. Sci. Technol. *32*, 27–33.

Foley, R.C., Liang, Z.M., and Singh, K.B. (1997). Analysis of type 1 metallothionein cDNAs in *Vicia faba*. Plant Mol. Biol. *33*, 583–591.

Fomina, M., and Gadd, G.M. (2002). Metal sorption by biomass of melanin-producing fungi grown in clay-containing medium. J. Chem. Technol. Biotechnol. *78*, 23–34.

Fomina, M., Ritz, K., and Gadd, G.M. (2003). Nutritional influence on the ability of fungal mycelia to penetrate toxic metal-containing domains. Mycol. Res. *107*, 861–871.

Fomina, M., Gadd, G.M.(2014). Biosorption: current perspective on concept, definition and application. Biores. Technol. *160*, 3–14.

Fu, F., and Wang, Q. (2011). Removal of heavy metal ions from wastewaters: a review. J Environ. Manag. *92*, 407–418.

Gabr, R.M., Hassan, S.H.A., and Shoreit, A.A.M. (2008). Biosorption of lead and nickel by living and non-living cells of *Pseudomonas aeruginosa* ASU 6a. Int. Biodeterior. Biodegrad. *62*, 195–203.

Gadd, G.M. (1993). Interactions of fungi with toxic metals. New Phytol. *124*, 25–60.

Gadd, G.M. (1999). Fungal production of citric and oxalic acid: importance in metal speciation, physiology and biogeochemical processes. Adv. Microb. Physiol. *41*, 47–92.

Gadd, G.M. (2001). Microbial metal transformations. J. Microbiol. *39*, 83–88.

Gadd, G.M. (2004). Microbial influence on metal mobility and application for bioremediation. Geoderma *122*, 109–119.

Gadd, G.M. (2007). Geomycology: biogeochemical transformations of rocks, minerals, metals and radionuclides by fungi, bioweathering and bioremediation. Mycol. Res. *111*, 3–49.

Gadd, G.M. (2009). Biosorption: critical review of scientific rationale, environmental importance and significance for pollution treatment. J. Chem. Technol. Biotechnol. *84*, 13–28.

Gadd, G.M. (2010). Metals, minerals and microbes: geomicrobiology and bioremediation. Microbiol. *156*, 609–643.

Gadd, G.M., Rhee, Y.J., Stephenson, K., and Wei, Z. (2012). Geomycology: metals, actinides and biominerals. Environ. Microbiol. Reports 4, 270–296.

Gadd, G.M., Bahri-Esfahani, J., Li, Q., Rhee, Y.J., Wei, Z., 1, Fomina, M., and Liang, X. (2014). Oxalate production by fungi: significance in geomycology, biodeterioration and bioremediation. Fungal Biol. Rev. *28*, 36–55.

Garg, N., and Aggarwal, N. (2011). Effects of interactions between cadmium and lead on growth, nitrogen fixation, phytochelatin, and glutathione production in mycorrhizal *Cajanus cajan* (L.) Millsp. J. Plant Growth Regul. *30*, 286–300.

Glick, B.R. (2010). Using soil bacteria to facilitate phytoremediation. Biotechnol. Adv. *28*, 367–374.

González-Chávez, M.C., Carrillo-González, R., Wright, S.F., and Nichols, K.A. (2004). The role of glomalin, a protein produced by mycorrhizal fungi, in sequestering potentially toxic elements. Environ. Pollut. *130*, 317–323.

González-Chávez, M.C., Carrillo-González, R., and Gutierres-Castorena, M.C. (2009). Natural attenuation in a slagheap contaminated with cadmium: the role of plants and arbuscular fungi. J. Hazard. Mater. *161*, 1288–1298.

Griffith, G., Easton, G., Detheridge, A., Roderick, K., Edwards, A., Worgan, H.J., Nicholson, J., and Perkins, W.T. (2007). Copper deficiency in potato dextrose agar causes reduced pigmentation in cultures of various fungi. FEMS Microbiol. Lett. *276*, 165–171.

Gupta, M.K., Kiran, K., Amita, S., and Shikha, G. (2014). Bioremediation of heavy metal polluted environment using resistant bacteria. J. Environ. Res. Develop. *8*, 883–889.

Guzik, U., Hupert-Kocurek K., Sałek, K., and Wojcieszyńska, D. (2012). Influence of metal ions on bioremediation activity of prot-ocatechuate 3,4-dioxygenase from *Stenotrophomonas maltophilia* KB2. World J Microbiol. Biotechnol. *29*, 267–273.

Hall, J.L. (2002). Cellular mechanisms for heavy metal detoxification and tolerance. J. Exp. Bot. *53*, 1–11.

Hardoim, P.R., van Overbeek, L.S., and van Elsas, J.D. (2008). Properties of bacterial endophytes and their proposed role in plant growth. Trends Microbiol. *16*, 463–471.

Howlett, N.G., and Avery, S.V. (1997). Induction of lipid peroxidation during heavy metal stress in *Saccharomyces cerevisiae* and influence of plasma membrane fatty acid unsaturation. Appl. Environ. Microbiol. *62*, 3960–3966.

Hrynkiewicz, K., Haug, I., and Baum, C. (2008). Ectomycorrhizal community structure under willows at former ore mining sites. Eur. J. Soil Biol. *44*, 37–44.

Hrynkiewicz, K., Baum, C., Niedojadło, J., and Dahm, H. (2009). Promotion of mycorrhiza formation and growth of willows by the bacterial strain *Sphingomonas* sp. 23L on fly ash. Biol. Fertil. Soils *45*, 385–394.

Hrynkiewicz, K., Baum, C., and Leinweber, P. (2010). Density, metabolic activity, and identity of cultivable rhizosphere bacteria on *Salix viminalis* in disturbed arable and landfill soils. J. Plant Nutr. Soil Sci. *173*, 747–756.

Hrynkiewicz, K., and Baum, C. (2012). The potential of rhizosphere microorganisms to promote the plant growth in disturbed soils. In: Environmental Protection Strategies for Sustainable Development, Strategies for Sustainability, A. Malik, and E. Grohmann, eds. (Springer, Netherlands), pp 35–64.

Hrynkiewicz, K., Dąbrowska, G., Baum, C., Niedojadło, K., and Leinweber, P. (2012). Impact of *Bacillus cereus* on metal uptake and expression of metallothionein MT1 in non-mycorrhizal and mycorrhizal willows. Wat. Air Soil Pollut. *223*, 957–968.

Hrynkiewicz, K., and Baum, C. (2013). Selection of ectomycorrhizal willow genotype in phytoextraction of heavy metals. Environ. Technol. *34*, 225–230.

Hrynkiewicz, K., Złoch, M., Kowalkowski, T., Baum, C., Niedojadło, K., and Buszewski, B. (2015). Strain-specific bioaccumulation and intracellular distribution of Cd^{2+} in bacteria isolated from the rhizosphere, ectomycorrhizae, and fruitbodies of ectomycorrhizal fungi. Environ. Sci. Pollut. Res. *22*, 3055–3067.

Huang, G.Y., and Wang, Y.S. (2009). Expression analysis of type 2 metallothionein gene in mangrove species (*Bruguiera gymnorrhiza*) under heavy metal stress. Chemosphere 77, 1026–1029.

Hupert-Kocurek, K., Saczyńska, A., and Piotrowska-Seget, Z. (2013). Cadmium increases catechol 2,3-dioxygenase activity in *Variovorax* sp. 12S, a metal-tolerant and phenol degrading strain. Ant. Leeuwenhoek *104*, 845–853.

Iqbal, M., Ahmad, A., Ansari, M.K.A., Qureshi, M.I., Aref, I.M., Khan, P.R., Hegazy, S.S., El-Atta, H., Husen, A., and Hakeem, K.R. (2015). Improving the phytoextraction capacity of plants to scavenge metal(loid)-contaminated sites. Environ. Rev. *23*, 44–65.

Iram, S., Zaman, A., Iqbal, Z., and Shabbir, S. (2013). Heavy metal tolerance of fungus isolated from soil contaminated with sewage and industrial wastewater. Pol. J. Environ. Stud. *22*, 691–697.

Jain, S., and Bhatt, A. (2013). Proteomic analysis of diversified of extremophilic strains of *Pseudomonas* in the presence of cadmium. Agric. Res. *2*, 354–359.

Jamal, A., Ayub, N., Usman, M., and Khan, A.G. (2002). Arbuscular mycorrhizal fungi enhance zinc and nickel uptake from contaminated soil by soybean and lentil. Int. J. Phytoremed. *4*, 205–221.

Janoušková, M., Pavlíková, D., Macek, T., and Vosátka, M. (2005). Arbuscular mycorrhiza decrease cadmium extraction by transgenic tabacco with inserted metallothionein. Plant Soil *272*, 29–40.

Jarosławiecka, A., and Piotrowska-Seget, Z. (2014). Lead resistance in micro-organisms. Microbiology *160*, 12–25.

Jefferson, K.K. (2004). What drives bacteria to produce a biofilm? FEMS Microbiol. Lett. *236*, 163–173.

Jentschke, G., and Godbold, D.L. (2000). Metal toxicity and ectomycorrhizas. Physiol. Plant. *109*, 107–116.

Joner, E.J., and Leyval, C. (2001). Time-course of heavy metal uptake in maize and clover as affected by root density and different mycorrhizal inoculation regimes. Biol. Fertil. Soils *33*, 351–357.

Joo, J.-H., Hassan, S.H.A., and Oh, S.-E. (2010). Comparative study of biosorption of Zn^{+2} by *Pseudomonas aeruginosa* and *Bacillus cereus*. Int. Biodeterior. Biodegrad. *64*, 734–741.

Juwarkar, A.A., Nair, A., Dubey, K.V., Singh, S.K., and Devotta, S. (2007). Biosurfactant technology for remediation of cadmium and lead contaminated soils. Chemosphere, *10*, 1996–2002.

Keller, L., and Surette, M.G. (2006). Communication in bacteria: an ecological and evolutionary perspective. Nat. Rev. Microbiol. *4*, 249–258.

Koedrith, P., and Seo, Y.R. (2011). Advances in carcinogenic metal toxicity and potential molecular markers. Int. J. Mol. Sci. *12*, 9576–9595.

Krauss, G.-J., Solé, M., Krauss, G., Schlosser, D., Wesenberg, D., and Bärlocher, F. (2011). Fungi in freshwaters: ecology, physiology and biochemical potetntial. FEMS Microbiol. Rev. *35*, 620–651.

Kuffner, M., Puschenreiter, M., Wieshammer, G., Gorfer, M., and Sessitsch, A. (2008). Rhizosphere bacteria affect growth and metal uptake of heavy metal accumulating willows. Plant Soil *304*, 35–44.

Kumar, K.S., Dahms, H.-U., Won, E.-J., Lee, J.-S., and Shin, K.-H. (2015). Microalgae – a promising tool for heavy metals remediation. Ecotoxicol. Environ. Saf. *113*, 329–352.

Kuroda, K., and Ueda, M. (2010). Engineering of microorganisms towards recovery of rare metal ions. Appl. Microbiol. Biotechnol. *87*, 53–60.

Labrenz, M., Druschel, G.K., and Thomsen-Ebert, T. (2000). Formation of sphalerite (ZnS) deposits in natural biofilms of sulfate-reducing bacteria. Science *290*, 1744–1747.

Lacerda, C.M.R., Choe, L.H., and Reardon, K.F. (2007). Metaproteomic analysis of bacterial community response to cadmium exposure. J. Proteom Res. *6*, 1145–1152.

Lata, S., Singh, P.K., and Samadder, S.R. (2015). Regeneration of adsorbents and recovery of heavy metals: a review. Int. J. Sci. Technol. *12*, 1461–1478.

Lee, S.W., Glickmann, E., and Cooksey, D.A. (2001). Chromosomal locus for cadmium resistance in *Pseudomonas putida* consisting of a cadmium-transporting ATPase a MerR family response regulator. Appl. Environ. Microbiol. *67*, 1437–1444.

Li, H., Cong, Y., Lin, J., and Chang, Y.H. (2015). Enhanced tolerance and accumulation of heavy metal ions by engineered *Escherichia coli* expressing *Pyrus calleryana* phytochelatin synthetase. J. Basic Microbiol. *55*, 398–405.

Li, Q., Scetenyi, L., and Gadd, G.M. (2014). Biomineralization of metal carbonates by *Neurospora crassa*. Environ. Sci. Technol. *48*, 14409–14416.

Li, Q., Csetenyi, L., Paton, G.I., and Gadd, G.M. (2015). $CaCO_3$ and $SrCO_3$ bioprecipitation by fungi isolated from calcareous soil. Environ. Microbiol. *17*, 3082–3097.

Loebus, J., Leitenmaier, B., Meissner, D., Braha, B., Krauss, G.-J., Dobritzsch, D., and Freisinger, E. (2013). The major function of metallothionein from the aquatic fungus *Heliscus lugdunensis* in cadmium detoxification. J. Inorganic Biochem. *127*, 253–260.

Luque-Garcia, J.L., Cabezas-Sanchez, P.C., and Camara, C. (2011). Proteomics as a tool for examining the toxicity of heavy metals. Trends Anal. Chem. *30*, 703–716.

Ma, Y., Prasad, M.N.V., Rajkumar, M., and Freitas, H. (2011). Plant growth promoting rhizobacteria and endophytes accelerate phytoremediation of metalliferous soils. Biotechnol. Adv. *29*, 248–258.

Marques, A.P.G.C., Oliveira, R.S., Rangel, A.O.S.S., and Castro, P.M.L. (2006). Zinc accumulation in *Solanum nigrum* is enhanced by different arbuscular mycorrhizal fungi. Chemosphere *65*, 1256–1263.

Massaccesi, G., Romero, M.C., Cazau, M.C., and Bucsinszky, A.M. (2002). Cadmium removal capacities of filamentous soil fungi isolated from industrially polluted sediments, in La Plata (Argentina). World J. Microb. Biotechnol. *18*, 817–820.

Mehta, S.K., and Gaur, J.P. (2005). Use of algae for removing heavy metal ions from wastewater: progress and prospects. Crit. Rev. Biotechnol. *113*, 113–152.

Michalak, I., Chojnacka, K., and Witek-Krowiak, A. (2013). State of the art for the biosorption process – a review. Appl. Biochem. Biotechnol. *170*, 1389–1416.

Monteiro, C., Castro, P.M.L., and Malcata, F.X. (2012). Metal uptake by microalgae, underlying and practical application. Biotechnol. Prog. *28*, 299–311.

Mukherjee, A., Das, D., Mondal, S.K., Biswas, R., Das, T.K., Boujedaini, N., and Khuda-Bukhsh, A.R. (2010). Tolerance of arsenate-induced stress in *Aspergillus niger*, a possible candidate for bioremediation. Ecotoxicol. Environ. Saf. 73, 172–182.

Mukhopadhyay, M., Noronha, S.B., and Suraishkumar, G.K. (2011). A review of experimental studies of biosorption of heavy metals by *Aspergillus niger*. Can. J. Chem. Eng. *89*, 889–900.

Njikam, E., and Schiewer, S. (2012). Optimization and kinetic modeling of cadmium desorption from citrus peels: a process for biosorbent regeneration. J. Hazard. Mater. *30*, 242–248.

Obertello, M., Wall, L., Laplaze, L., Nicole, M., Auguy, F., Gherbi, H., Bogusz, D., and Franche, C. (2007). Functional analysis of the metallothionein gene cgMT1 isolated from the actinorhizal tree *Casuarina glauca*. Mol. Plant-Microbe Interact. *20*, 1231–1240.

Orell, A., Remonsellez, F., Arancibia, R., and Jerez, C.A. (2013). Molecular characterization of copper and cadmium resistance determinants in the biomining thermoacidophilic archaeon *Sulfolobus metalicus*. Archaea, *2013*, Art. ID 289236.

Ozkuthu, F., Sekeroglu, N., and Kara, S.M. (2006). Monitoring of cadmium and micronutrients in spices commonly consumed in Turkey. Res. J. Agric. Biol. Sci. *2*, 223–226.

Paraszkiewicz, K., Frycie, A., Słaba, M., and Długoński, J. (2007). Enhancement of emulsifier production by *Curvularia lunata* in cadmium, zinc and lead presence. Biometals, *20*, 797–805.

Paraszkiewicz, K., Bernat, P., Naliwajski, and Długoński, J. (2010). Lipid peroxidation in the fungus *Curvularia lunata* exposed to nickel. Arch. Microbiol. *192*, 135–141.

Park, D., Yun, Y.-S., Kim, J.Y., and Park, J.M. (2008). How to study Cr(VI) biosorption: use of fermentation waste for detoxifying Cr(VI) in aqueous solution. Chem. Engin. J. *136*, 173–179.

Park, D., Yun, Y.S., Moon, and Park, J. (2010). The past, present and future trends of biosorption. Biotechol. Bioproc. Eng. *15*, 86–102.

Parker, D.L., Mihalick, J.E., Plude, J.L., Plude, M.J., Clark, T.P., Lynn, E., Flom, J.J., Rai, L.C., and Kumar, H.D. (2000). Sorption of metals by extracellular polymers from the cyanobacterium *Microcystis aeruginosa* fo. *flos-aquae* strain C3-40. J. Appl. Phycol. *12*, 219–240.

Peng, K.J., Luo, C.L., and, Chen, Y.H. (2009). Cadmium and other metal uptake by *Lobelia chinensis* and *Solanum nigrum* from contaminated soils. Bull. Environ. Contam. Toxicol. *83*, 260–264.

Pillay, V.K., and Nowak, J. (1997). Inoculum density, temperature, and genotype effects on in vitro growth promotion and epiphytic and endophytic colonization of tomato (*Lycopersicon esculentum* L) seedlings inoculated with a pseudomonad bacterium. Can. J. Microbiol. *43*, 354-361.

Rahman, Z., and Singh, V.P. (2014). Cr(VI) reduction by *Enterobacter* sp. DU17 isolated from the tannery waste dump site and characterization of the bacterium and the Cr(VI) reductase. Int. Biodeterior. Biodegr. *91*, 97–103.

Rajkumar, M., Ae, N., Prasad, M.N.V., and Freitas, H. (2010). Potential of siderophore-producing bacteria for improving heavy metal phytoextraction. Trends Biotechnol. *28*, 142–149.

Rhee, Y.J., Hillier, S., and Gadd, G.M. (2012). Lead transformation to pyromorphite by fungi. Curr. Biol. *22*, 237–241.

Rhee, Y., Hillier, S., Pendlowski, H., and Gadd, G.M. (2014a). Pyromorphite formation in a fungal biofilm community growing on lead metal. Environ. Microbiol. *16*, 1441–1451.

Rhee, Y.J., Hillier, S., Pendlowski, H., and Gadd, G.M. (2014b). Fungal transformation of metallic lead to pyromorphite in liquid medium. Chemosphere *113*, 17–21.

Rillig, M.C., Caldwell, B.A., Wösten, H.A.B., and Sollins, P. (2007). Role of proteins in soil carbon and nitrogen storage: controls on persistence. Biogeochem. *85*, 25–44.

Rodriguez, L., Lopez-Bellido, F.J., Carnicer, A., Recreo, F., Tallos, A., and Monteagudo, J.M. (2005). Mercury recovery from soils by phytoremediation. In: Book of Environmental Chemistry, E. Lichtfouse, J. Schwarzbauer and R. Didier, eds. (Springer: Berlin,), pp. 197–204.

Ron, E.Z., Minz, D., Finkelstein, N.P., and Rosenberg, E. (1992). Interactions of bacteria with cadmium. Biodegradation 3, 161–170.

Saha, B., and Orvig, C. (2010). Biosorbents for hexavalent chromium elimination from industrial and municipal effluents. Coordin. Chem. Rev. *254*, 2959–2972.

Sanyal, A., Rautaray, D., Bansal, V., and Sastry, M. (2005). Heavy metal remediation by a fungus as a means of production of lead and cadmium carbonate crystals. Langmuir *21*, 7220–7224.

Schulz, B., and Boyle, C. (2006). What are endophytes? In: Microbial Root Endophytes, B.J.E. Schulz, C.I.C. Boyle, and T.N. Sieber (eds.) Microbial Root Endophytes. (Springer: Berlin), pp. 1r:B1–13.

Scott, J.A., and Palmer, S.J. (1990). Sites of cadmium uptake in bacteria used for biosorption. Appl. Microbiol. Biotechnol. *33*, 221–225.

Sessitsch, A., Kuffner, M., Kidd, P., Vangronsveld, J., Wenzel, W.W., Fallmann, K., and Puschenreiter, M. (2013). The role of plant-associated bacteria in the mobilization and phytoextraction of trace elements in contaminated soils. Soil Biol. Biochem. *60*, 182–194.

Sharma, S., Sundaram, C.S., Luthra, P.M., Singh, Y., Sirdeshmukh, R., and Gade, W.N. (2006). Role of proteins in resistance mechanism of *Pseudomonas fluorescens* against heavy metal induced stress with proteomics approach. J. Biotechnol. *126*, 374–382.

Shine, A.M., Shakya, V.P.S., and Indnurm, A. (2015). Phytochelatin synthase is required for tolerating metal toxicity in a basidiomycete yeast and is a conserved factor involved in metal homeostasis in fungi. Fungal Biol. Biotechnol. 2, 3.

Singh, B.K., Walker, A., and Wright, D.J. (2006). Bioremedial potential of fenamiphos and chlorpyrifos degrading isolates: Influence of different environmental conditions. Soil Biol. Biochem. *38*, 2682–2693.

Singh, R., Singh, P., and Sharma, R. (2014). Microorganism as a tool of bioremediation technology for cleaning environment: A review. Proceedings of the Int. Academy Ecol. Environ. Sci. *4*, 1–6.

Siripornadulsil, S., Thanwisai, L., and Siripornadulsil, W. (2014). Changes in the proteome of the cadmium-tolerant bacteria *Cupriavidus taiwanensis* KKU2500-3 in response to cadmium toxicity. Can. J. Microb. *60,* 121–131.

Słaba, M., Szewczyk, R., Bernat, P., and Długoński, J. (2009). Simultaneous toxic action zinc and alachlor resulted in enhancement of zinc uptake by *Paecilomyces marquandii*. Sci. Total Environ. *407,* 4127–4133.

Słaba, M., and Długoński, J. (2011). Efficient Zn^{2+} and Pb^{2+} uptake by filamentous fungus *Paecilomyces marquandii* with engagement of metal hydrocarbonates precipitation. Int. Biodeterior. Biodegrad. *65,* 954–960.

Słaba, M., Bernat, P., Różalska, S., Nykiel, J., and Długoński, J. (2013a). Comparative study of metal induced phospholipid modifications in the heavy metal tolerant filamentous fungus *Paecilomyces marquandii* and the implication for membrane integrity. Acta Biochim. Pol. *60,* 695–700.

Słaba, M., Gajewska, E., Bernat, P., Fornalska, M., and Długoński, J. (2013b). Adaptive alteration in the fatty acids composition under induced oxidative stress in heavy metal-tolerant filamentous fungus *Paecilomyces marquandii* cultured in ascorbic acid presence. Environ. Sci. Pollut. Res. *20,* 3423–3434.

So, N., Rho, J., Lee, S., Hancock, I.C., and Kim, J. (2001). A lead-absorbing protein with superoxide dismutase activity from *Streptomyces subrutilus*. FEMS Microbiol. Lett. *194,* 91–98.

Sriprang, R., Hayashi, M., Ono, H., Takagi, M., Hirata, K., and Murooka, Y. (2003). Enhanced accumulation of Cd^{2+} by a *Mesorhizobium* sp. transformed with a gene from *Arabidopsis* thaliana coding for phytochelatin synthase. Appl. Environ. Microbiol. *69,* 1791–1796.

Srivastava, S., and Goyal, P. (2010). Reusability of biomaterial: a cost effective approach. In: Novel Biomaterials. Decontamination of Toxic Metals from Wastewater. Environ. Sci. Eng. (Springer: Heidelberg), pp. 93–96.

Sturz, A.V., Christie, B.R., and Nowak, J. (2000). Bacterial endophytes: potential role in developing sustainable systems of crop production. Crit. Rev. Plant Sci. *19,* 1–30.

Sun, J., Zou, X., Ning, Z., Sun, M., Peng, J., and Xiao, T. (2012). Culturable microbial groups and thallium – tolerant fungi in soils with high thallium contamination. Sci. Total Environ. *441,* 258–264.

Sutherland, I.W. (2001). The biofilm matrix – an immobilized but dynamic microbial environment. Trends Microbiol. *9,* 222–227.

Tabak, H.H., Lens, P., van Hullebusch, E.D., and Dejonghe, W. (2005). Developments in bioremediation of soils and sediments polluted with metals and radionuclides – 1. Microbial processes and mechanisms affecting bioremediation of metal contamination and influencing metal toxicity and transport. Rev. Environ. Sci. Biotechnol. *4,* 115–156.

Takiquchi, M., and Yoshihara, S. (2006). New aspects of cadmium as endocrine disruptor. Environ. Sci. *13,* 107–116.

Tan, K.H. (2009). Environmental Soil Science, 3rd edn (CRC Press: New York).

Tang, X., Zhu, Y., and Meng, Q. (2007). Enhanced cruder oil biodegradability of *Pseudomonas aeruginosa* ZJU after preservation in crude oil-containing medium. World J. Microbiol. Biotechnol. *23,* 7–14.

Tangahu, B.V., Abdullah, S.R.S., Basri, H., Idris, M., Anuar, N., and Mukhlisin, M. (2011). A review on heavy metals (As, Pb, and Hg) uptake by plants through phytoremediation. Int. J. Chem. Engin. Res., doi:10.1155/2011/939161.

Tibbett, M., and Sanders, F.E. (2002). Ectomycorrhizal symbiosis can enhance plant nutrition through improved access to discrete organic nutrient patches of high resource quality. Ann. Bot. *89,* 783–789.

Tibbett, M., Sanders, F.E., and Cairney, J.W.G. (1998). The effect of temperature and inorganic phosphorus supply on growth and acid phosphatase production in arctic and temperate strains of ectomycorrhizal *Hebeloma* spp. in axenic culture. Mycol. Res. *102,* 129–135.

Todorova, D., Nedeva, D., Abrashev, A., and Tsekova, K. (2008). Cd (II) stress response during the growth of *Aspergillus niger* B 77. J. Appl. Microb. *104,* 178–184.

Tsyganov, V.E., Borisov, A.Y., and Tikhonovich, I.A. (2001). Characterisation of plant growth-promoting rhizobacteria isolated from polluted soils and containing 1-aminocyclopropane-1-carboxylate deaminase. Can. J. Microbiol. *47,* 642–652.

Ullah, A., Heng, S., Munis, M.F.H., Fahad, S., and Yang, X. (2015). Phytoremediation of heavy metals assisted by plant growth promoting (PGP) bacteria: A review. Environ. Exp. Bot. *117,* 28–40.

Van der Heggen, M., Martins, S., Flores, G., and Soares, E. (2010). Lead toxicity in *Saccharomyces cerevisiae*. Appl. Microbiol. Biotechnol. *88,* 1355–1361.

Velmurugan, P., Shim, J., You, Y., Choi, S., Kamala-Kannan, S., Lee, K.-J., Kim, H.J., and Oh, B.-T. (2010). Removal of zinc by live, dead, and dried biomass of *Fusarium* spp. isolated from the abandoned-metal mine in South Korea and its perspective of producing nanocrystals. J. Hazard. Mater. *182,* 317–324.

Vido, K., Spector, D., Lagniel, G., Lopez, S., Toledano, M.B., and Labarre, J. (2001). A proteome analysis of the cadmium response in *Saccharomyces cerevisae*. J. Biol. Chem. *276,* 8469–8474.

Vijayaraghavan, K., Yan, Y.-S. (2008). Bacterial biosorbents and biosorption. Biotechnol. Adv. *26,* 266–291.

Voiblet, C., Duplessis, S., Encelot, N., and Martin, F. (2001). Identification of symbiosis-regulated genes in *Eucalyptus globulus–Pisolithus tinctorius* ectomycorrhiza by differential hybridization of arrayed cDNAs. Plant J. *25,* 181–191.

Wang, J., and Chen, C. (2009). Biosorbents for heavy metals removal and their future. Biotechnol. Adv. *27,* 195–226.

Welti, R., Li, W., Li, M., Sang, Y., Biesiada, H., Zhou, H.E., Rajashekar, C.B., Williams, T.D., and Wang, X. (2002). Profiling membrane lipids in plant stress responses. Role of phospholipase D alpha in freezing-induced lipid changes in *Arabidopsis*. J. Biol. Chem. *277,* 31994–32002.

White, C., and Gadd, G.M. (2000). Copper accumulation by sulfate-reducing bacterial biofilms. FEMS Microbiol. Lett. *183,* 313–318.

Won, S.W., Kotte, P., Wei, W., Lim, A., and Yan, Y.-S. (2014). Biosorbents for recovery of precious metals. Biores. Technol. *160,* 203–212.

Wu, S.C., Cheung, K.C., Luo, Y.M., and Wong, M.H. (2006). Effects of inoculation of plant growth-promoting rhizobacteria on metal uptake by *Brassica juncea*. Environ. Pollut. *140,* 124–135.

Wuana, R.A., and Okieimen, F.E. (2011). Heavy metals in contaminated soils: a review of sources, chemistry, risks and best available strategies for remediation. International Scholarly Research Network, ISRN Ecology, Volume 2011, Article ID 402647, 20 pages, doi:10.5402/2011/402647.

Wunschmann, J., Beck, A., Meyer, L., Letzel, T., Grill, E., and Lendzian, K., J. (2007). Phytochelatins are synthesized by two vacuolar serine carboxypeptidases in *Saccharomyces cerevisie*. FEBS Lett. *581*, 1681–1687.

Xiao, X., Luo, S.L., Zeng, G.M., Wei, W.Z., Wan, Y., Chen, L., Guo, H.J., Cao, Z., Yang, L.X., Chen, J.L., and Xi, Q. (2010). Biosorption of cadmium by endophytic fungus (EF) *Microsphaeropsis* sp. LSE10 isolated from cadmium hyperaccumulator *Solanum nigrum* L. Biores. Technol. *101*, 1668–1674.

Yan, G., and Viraraghavan, T. (2001). Heavy metal removal in a biosorption column by immobilized *Mucor rouxii* biomass. Biores. Technol. *78*, 243–249.

Yang, J., Cao, J., Xing, J., and Yuan, H. (2015). Lipid production combined with biosorption and bioaccumulation of cadmium, copper, manganese and zinc by oleaginous microalgae *Chlorella minutissima* UTEX2341. Biores. Technol. *175*, 537–544.

Zhu, J., Zhang, Q., Wu, R., and Zhang, Z. (2010). HbMT2, an ethephon-induced metallothionein gene from *Hevea brasilienis* responds to H_2O_2 stress. Plant Physiol. Biochem. *48*, 710–715.

Zimmer, D., Baum, C., Leinweber, P., Hrynkiewicz, K., and Meissner, R. (2009). Associated bacteria increase the phytoextraction of cadmium and zinc from a metal-contaminated soil by mycorrhizal willows. Int. J. Phytorem. *11*, 200–213.

Application of Recent Omics Achievements in Bioremediation Processes Illustrated by Progress in Microbial Surfactants Commercialization

13

Katarzyna Paraszkiewicz, Jerzy Długoński and Dariusz Trzmielak

Contents

Abstract

This section comprises a critical analysis of microbial surfactant commercialization. Petroleum-derived active surface agents (synthetic surfactants) are applicable in various industrial sectors but pose significant environmental risks. In comparison with synthetic equivalents, microbial surfactants are environmentally friendly, biodegradable, and less toxic, as well as exhibit higher surface activity and selectivity; however, they are still too expensive to compete on the market. However, available reports forecast an increased share of biosurfactants as market products. Progress in biosurfactants commercialization (coupled with a better understanding of the roles of these compounds in microbial physiology and in biodegradation processes) results from advances in cell and molecular biology (e.g. genomics, proteomics, metabolomics and lipidomics).

13.1 Introduction

The new biotechnology knowledge, including omics sciences, that is generated by expenditures on research in pure science and industrial research and development cannot rely on intellectual resources alone. It should affect the viability of individual organizations and entire industries (Robertson and Jacobson, 2011). Major challenges in the commercialization of biotechnology involve market acceptance and adoption (Valpatti and Yetisen, 2014). In assessing the commercialization benefits, there are three types of risks considered in biotechnology research development: (1) risks inherent in the technology itself; (2) social – cultural context risks, i.e., which societal group benefits from the emerging technology and which group is harmed; and (3) diffusion risk – the spread of knowledge from an origin source to one or more recipients. In the context of biotechnology, diffusion risk is the probability that the new technology reaches its greatest penetration of the biotechnology marketplace and either attains widespread acceptance by producers and product consumers (Cohen, 2007b). Applications of biotechnology are broad, and the potential benefit to societies is great. However, the acceptance of biotechnological growth within a particular nation and industry will depend on the balance between the commercial benefits and the evaluation of the necessary types of risks for particular entrepreneurs, consumers and all society. As the world gains an increased understanding of biotech's benefits versus risks evaluated, its commercialization will accelerate (Azzone and Pozza, 2003).

Zehner *et al.* (2013) reported that the time line to develop and commercialize a new biotech product is measured in decades rather than years or months for most technology products. Consequently the commercialization process often starts from basic research (Fig. 13.1). Basic research is very important to biotechnology innovation, and the commercialization decision is whether to proceed with, modify, or terminate a particular innovation. Because the biotechnology commercialization process is relatively long and risky for investors calculating the net benefits, the tendency of practical innovation commercialization is the justification for allocating public and private spending on each research and commercialization stage. Estimating the value of an early stage biotechnology, which can be monetized for longer than decades, is significantly difficult. At the time that the R&D investment must be considered, the commercialization process should be shaped and recognized.

The lengthy process imposes a number of critical and time-sensitive decisions on biotech entrepreneurs, which they may or may not be capable of handling. For example, because biotech is heavily grounded in basic scientific research, which is mostly undertaken in university research labs, the scientist-entrepreneur may have to devote an inordinate amount of time to finding grants to support the basic research – idea generation stage [stage 1]. Basic research focuses on the discovery and development of a concept. Stage 2 is the laboratory proof of concept and the conceptualized phase. This early stage involves genetic engineering techniques. Stage 3 is invention conversion into innovation of the potential product. Research is initially subjected to a series of biological research tests or screens, which are designed to demonstrate the biological activity of the potential trait (McDougall, 2012). Additionally, an activity that must be considered beyond stage 3 is patenting. Scientists are seeking to expand inventions. Entrepreneurs are also seeking to minimize new technology development costs, reduce business risk and create value for investors. All of this can be accomplished by exploiting patents. Access to intellectual property is the key to creating market value (Smith and Parr, 2005). Stage 4 involves trials or the establishment of effectiveness and efficacy product characteristics. Stage 5 is the conversion of the invention into an innovation located in the marketplace. The final stage focuses

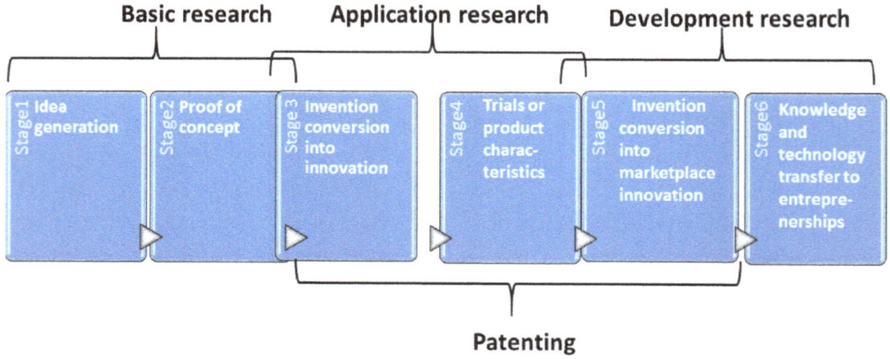

Figure 13.1 Commercialization process.

on knowledge and technology transfer from labs to entrepreneurships [stage 6]. The transition from invention to innovation involves interactions among the scientists, the university technology transfer office, the commercialization organization, science parks, technology incubators and business elements such legal and venture capital organizations (Gallwas, 2005; Ganguli *et al.*, 2008; Sinha, 2012).

As the scientist-entrepreneurs progress through the invention phase, they will spend time dealing with the technology transfer office as well as prospective non-university organizations such as legal and investment organizations. Finally, when the molecule is in the innovation – complete product stage [stage 5 and 6], the scientist-entrepreneur must address venture firms. The funding process is time-consuming and requires the scientist entrepreneur to build multiple relationships across many industries or potential applications throughout the process from idea to proof of concept to prototype to market launch. Individuals must be educated regarding the merits of the molecule at each step. Equally challenging is the fact that for the project to progress, funding must be obtain from different sources – each with a requirement for different scientific data as well as a different time frame. Basic research could be funded through a 4- to 7-year research grant, but the time horizon of a venture firm to achieve commercial success is 3 to 5 years. The cost of developing and launching new technology products is significant. A rough rule of thumb is that for every one Euro invested in the cost of discovery of the principle, the cost of developing a prototype is ten times the cost of discovery, and the cost of market introduction is tenfold the cost of the prototype (Jolly, 1997). Developing new biotech products is expensive and takes longer than pharma products but has a success rate of 30.2% compared with pharma's 21.5% (McDougall, 2012).

Reports from the US Environmental Protection Agency (US EPA) providing overviews of certain non-combustion technologies designed to remove persistent organic pollutants (POPs) from contaminated soil and sediments, as well as technologies for treating mining-influenced water (MIW), were published in 2010 and 2014, respectively. These technologies were identified by the US EPA by reviewing the technical literature (e.g. EPA reports, site-specific reports, industry publications, and several databases) and by contacting subject-matter experts. As described in the reports, the technologies were characterized in detail, including a description of the commercial availability, advantages and limitations, type of treated pollutants, sites where the treatment was (*ex situ* or *in situ*), pre-treatment needs, scale, power requirements, effectiveness and costs.

The market value of synthetic surfactants is predicted to reach \$2.8 billion in 2023 (Grand View Research, Inc., 2014). Biosurfactants, which are surface-active agents synthesized by microorganisms, are a worthy alternative for the classic petrochemical surfactants and can be successfully exploited in various applications designed to clean the environment (bioremediation, phytoremediation, and contaminant removal by washing/flushing techniques). Nevertheless, the commercial potential of biosurfactants is still limited by relatively high costs of production (Mao *et al.*, 2015). The effort to achieve the marketable production of biosurfactants (exhibiting eco-friendly properties and often described as a 'green solution' or 'green chemistry') is in agreement with the European Surfactant Directive (Regulation EC No. 648/2004) requiring a description of a surfactant's biodegradability to protect the environment and human health.

13.2 Bio-based and microbial surfactants market analysis

A report on the present global bio-based and microbial surfactants market and this segment's prospects (to 2020) has recently been published in the CAGR report (2014). Primary sources (opinions of scientific and industrial experts and suppliers who were interviewed) and various secondary sources (such as databases) were used to prepare the report. The global biosurfactant market has been analysed by product type, end-use, geographical insight and producing companies. Notably, the CAGR evaluation was made for all bio-based surfactants, including not only microbial surfactants but also predominantly chemically prepared derivatives (with hydrophobic or hydrophilic portions obtained from plant-based compounds).

According to the CAGR report (2014), the world production of surfactants is currently estimated at 15 Mton/year about half of which are soaps. In 2013, the global bio-based surfactants market was 344,068.40 tons and is expected to reach 461,991.67 tons by 2020. The market segmented by product types comprises mainly plant-based surfactants with methyl ether sulfonates (MESs) and alkyl polyglucosides (APGs) as the largest consumed bio-based surfactants, accounting for 33 and 25% of the global market, respectively. Economical production of microbial surfactants has been achieved in a case of microorganisms belong mainly to *Bacillus*, *Pseudomonas* and *Candida* genera synthesizing lipopeptides, rhamnolipids and sophorolipids, respectively. Among the microbial biosurfactants, the most important group of commercialized compounds are rhamnolipids, with a relatively small market in 2013 but expected to grow to 5.4% from 2014 to 2020.

Based on the end-use, the global biosurfactants market was divided into textiles, personal care, food processing, household detergents, oilfield chemicals, agricultural chemicals, industrial and institutional cleaners, and others. The personal care and household detergents segment was estimated as dominant, expected to account for over 56.8% (50.7 kilo tons) of the market share by 2020, followed by food processing applications (11%), oil field chemicals, agricultural chemicals and agents used in the textile industry.

According to the regional (geographical) insight, Europe is the largest consumer of biosurfactants (with over 50% of the global consumption in 2013), probably due to a high awareness among European consumers regarding the health hazards associated with synthetic surfactants. North America is the second largest consumer of biosurfactants.

BASF (with a market share of over 20%) is reportedly the top bio-based surfactant company, followed by Ecover, Urumqui Unite, MG Intobio and Saraya (together covering a market share of over 60%). The other leading players are Akzonobel N.V. (Netherlands), Air Products and Chemicals (US), Clariant AG (Switzerland), DuPont EI DE Nemours & Co. (US), Evonik Industries AG (Germany), Henkel AG & Co. KGaA (Germany), Huntsman Corporation (US), Kao Corporation (Japan), and The Dow Chemical Company (US).

Soliance Company (France) manufactures sophorolipids (under the trade name Sopholiance) used in cosmetic and skin products with antimicrobial activity. Iwata Chemical Co., Ltd, produces spiculosporic acid and rhamnolipids. Jeneil Biosurfactant Company (USA) produces rhamnolipids with good emulsification, wetting, detergency, and foaming properties. Rhamnolipids produced by this company are marketed for agricultural purposes as an EPA-approved biofungicide with the trade name ZONIX Biofungicide. Moreover, its RECO product line is used to clean and recover oils from storage tanks. Wako Pure Chem. Industries is a manufacturer of surfactin. In addition, Sigma – Aldrich offers surfactin (with a CMC of 7.5–20 μmol/l) iturin and fengycin suitable for chromatographic analysis but also with a high price (for example, the cost of iturin is approximately 200 EUR for a 5 mg vial). The cost of the RAG-1 emulsan containing broth is approximately 50 dollars/kg. To compete with the synthetic surface agents, the cost of biosurfactant production must be brought down to £1.70 per litre (Sekhon Randhawa and Rahman, 2014). The high-molecular-weight polymer RAG-1 emulsan, produced by *Acinetobacter calcoaceticus*, is the only commercially available microbial bioemulsifier at present (Suthar *et al.*, 2008; Banat *et al.*, 2010). Sakhon Randhavwa and Rahman (2014) stated that the cost of biosurfactant production must be reduced to around £1.7 per litre to successfully compete with synthetic counterparts in the market. The summary (prepared by these authors) of companies producing biosurfactants on an industrial scale includes 17 companies located mostly in the US and Japan.

13.3 Bottlenecks of microbial surfactant commercialization

The commercialization application of microbial surfactants is dependent on three main factors: economics, commercial utilization and regulatory issues. Certainly, the primary driver in all technologies is return versus cost (Faber *et al.*, 2003). The market value must be considered, neither the future return (benefits for investors, enterprises) nor the costs of bringing them to market. As an example, the main microbial surfactant disadvantage limiting their use on an industrial scale is the relatively high production cost and low microbial productivity, caused by many bottlenecks, including (among others) problems with enlarging the production scale, obtaining compounds of high purity (particularly required in pharmaceutical, food and cosmetic applications), and achieving overproducing, non-pathogenic microbial strains exhibiting high biosurfactant production during growth in inexpensive media containing by-products or wastes from the various industries. The regulation of biosurfactant synthesis is in many cases barely understood (rendering the genetic improvement of microorganisms difficult); biosurfactants are produced as a mixture of many isomers, and the interactions between biosurfactant cells/organisms and the abiotic environment are not fully recognized (Singh *et al.*, 2007; Banat *et al.*, 2010; Raaijmakers *et al.*, 2010).

Regulation involves the control of technology adoption and use through direct intervention in the inventors', producers' and users' decision-making process, although regulatory authority generally derives from the public sector and public and private insurers. The principal legislative agencies focus on knowledge-processing activities that have included studies of the relations between new technology and safety and ethical implications (Cohen and Hanft, 2007a). Commercial utilization may be assessed by the following evaluative purposes (Cohen and Hanft, 2007b): (1) to assess the safety, efficacy, and effectiveness of technology or products when applied for a specific industrial purpose; (2) to provide information to entrepreneurs, investors, and consumers regarding the added value of technology compared with existing alternatives; (3) to provide information on appropriate strategies for using a technology to achieve favourable outcomes and to assure quality of products; and (4) to illuminate issues related to the social, ethical, and legal implications of adoption and the commercial usage.

The laboratory proof of concept and the conceptualized phase for microbial surfactants should take into consideration future large-scale production and penetration of products into the market. The magnitude of future benefits depends upon the effects of the innovation on the cost of production (Raghukumar, 2014).

Currently, the most advanced level of research focused on wider biosurfactant commercialization has been achieved for rhamnolipids (Müller *et al.*, 2012; Lovaglio *et al.*, 2015), sophorolipids, MELs and cellobiose lipids (mostly due to the progress in the knowledge of biosurfactants gene regulatory pathways and metabolic engineering) (Roelants *et al.*, 2014; Dobler *et al.*, 2016).

According to Müller *et al.* (2012), a sustainable bioeconomy approach should combine a holistic X-omics strategy with metabolic engineering to achieve the next step in biosurfactant production based on non-food renewable resources. A great intensity of ongoing research programmes gives real hope to overcome commercialization problems and increase the penetration of microbial surfactants into the market (Fig. 13.2).

The main challenges that should be met to attain the economical large-scale production of biosurfactants include (1) an effective microorganism (non-toxic, non-pathogenic biosurfactant producer capable of growth and biosurfactant synthesis during the industrial by-products or waste utilization, obtained by efficient and modern screening techniques – e.g. metanomics – as well as resulting from progress in genomics, metabolomics, proteomics and lipidomics studies); (2) optimized production conditions (achieved by the selection of cheap culture medium components, fermentation process type and other operating parameters obtained through metabolomics study, experimental design techniques, surface response methodology and artificial neutral network coupled with a genetic algorithm); (3) inexpensive downstream processes (biosurfactant recovery and purification, if necessary); and (4) easy and

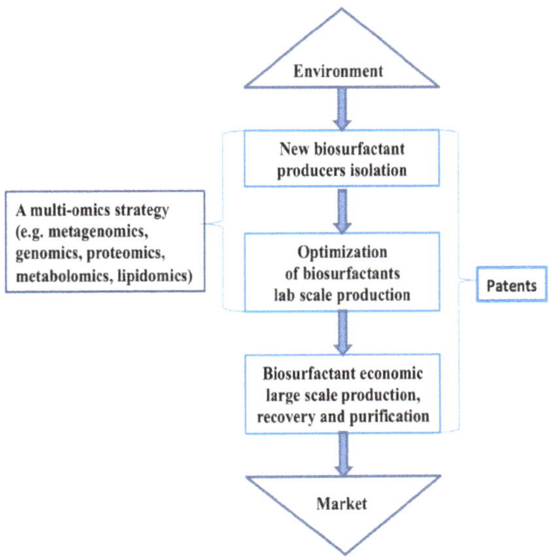

Figure 13.2 Biosurfactant milestones from environment to market.

inexpensive management of fermentation process wastes (Banat *et al.*, 2010).

13.4 Acceleration of microbial surfactant application by progress in basic sciences

The middle stages of the commercialization process have a place if the invention conversion into innovation and establishment of effectiveness and the efficacy products characteristics are recognized. The structures of synthesized biosurfactants and the ability to produce a mixture of biosurfactant variants first depend on microorganism genomics but can also be influenced by the culture medium composition and conditions used during the cultivation. Although many patents concerning microbial surfactants are available, only a few such compounds are produced on an industrial scale. Among several approaches useful to improve biosurfactant productivity, combined omics analysis and computational modelling is especially promising (Lee *et al.*, 2005; Reis *et al.*, 2013).

13.4.1 Isolation of biosurfactant producers

Presently, many rapid screening tests (e.g. the drop-collapse test, assays of emulsifying activity and haemolysis, and the microplate technique) are available. These tests are very useful for quickly obtaining a large collection of biosurfactant-producing microorganisms (Persson and Molin, 1987; Chen *et al.*, 2007; Satpute *et al.*, 2010; Walter *et al.*, 2010; Płaza *et al.*, 2015).

The search for new biosurfactant-producing strains from uncommon or extreme microbial habitats (extremophiles such as thermophilic and halophilic microorganisms) allowing the isolation of new compounds with novel properties is still important due to its biotechnological and environmental significance (Kumar *et al.*, 2008; Banat *et al.*, 2010). For example, Jadhav *et al.* (2013) established that among 14 bacterial strains isolated from Antarctica sea water samples, one isolate identified as *Oceanobacillus* sp. was a producent of a glycolipoprotein biosurfactant exhibiting emulsifying properties and antimicrobial activity against *E. coli* but no cytotoxicity against a normal cell line. According to results published by Plante *et al.* (2008), polluted marine habitats rich in surfactants are a source of microorganisms possessing both degradative capabilities and the ability to thrive in the presence of surfactants. From the guts of *Nereis succinea* and *Amphitrite ornata* (marine deposit feeder invertebrates) were isolated bacterial strains exhibiting a high resistance to both cationic and anionic surfactants.

There are significant interests in the search for novel biosurfactant producers, but it is worth mentioning that the numbers of cultured organisms from the environmental samples are estimated to be only a tiny fraction (0.001%) of the actual microbial diversity (Sekhon Randhawa and Rahman, 2014). Metagenomics (the culture-independent genomic analysis) of microbial communities is indicated to be a powerful tool for exploring novel compounds from uncultured bacteria associated with natural ecosystems. According to Sachdev and Cameotra (2010), the functional metagenomic technique required to obtain the novel biosurfactants from uncultured bacteria contains the following steps: (1) the extraction of total metagenomic DNA; (2) DNA fragmentation by restriction enzymes followed by the insertion to suitable expression vectors; (3) transformation of DNA constructs in hosts such as *E. coli*; and (4) screening of the metagenomic clones for biosurfactant production by known conventional and/or high throughput techniques. An immunological detection by competitive ELISA assay is also reported to be useful for monitoring *in situ* biosurfactant production (Raaijmakers *et al.*, 2010).

Progress in research focused on biosurfactant cost-efficient production is associated with the development of new analytical methods, especially matrix assisted laser desorption/ionization time of flight (MALDI-TOF) mass spectrometry and tandem mass spectrometry (MS/MS) (Satpute *et al.*, 2008; Heyd *et al.*, 2008; Baker and Chen, 2010; Janek *et al.*, 2010; Pecci *et al.*, 2010; Raaijmakers *et al.*, 2010). For example, a recently combined approach of a molecular tool based on the PCR method and LC-MS/MS technique was proposed to distinguish potent surfactin, iturin and fengycin producers among screened *Bacillus* isolates (Płaza *et al.*, 2015).

13.4.2 Multi-omics studies on biosurfactant producing microorganisms

The development of biosurfactant over/hyper producers requires genetic manipulations of selected robust wild-strains (Dhanarajan and Sen, 2014). The most advanced studies are conducted on developing overproducers of low-molecular weight biosurfactants (including lipopeptides and glycolipids such as rhamnolipids, sophorolipids, cellobiose lipids and mannosylerythritol lipids). Rhamnolipids (containing one or two L-rhamnose molecules coupled to a mono or dimer of β-hydroxy fatty acids) are produced by many *Pseudomonas* species, usually as heterogeneous mixtures of various mono- and dirhamnolipids. The biosynthetic pathways and the main gene regulatory factors involved in rhamnolipid production within *P. aeruginosa*, as well as strategies applied to metabolic engineering, were recently described by Dobler *et al* (2016). In *P. aeruginosa*, three genes (carried on two distinct operons) encode the enzymes responsible for the final steps of rhamnolipid synthesis: one operon carries the rhlAB genes and the other, rhlC (Abdel-Mawgoud *et al.*, 2010). Additionally, direct and indirect factors such as quorum sensing, nutritional status and stress response can influence rhamnolipid production (Reis *et al.*, 2011; Geis *et al.*, 2014). Strains of *P. aeruginosa* are opportunistic pathogens, and this fact limits (to a large extent) industrial-scale production due to the health hazard of the process. Successful limitation of the secreted biosurfactant mixture to only mono-rhamnolipids was obtained through cloning the rhlABRI operon from *P. aeruginosa* into *E. coli* and non-pathogenic *P. putida* strains (Cabrera-Valladares *et al.*, 2006; Cha *et al.*, 2008). Wang *et al.* (2007) obtained rhamnolipid-producing *E. coli* and *P. aeruginosa* strains after transposon-mediated chromosome integration of the rhamnosyltransferase 1 complex. Approaches that have been taken to produce rhamnolipids using non-pathogenic microorganisms by genetic engineering of different bacteria have been summarized by Toribio *et al.* (2010).

Sophorolipids, produced by yeast strains such as *Starmerella bombicola* (former *Candida bombicola*) are glycolipid biosurfactants constituted from a sophorose molecule (2-*O*-β-D-glucopyranosyl-D-gluco-pyranose) linked to a terminal or subterminal hydroxylated fatty acid. A sophorolipid gene cluster consisting of five genes is directly involved in sophorolipid synthesis (Van Bogaert *et al.*, 2013). The first step in the biosynthesis of sophorolipids is the terminal or subterminal hydroxylation of a fatty acid by the cytochrome P-450 monooxygenase enzyme CYP52M1 (Geys *et al.*, 2014). To date, several knock-out strains have been created for improving production yields and for obtaining various modified sophorolipids (Van Bogaert *et al.*, 2009, Saerens *et al.*, 2011; Van Bogaert *et al.*, 2013).

Despite the growing knowledge of *Bacillus* and *Pseudomonas* genomes, the preparation of biosurfactant-over producing recombinants is very difficult because of the high complexity of the transcriptional regulatory network of genes involved in lipopeptides and rhamnolipids synthesis (Das *et al.*, 2008; Raaijmakers *et al.*, 2010; Reis *et al.*, 2011; Muligan *et al.*, 2014). For example, the peptide portion of the cyclic lipopeptide – surfactin, is synthesized non-ribosomally by a large multienzyme peptide synthetase complex containing four enzymatic subunits and catalyses the incorporation of amino acids into the peptide portion of surfactin by a thiotemplate mechanism. Then, the peptide part is linked to a hydroxyl fatty acid residue by acyltransferase (thioesterase) Synthesis of other lipopeptides (e.g. iturin, fengycin, lichenysin and arthrofactin) is mediated by similar enzyme complexes, called non-ribosomal peptide synthetases (NRPSs), exhibiting a high degree of structural similarity among themselves even from distant microbial species (Sen, 2010; Dhanarajan and Sen, 2014; Qiu *et al.*, 2014).

Recently, genome sequences of two strains of *Rhodococcus* able to produce trihalolipid biosurfactants have been published (Kuyukina *et al.*, 2015; Sambles and White, 2015). Analysis of the draft genome of the *Pseudomonas putida* member confirmed the presence of putisolvin biosynthesis genes, and the corresponding lipopeptide biosurfactants were found to contribute to the antimicrobial activity against *P. aeruginosa* (Ye *et al.*, 2014).

The development of lipidomics studies allows, among other aspects, recognizing in biosurfactant secreting cells the mechanisms responsible for managing the stress caused by their own membrane-destabilizing agents. Many cyclic lipopeptide biosurfactants easily interfere with phospholipids, which are important components of cytoplasmic membranes. Surfactin displays a particularly strong membrane-destabilizing action. Many studies on the identification of the molecular mechanisms of surfactin-induced membrane destabilization have revealed that this compound can interact with the lipid membrane by (1) acting as a mobile cation carrier; (2) forming cationic channels; and/or (3) destroying the membrane via a detergent action (Heerklotz and Seelig 2007; Shen *et al.*, 2010; Seydlová *et al.*, 2013). These effects can be observed at surfactin concentrations far below its critical micelle concentration (CMC). The presence of surfactin in higher concentrations leads to complete disruption and solubilization of the lipid bilayer and the formation of mixed micelles (Ongena and Jacques, 2008).

For example, Seydlovà and Svobodovà (2008) used a surfactin-producing strain of *B. subtilis* to demonstrate a significant increase in the levels of high-melting fatty acids and cardiolipin (regarded as a stress phospholipid stabilizing the lipid bilayer) in the bacilli cytoplasmic membrane. The most significant changes in the membrane polar head groups (resulting in the progressive rigidization of the membrane interior) occurred during the stationary phase, when surfactin was accumulated in the culture medium. Seydlová *et al.* (2013) when comparing the phospholipid composition of the surfactin producer's membrane (Sf+) to that of its non-producing mutant (Sf–), determined a progressive accumulation (up to 22%) of the stress phospholipid cardiolipin in the Sf+ membrane and suggested that cardiolipin, due to its negative charge, may prevent the surfactin–membrane interaction or that the surfactin pore formation activity thus plays a role as a key factor in the membrane protection.

Recently developed high-throughput experimental techniques supported by bioinformatics has resulted in the accumulation of data providing a basis for a deeper understanding of microbial biosurfactant synthesis and secretion (Sachdev and Cameotra, 2010; Reis *et al.*, 2013) The genomic and proteomic approaches for biotechnology are still limited, despite *Pseudomonas* and *Bacillus* strains being extensively studied due to their versatility as well as their ecological and industrial importance. High-throughput 'omics' analysis, predictive computational modelling or simulation and experimental perturbation can be combined to generate new knowledge of the cellular physiology and metabolism, to design strategies for metabolic and cellular engineering of organisms (Reis *et al.*, 2013).

13.4.3 Optimization of bioprocess conditions

The development of low-cost processes involving the use of raw materials can account for 10–30% of the final product cost. Because inexpensive raw materials reported as growth substrates for biosurfactant production contain various starchy substrates, vegetable oils and oil wastes, olive oil mill effluent, whey, distillery wastes, animal fats, soap stocks and molasses, the choice of appropriate carbon and nitrogen sources (e.g. the usage of nitrate instead of ammonium for rhamnolipid production by *Pseudomonas*), the C/N ratio, and the concentration of particular mineral salts are important factors that may greatly promote biosurfactant productivity (Santos *et al* 2002; Müller, *et al* 2012). Some microbial surfactants (e.g. lipopeptides and rhamnolipids) are secondary metabolites. Therefore, in many cases, it has been proven that growth-limiting conditions, a high carbon to nitrogen ratio (C/N), and the limitation of certain microelements improve biosurfactant yield (Reis *et al.*, 2013). The successful use of inexpensive, renewable resources as culture media components has been described in many reviews and experimental papers (Mukherjee *et al.*, 2006; Das and Mukherjee, 2007; De Sousa *et al.*, 2011; Banat *et al.*, 2014; Henkel *et al.*, 2012; Khondee *et al.*, 2015, Marti *et al.*, 2015; Moya *et al.*, 2015; Ramírez *et al.*, 2015).

Among various cultivation strategies, the culturing in shake flasks, due to its simplicity, is used in the preliminary step of the biosurfactant optimization study. Bioreactors provide high control over the relevant parameters, such as oxygen transfer and pH, during the entire cultivation process and allows for up-scaling studies (Reis *et al.*, 2013). The important problem associated with production of some biosurfactants in bioreactors is the intense foaming because neither mechanic foam breakers nor chemical antifoam agents are recommended. Strategies proposed to overcome this problem involve using a non-dispersive oxygenated device controlled by a programmable logic controller (PLC) (Kronemberger *et al.*, 2010).

The application of experimental design techniques accelerates the development of optimal and economical conditions for biosurfactant production (Banat *et al.*, 2010). For example, the use of the response surface methodology (RSM) caused an increase in biosurfactant production by *Rhodococcus* spp. MTCC 2574 and *Gordonia* sp. BS29 (Mutalik *et al.*, 2008; Franzetti *et al.*, 2009). Kronemberger and co-workers (2008) reported a significant increase in yield by optimizing cultural conditions using statistical tools. The improvement in biosurfactant production using an artificial neural network coupled with a genetic algorithm was described by Pal *et al.* (2009) and Sivapathasekaran *et al.* (2010). The application of RSM for optimizing the biosurfactant production has recently been described by Coronel-León *et al.* (2015) and Ebadipour *et al.* (2015).

The next commercialization process stage is the conversion of the invention into an innovation located in the marketplace. This stage also reduces the uncertainty affecting R&D projects. The innovation must meet performance parameters in the R&D process or in the marketplace. The innovation incentives must be approved for marketplace release and may not infringe upon existing patent rights. Finally, end users will exhibit willingness to pay for an innovation or the emergence of competing products (Graff *et al.*, 2010).

13.4.4 Downstream processing methods

The cost of downstream processes (approximately constituting from 60 to even 80% of the total production costs) depends to a great extent on the planned biosurfactant application. For example, in many environmental applications, biosurfactants of high purity are not required. Methods most often used for biosurfactant recovery comprise acid precipitation, solvent extraction, crystallization, ammonium sulfate precipitation, foam fractionation, ultrafiltration and ion exchange chromatography (Desai and Banat, 1997; Mukherjee *et al.*, 2006; Satpute *et al.*, 2010; Reis *et al.*, 2013). The choice of the isolation method is strongly dependent on the biosurfactant structure and properties. Toxic solvents (e.g. acetone, methanol and chloroform) commonly used to extract many biosurfactants on a laboratory scale are not suitable on the industrial scale due to the cost and harmful effects (Banat *et al.*, 2010).

Novel techniques proposed for the recovery and purification of various biosurfactants were summarized by Banat *et al.* (2010). The above-mentioned methods comprise, among other techniques, the use of LC combined with the manipulation of a gradient and flow rate of solvents, liquid membrane extraction, a two-stage ultrafiltration followed by adsorption or ion exchange (Chen *et al.*, 2008a,b; Chen and Juang, 2008; Dimitrov *et al.*, 2008; Sivapathasekaran *et al.*, 2010).

According to a classic analysis of innovation approvals, marketplace pressure affects the use of alternative substrates such as agro-based industrial wastes, which is one of the attractive strategies for economical biosurfactants production. The development of low-cost processes and raw materials can account for 10–30% of the final product cost. Further optimization of culture medium and growth conditions can significantly increase the yield.

The final stage of commercialization process focuses on knowledge and technology transfer from labs to entrepreneurships.

13.5 Patents on microbial surfactants

Currently, there are more than 250 available patents on microbial biosurfactants (Shete *et al.*, 2006; Rahman and Gakpe, 2008). For example, Shete and co-workers (2006) described 255 patents related to microbial low- and high-molecular-weight biosurfactants, by scanning eight international as well as five Indian patent sites (e.g. US Patent and Trademark Office; European Patent Office; Canadian Patents Database; Japanese Patent Office; Indian Patent Office; Indian Patent Database). Most of those patents concern biosurfactants produced by microorganisms from rather few genera (*Acinetobacter, Pseudomonas, Bacillus* and *Candida*) and characterize conditions of microbial production, chemical preparation of biosurfactant derivatives or methods for the practical use of these compounds. Additionally, in some patents, the biosurfactant-producing organism is not specified, or the use of microbial consortia is described. As was concluded by Shete *et al.* (2006), the highest number of patents was issued for the use of biosurfactants in the petroleum industry, followed by the cosmetic industry, for bioremediation and as antimicrobial agents. Although the production of biosurfactants is usually higher compared with that of their synthetic counterparts, in many patents, this problem has been solved by describing the use of the crude (non-purified) product or even the fermentation broth (obtained after the removal of microbial cells).

Among polymers, mainly emulsan (polyionic lipopolysaccharide) synthesized by *Acinetobacter calcoaceticus* RAG-1 ATCC 31012 (first marketed by Petroleum Fermentations, Netherlands) was broadly patented, especially for environmental use, in cleaning oil-contaminated vessels, during oil spills, in the MEOR technique and to facilitate pipeline transportation of heavy crude oil.

Bacilli lipopeptides, due to their strong antifungal activity, are patented mainly for preventing microbial infections. Nevertheless, the lipopeptide lichenysin from *B. licheniformis* was patented for use in MEOR. By contrast, rhamnolipids (produced by various members of *Pseudomonas*) have a broad environmental usage both in the petroleum industry and in contaminated biota remediation techniques and, compared with other microbial biosurfactants, have the highest number of patents. Sekhon Randhawa and Rahman (2014) have summarized 42 major patents obtained on rhamnolipids issued between 1986 and 2014.

13.6 Conclusions

Petroleum-derived surfactants still dominate the market, but there is a growing effort to replace them with microbial biosurfactants. The currently observed acceleration of biosurfactant commercialization has mainly resulted from the progress in basic sciences (multi-omics studies on biosurfactant-producing microorganisms). The development of heterologous and non-pathogenic production systems with an improved ability for waste utilization (as growth substrates) will increase further market penetration by these versatile microbial metabolites.

References

Abdel-Mawgoud, A.M., Lépine, F., and Déziel, E. (2010). Rhamnolipids: diversity of structures, microbial origins and roles. Appl. Microbiol. Biotechnol. *86*, 1323–1336.

Azzone, G., and Pozza, I.D. (2003). An integrated strategy for launching a new product in the biotech industry. Management Decision. *9*, 832 – 843.

Baker, S.C., and Chen, C.Y. (2010). Enrichment and purification of lipopeptide biosurfactants. Adv. Exp. Med. Biol. *672*, 281–288.

Banat, I.M., Franzetti, A., Gandolfi, I., Bestetti, G., Martinotti, M.G., Fracchia, L., Smyth, T.J., and Marchant, R. (2010). Microbial biosurfactants production, applications and future potential. Appl. Microbiol. Biotechnol. *87*, 427–44.

Banat, I.M., Satpute, S.K., Cameotra, S.S., Patil, R., and Nyayanit, N.V. (2014). Cost effective technologies and renewable substrates for biosurfactants' production. Front Microbiol. *5*, 697.

Cabrera-Valladares, N., Richardson, A.P., Olvera, C., Trevino, L.G., Déziel, E., Lépine, F., and Soberon-Chavez, G. (2006). Monorhamnolipids and 3-(3-hydroxyalkanoyloxy) alkanoic acids (HAAs) production using *Escherichia coli* as a heterologous host. Appl. Microbiol. Biot. 73, 187–194.

CAGR report. (2014). Biosurfactants market by product (Market Research Report Code, GVR20). Grand View Research Inc. Available online at: http// www.grandviewresearch.com/industry-analysis/biosurfactants-industry.

Cha, M., Lee, N., Kim, M., and Lee, S. (2008). Heterologous production of *Pseudomonas aeruginosa* EMS1 biosurfactant in *Pseudomonas putida*. Bioresour. Technol. *99*, 2192–2199.

Chen, C.Y., Baker, S.C., and Darton, R.C. (2007). The application of a high throughput analysis method for the screening of potential biosurfactants from natural sources. J. Microbiol. Methods. *70*, 503–10.

Chen, H.L., and Juang, R.S. (2008). Recovery and separation of surfactin from pretreated fermentation broths by physical and chemical extraction. Biochem. Eng. J. *38*, 39–46.

Chen, H.L., Chen, Y.S., and Juang, R.S. (2008). Recovery of surfactin from fermentation broths by a hybrid salting-out and membrane filtration process. Sep. Purif. Technol. *59*, 244–252.

Cohen, A.B. (2007a). Biomedical innovation and the development of medical technology, In Technology in American health care, A.B. Cohen, and R.S. Hanft, eds. (The University of Michigan Press, Ann Arbior, USA), pp. 45–79.

Cohen, A.B. (2007b), The diffusion of new medical technology, In Technology in American health care, A.B. Cohen, and R.S. Hanft, eds. (The University of Michigan Press, Ann Arbior, USA), pp. 79–104.

Cohen, A.B., and Hanft, R.S. (2007a). Medical technology evaluation in the United State and sected other coutries. In Technology in American health care, A.B. Cohen, and R.S. Hanft, eds. (The University of Michigan Press, Ann Arbior, USA), pp. 219–249.

Cohen, A.B., and Hanft, R.S. (2007b). Evaluation Concerns: Medical technology evaluation in the future. In Technology in American health care, A.B. Cohen, and R.S. Hanft, eds. (The University of Michigan Press, Ann Arbior, USA), pp. 364–382.

Coronel-León, J., Marqués, A.M., Bastida, J., and Manresa, A. (2015). Optimizing the production of the biosurfactant lichenysin and its application in biofilm control. J. Appl. Microbiol. [Epub ahead of print].

Das, K., and Mukherjee, A.K. (2007). Comparison of lipopeptide biosurfactants production by *Bacillus subtilis* strains in submerged and solid state fermentation systems using a cheap carbon source: some industrial applications of biosurfactants. Process Biochem. *42*, 1191–1199.

Das, P., Mukherjee, S., and Sen, R. (2008). Genetic regulations of the biosynthesis of microbial surfactants: an overview. Biotechnol. Genet. Eng. Rev. *25*, 165–85.

De Sousa, J.R., Da Costa Correia, J.A., De Almeida, J.G.L., Rodrigues, S., Pessoa, O.D.L., Melo, V.M.M., and Gonçalves, L.R.B. (2011). Evaluation of a co-product of biodiesel production as carbon source in the production of biosurfactant by *P. aeruginosa* MSIC02. Process Biochem. *46*, 1831–1839.

Desai, J.D., and Banat, I.M. (1997). Microbial production of surfactants and their commercial potential. Microbiol Mol Biol Rev. *61*, 47–64.

Dhanarajan, G., and Sen, R. (2014). Amphiphilic molecules of microbial origin: classification, characteristics, genetic regulations and pathways for biosynthesis. In Biosurfactants: Research Trends and Applications, C.N. Mulligan, S.K. Sharma and A. Mudhoo, eds. (CRC Press Taylor andFrancis Group), pp. 31–48.

Dimitrov, K., Gancel, F., Montastruc, L., and Nikov, I. (2008). Liquid membrane extraction of bio-active amphiphilic substances: recovery of surfactin. Biochem. Eng. J. *42*, 248–253.

Dobler, L., Vilela, L.F., Almeida, R.V., and Neves, B.C. (2016). Rhamnolipids in perspective: gene regulatory pathways, metabolic engineering, production and technological forecasting. N. Biotechnol. *25*, 123–135.

Ebadipour, N., Lotfabad, T.B., Yaghmaei, S., and RoostaAzad, R. (2015). Optimization of low-cost biosurfactant production from agricultural residues through response surface methodology. Prep. Biochem. Biotechnol. *6*, 1–9.

Faber, D.C., Molina, J.A., Ohlrichs, Vander Zwaag, D.F., and Ferré, L.B. (2003). Commercialization of animal biotechnology. Theriogenology. *59*, 125–138.

Franzetti, A., Caredda, P., Ruggeri, C., La Colla, P., Tamburini, E., Papacchini, M., and Bestetti, G. (2009). Potential applications of surface active compounds by *Gordonia* sp. strain BS29 in soil remediation technologies. Chemosphere. *75*, 810–807.

Gallwas, G. (2005). Lesson in Innovation and Commercialization from the Biotechnology Revolution, In Nanotechnology, Science, Innovation, and Opportunity, L.E. Foster, ed. (Prentice Hall, Massachusetts, USA), pp. 3–12.

Ganguli, P., Khanna, R., Prickril, B. (2008). Defining the future: Emerging issues in biotechnology, intellectual property rights and technology transfer, In Technologty Transfer in Biotechnology. A global perspective, P. Ganguli, R. Khanna, and B. Prickril, eds. (Wiley-Vch, Weinheim, Germany), pp. 1–12.

Geys, R., Soetaert, W., and Van Bogaert, I. (2014). Biotechnological opportunities in biosurfactant production. Curr. Opin. Biotechnol. *30*, 66–72.

Graff, G.D., Zilberman, D., and Bennett, A. (2010). The commercialization of biotechnology traits. Plant Science. *179*, 635–644.

Heerklotz, H., and Seeling, J. (2007). Leakage and lysis of lipid membranes induced by the lipopeptide surfactin. Eur. Biophys. J. *36*, 305–314.

Henkel, M., Müller, M.M., Kügler, J.H., Lovaglio, R.B., Contiero, J., Syldatk, C., and Hausmann, R. (2012). Rhamnolipids as biosurfactants from renewable resources: Concepts for next-generation rhamnolipid production. Process Biochem. *47*, 1207–19.

Heyd, M., Kohnert, A., Tan, T.H., Nusser, M., Kirschhöfer, F., Brenner-Weiss, G., Franzreb, M., and Berensmeier, S. (2008). Development and trends of biosurfactant analysis and purification using rhamnolipids as an example. Anal. Bioanal. Chem. *391*, 1579–90.

Jadhav, V.V., Yadav, A., Shouche, Y.S., Aphale, S., Moghe, A., Pillai, S., Arora, A., and Bhadekar, R.K. (2013). Studies on biosurfactant from *Oceanobacillus* sp. BRI 10 isolated from Antarctic sea water. Desalination. *318*, 64–71.

Janek, T., Łukaszewicz, M., Rezanka, T., and Krasowska, A. (2010). Isolation and characterization of two new lipopeptide biosurfactants produced by *Pseudomonas fluorescens* BD5 isolated from water from the Arctic Archipelago of Svalbard. Bioresour. Technol. *101*, 6118–23.

Jolly, V.K. (1997). Commercialization new Technology. Getting from Mind to Market Harvard Business School Press (Boston, USA).

Keener, K., Hoban, T., and Balasubramanian, R. (2013). Biotechnology and it Applications (North Carolina, US: North Carolina State University, College of Agriculture and Life Science), FSR0031 working paper, March.

Khondee, N., Tathong, S., Pinyakong, O., Müller, R., Soonglerdsongpha, S., Ruangchainikom, C., Tongcumpou, C., and Luepromchai, E. (2015). Lipopeptide biosurfactant production by chitosan-immobilized *Bacillus* sp. GY19 and their recovery by foam fractionation. Biochem. Eng. J. *93*, 47–54.

Kronemberger Fde, A., Santa Anna, L.M., Fernandes, A.C., Menezes, R.R., Borges, C.P., and Freire, D.M. (2008). Oxygen-controlled biosurfactant production in a bench scale bioreactor. Appl. Biochem. Biotechnol. *147*, 33–45.

Kronemberger, F.D.A., Borges, C.P., and Freire, D.M.G. (2010). Fed-batch biosurfactant production in a bioreactor. Int. Rev. Chem. Eng. *2*, 513–8.

Kumar, A., Mody, K., and Jha, B. (2008). Evaluation of biosurfactant/bioemulsifier production by a marine bacterium. Bull. Environ. Contam. Toxicol. *79*, 617–621.

Lee, S.Y., Lee, D.Y., and Kim, T.Y. (2005). Systems biotechnology for strain improvement. Trends Biotechnol. *23*, 349–58.

Lovaglio, R.B., Silva, V.L., Ferreira, H., Hausmann, R., and Contiero, J. (2015). Rhamnolipids know-how: Looking for strategies for its industrial dissemination. Biotechnol Adv. *33*, 1715–26.

Mao, X., Jiang, R., Xiao, W., and Yu, J. (2015). Use of surfactants for the remediation of contaminated soils: a review. J. Hazard. Mater. *285*, 419–35.

Marti, M.E., Colonna, W.J., Reznik, G., Pynn, M., Jarrell, K., Lamsal, B., and Glatz, C.E. (2015). Production of fatty-acyl-glutamate biosurfactant by *Bacillus subtilis* on soybean co-products. Biochem. Eng. J. *95*, 48–55.

McDougall, P. (2012). The Cost and Time Involved in the Discovery, Development, and Authorization of a New Plant Biotechnology Derived Trait. A Consultancy

Study for Crop Life International. (Vineyard Business Center, Midlothian, UK), September.

Moya, R.I., Tsaousi, K., Rudden, M., Marchant, R., Jurado, Alameda, E., García, Román.M., and Banat, I.M. (2015). Rhamnolipid and surfactin production from olive oil mill waste as sole carbon source. Bioresour Technol. *198*, 231–6.

Mukherjee, S., Das, P., and Sen, R. (2006). Towards commercial production of microbial surfactants. Trends Biotechnol. *24*, 509–15.

Müller, M.M., Kügler, J.H., Henkel, M., Gerlitzki, M., Hörmann, B., Pöhnlein., M., Syldatk, C., and Hausmann, R. (2012). Rhamnolipids-Next generation surfactants? J. Biotechnol. *31*, 366–80.

Mulligan, C.N. (2014). Characterization, production and applications of lipopeptides. In Biosurfactants: Research Trends and Applications, C.N. Mulligan, S.K. Sharma and A. Mudhoo, eds. (CRC Press, Boca Raton, FL), pp. 147–176.

Mutalik, S.R., Vaidya, B.K., Joshi, R.M., Desai, K.M., and Nene, S.N. (2008). Use of response surface optimization for the production of biosurfactant from *Rhodococcus* spp. MTCC 2574. Bioresour. Technol. *99*, 7875–7880.

Ongena, M., and Jacques, P. (2008). *Bacillus* lipopeptides: versatile weapons for plant disease biocontrol. Trends Microbiol. *16*, 115–25.

Pal, M.P., Vaidya, B.K., Desai, K.M., Joshi, R.M., Nene, S.N., and Kulkarni, B.D. (2009). Media optimization for biosurfactant production by *Rhodococcus erythropolis* MTCC 2794: artificial intelligence versus a statistical approach. J. Ind. Microbiol. Biotechnol. *36*, 747–56.

Pecci, Y., Rivardo, F., Martinotti, M.G., and Allegrone, G. (2010). LC/ESI-MS/MS characterisation of lipopeptide biosurfactants produced by the *Bacillus licheniformis* V9T14 strain. J. Mass Spectrom. *45*, 72–8.

Persson, A., and Molin, G. (1987). Capacity for biosurfactant production of environmental *Pseudomonas* and *Vibrionaceae* growing on carbohydrates. Appl. Microbiol. Biotechnol. *26*, 439–442.

Plante, C.J., Coe, K.M., and Plante, R.G. (2008). Isolation of surfactant-resistant bacteria from natural, surfactant-rich marine habitats. Appl. Environ. Microbiol. *74*, 5093–9.

Płaza, G., Chojniak, J., Rudnicka, K., Paraszkiewicz, K., and Bernat, P. (2015). Detection of biosurfactants in *Bacillus* species: genes and products identification. J. Appl. Microbiol. *119*, 1023–34.

Qiu, Y., Xiao, F., Wei, X., Wen, Z., and Chen, S. (2014). Improvement of lichenysin production in *Bacillus licheniformis* by replacement of native promoter of lichenysin biosynthesis operon and medium optimization. Appl. Microbiol. Biotechnol. *98*, 8895–903.

Raaijmakers, J.M., De Bruijn, I., Nybroe, O., and Ongena, M. (2010). Natural functions of lipopeptides from *Bacillus* and *Pseudomonas*: more than surfactants and antibiotics. FEMS Microbiol. Rev. *34*, 1037–1062.

Raghukumar, S. (2014). The biotechnology pyramid: basic science to application of science. Current Science. *11*, 1789–1791.

Rahman, P.K.S.M., and Gakpe, E. (2008). Production, characterization and applications of biosurfactants – review. Biotechnol. *7*, 360–370.

Regulation EC No. 648/2004 of the European Parliament and of the Council on detergents.(2004). Official Journal of the European Union L 104, pp. 1–35. Type of document: Legislation. Link to full text: eur42319.pdf.

Reis, R.S., Pacheco, G.J., Pereira, A.G., and Freire, D.M.G. (2013). Biosurfactants: production and applications. In Biodegradation – Life of Science. R. Chamy, ed. (InTech, Rijeka, Croatia), pp. 31–61.

Reis, R.S., Pereira, A.G., Neves, B.C., and Freire, D.M.G. (2011). Gene regulation of rhamnolipid production in *Pseudomonas aeruginosa* – a review. Bioresour. Technol. *102*, 6377–6384.

Robertson, P.L., and Jacobson, D. (2011). Knowledge transfer and technology diffusion: an introduction. In Knowledge transfer and technology diffusion, P.L. Roberdson and D. Jacobson, eds. (Edward Elgar Publishing, Massachusetts, USA), pp. 1–34.

Roelants, S.L., De Maeseneire, S.L., Ciesielska, K., Van Bogaert, I.N., and Soetaert, W. (2014). Biosurfactant gene clusters in eukaryotes: regulation and biotechnological potential. Appl. Microbiol. Biotechnol. *98*, 3449–61.

Sachdev, D.P., and Cameotra, S.S. (2013). Biosurfactants in agriculture. Appl. Microbiol. Biotechnol. *97*, 1005–1016.

Saerens, K.M.J., Zhang, J., Saey, L., Van Bogaert, I.N.A., and Soetaert, W. (2011). Cloning and functional characterization of the UDP- glucosyltransferase UgtB1 involved in sophorolipid production by *Candida bombicola* and creation of a glucolipid- producing yeast strain. Yeast. *28*, 279–292.

Sambles, C.M., and White, D.A. (2015). Genome sequence of *Rhodococcus* sp. strain PML026, a trehalolipid biosurfactant producer and biodegrader of oil and alkanes. Genome Announc. 3, pii:e00433-15.

Santos, A.S., Sampaio, A.P., Vasquez, G.S., Santa Anna, L.M., Pereira, N. Jr., and Freire, D.M. (2002). Evaluation of different carbon and nitrogen sources in production of rhamnolipids by a strain of *Pseudomonas aeruginosa*. Appl Biochem Biotechnol. *98–100*, 1025–35.

Satpute, S.K., Banpurkar, A.G., Dhakephalkar, P., Banat, I.M., and Chopade, B.A. (2010). Methods for investigating biosurfactants and bioemulsifiers: a review. Crit. Rev. Biotechnol. *30*, 127–44.

Sekhon Randhawa, K.K., and Rahman, P.K. (2014). Rhamnolipid biosurfactants – past, present, and future scenario of global market. Front. Microbiol. 5, 454.

Seydlovà, G., and Svobodovà, J. (2008). Development of membrane lipids in the surfactin producer *Bacillus subtilis*. Folia Microbiol. 53, 303–307.

Seydlová, G., Fišer, R., Čabala, R., Kozlík, P., Svobodová, J., and Pátek, M. (2013). Surfactin production enhances the level of cardiolipin in the cytoplasmic membrane of *Bacillus subtilis*. Biochim. Biophys. Acta. *1828*, 2370–2378.

Shen, H.H., Thomas, R.K., Penfold, J., and Fragneto, G. (2010). Destruction and solubilization of supported phospholipid bilayers on silica by the biosurfactant surfactin. Langmuir. *26*, 7334–7342.

Shete, A.M., Wadhawa, G., Banat, I.M., and Chopade, B.A. (2006). Mapping of patents on bioemulsifier and biosurfactant: A review. J. Sci. Ind. Res. *65*, 91–115.

Singh, A., van Hamme, J.D., and Ward, O.P. (2007). Surfactants in microbiology and biotechnology: Part 2. Applications aspects. Biotech. Adv. *25*, 99–121.

Sinha, G. (2012). One Patent for Europe. Nat Biotechnol. *3*, 200–201.

Sivapathasekaran, C., Mukherjee, S., Ray, A., Gupta, A., and Sen, R. (2010). Artificial neural network modeling and genetic algorithm based medium optimization for the improved production of marine biosurfactant. Bioresour. Technol. *101*, 2884–2887.

Smith, G.V., and Parr, R.L. (2005). Intellectual property. Valuation, Exploitation, and infridement damages. (Wiley, New Jersey, USA).

Suthar, H., Hingurao, K., Desai, A., and Nerurkar, A. (2008). Evaluation of bioemulsifier mediated microbial enhanced oil recovery using sand pack column. J. Microbiol. Methods *75*, 225–230.

Toribio, J., Escalante, A.E., and Soberón-Chávez G. (2010). Rhamnolipids: Production in bacteria other than *Pseudomonas aeruginosa*. Eur. J. Lipid Sci. Technol. *112*, 1082–1087.

US EPA (2010). Reference Guide to Non-combustion Technologies for Remediation of Persistent Organic Pollutants in Soil, Second Edition – 2010 (103 pp, 9 MB, 2010, 542-R-09–007).

US EPA (2014). Reference Guide to Treatment Technologies for Mining-Influenced Water (94 pp, 4 MB, 2014, 542-R-14–001).

Van Bogaert, I.N., Holvoet, K., Roelants, S.L., Li, B., Lin, Y.C., Van de Peer, Y., and Soetaert, W. (2013). The biosynthetic gene cluster for sophorolipids: a biotechnological interesting biosurfactant produced by *Starmerella bombicola*. Mol. Microbiol. *88*, 501–9.

Van Bogaert, I.N., Sabirova, J., Develter, D., Soetaert, W., and Vandamme, E.J. (2009). Knocking out the MFE-2 gene of *Candida bombicola* leads to improved medium-chain sophorolipid production. FEMS Yeast Res. *9*, 610–617.

Walter, V., Syldatk, C., and Hausmann, R. (2010). Screening concepts for the isolation of biosurfactant producing microorganisms. Adv. Exp. Med. Biol. *672*, 1–13.

Wang, Q., Fang, X., Bai, B., Liang, X., Shuler, P.J., Goddard, W.A. III, and Tang, Y. (2007). Engineering bacteria for production of rhamnolipid as an agent for enhanced oil recovery. Biotechnol. Bioeng. *98*, 842–853.

Ye, L., Hildebrand, F., Dingemans, J., Ballet, S., Laus, G., Matthijs, S., Berendsen, R., and Cornelis, P. (2014). Draft genome sequence analysis of a *Pseudomonas putida* W15Oct28 strain with antagonistic activity to Gram-positive and *Pseudomonas* sp. pathogens. PLoS ONE *4*, e110038.

Zehner, W.B., Trzmielak, D.M., Zehner, J.A. (2013). Commercialization of Biotechnology Science Molecules to Market – Global and American Perspective, In Innowacje i komercjalizacja w biotechnologii, D.M. Trzmielak, ed. (University of Lodz Publishing, Poznań-Łódź, Poland), pp. 177–189.

Index

R

S

www.ingramcontent.com/pod-product-compliance
Ingram Content Group UK Ltd.
Pitfield, Milton Keynes, MK11 3LW, UK
UKHW061959041026
15591UKWH00016B/1090

9 781910 190456